DIFFERENTIAL GEOMETRY FOR PHYSICISTS AND MATHEMATICIANS

Moving Frames and Differential Forms:
From Euclid Past Riemann

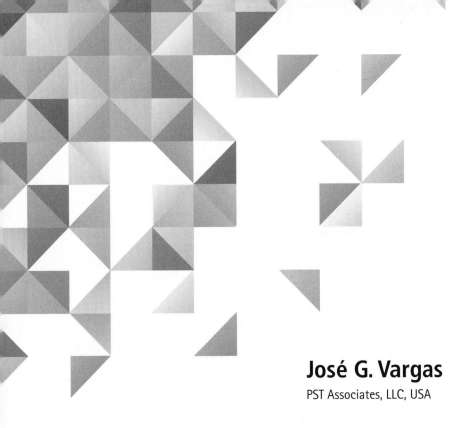

José G. Vargas

PST Associates, LLC, USA

DIFFERENTIAL GEOMETRY FOR PHYSICISTS AND MATHEMATICIANS

Moving Frames and Differential Forms: From Euclid Past Riemann

World Scientific

NEW JERSEY · LONDON · SINGAPORE · BEIJING · SHANGHAI · HONG KONG · TAIPEI · CHENNAI

Published by

World Scientific Publishing Co. Pte. Ltd.

5 Toh Tuck Link, Singapore 596224

USA office: 27 Warren Street, Suite 401-402, Hackensack, NJ 07601

UK office: 57 Shelton Street, Covent Garden, London WC2H 9HE

Library of Congress Cataloging-in-Publication Data
Vargas, José G.
  Differential geometry for physicists and mathematicians : moving frames and differential forms : from Euclid past Riemann / by José G Vargas (PST Associates, LLC, USA).
    pages cm
  Includes bibliographical references and index.
  ISBN 978-9814566391 (hardcover : alk. paper)
  1. Mathematical physics. 2. Geometry, Differential. I. Title.
  QC20.V27 2014
  516.3'6--dc23

                          2013048730

British Library Cataloguing-in-Publication Data
A catalogue record for this book is available from the British Library.

Printed in Singapore by World Scientific Printers.

To Ms. Gail Bujake for her contribution in making this world a better place through her support of science and technology

Thanks for your support, without which this work would not have been possible. I will always be grateful.

José

Acknowledgements

It would be difficult to acknowledge in detail the many persons who have contributed to the journey that the writing of this book has represented. In each new phase, new names had to be added. Not all contributions were essential, but all of them important.

Different forms of inspiration, mathematical or pedagogical, direct or indirect, personal or through their writings (in the case of the deceased) are due to famous mathematicians like É. Cartan and E. Kähler, but also less known figures like the late Professors Y. H. Clifton (a mathematician at the department of Physics of Utah State) and Fernando Senent, physicist at the University of Valencia in Spain.

Moral support and encouragement through the years are due to Professors Douglas G. Torr and Alwyn van der Merwe.

For a variety of reasons support is also due to Doctors and/or Professors Jafar Amirzadeh, Vladimir Balan, Howard Brandt, Iulian C. Bandac, Zbigniew Oziewicz, Marsha Torr and Yaohuan Xu, and to Ms. Luminita Teodorescu.

Phase Space Time Associates provided generous support.

And last but most, I acknowledge my wife Mayra for her patience and understanding.

Preface

The principle that informs this book. This is a book on differential geometry that uses the method of moving frames and the exterior calculus throughout. That may be common to a few works. What is special about this one is the following. After introducing the basic theory of differential forms and pertinent algebra, we study the "flat cases" known as affine and Euclidean spaces, and simple examples of their generalizations. In so doing, we seek understanding of advanced concepts by first dealing with them in simple structures. Differential geometry books often resort to formal definitions of bundles, Lie algebras, etc. that are best understood by discovering them in a natural way in cases of interest. Those books then provide very recondite examples for the illustration of advanced concepts, say torsion, even though very simple examples exist. Misunderstandings ensue.

In 1492 Christopher Columbus crossed the Atlantic using an affine connection in a simplified form (a connection is nothing but a rule to navigate a manifold). He asked the captains of the other two ships in his small flotilla to always maintain what he considered to be the same direction: West. That connection has torsion. Élie Cartan introduced it in the mathematical literature centuries later [13]. We can learn connections from a practical point of view, the practical one of Columbus. That will help us to easily understand concepts like frame bundle, connection, valuedness, Lie algebra, etc., which might otherwise look intimidating. Thus, for example, we shall slowly acquire a good understanding of affine connections as differential 1−forms in the affine frame bundle of a differentiable manifold taking values in the Lie algebra of the affine group and having such and such properties. Replace the term affine with the terms Euclidean, conformal, projective, etc. and you have entered the theories of Euclidean, projective, conformal ... connections.

Cartan's versus the modern approach to geometry. It is sometimes stated that É. Cartan's work was not rigorous, and that it is not possible to make it so. This statement has led to the development of other methods to do differential geometry, full of definitions and distracting concepts; not the style that physicists like.

Yeaton H. Clifton was a great differential topologist, an opinion of this author which was also shared by the well known late mathematician S.-S. Chern in private conversation with this author. Clifton had once told me that the only thing that was needed to make rigorous Cartan's theory of connections was to add a couple of definitions. A few years later, upon the present author's prodding, Clifton delivered on his claim. To be precise, he showed that just a major definition and a couple of theorems were needed. The proof is in the pudding. It is served in the last section of chapter 8 and in the second section of chapter 9.

Unfortunately, Cartan's approach has virtually vanished from the modern literature. Almost a century after his formulation of the theory of affine and Euclidean connections as a generalization of the geometry of affine and Euclidean spaces [11], [12], [14], an update is due on his strategy for the study of generalized spaces with the method of the moving frame [20]. We shall first study from the perspective of bundles and integrability of equations two flat geometries (their technical name is Klein geometries) and then proceed with their Cartan generalization. In those Klein geometries, affine and Euclidean, concepts like equations of structure already exist, and the mathematical expression of concepts like curvature and torsion already arise in full-fledged form. It simply happens that they take null values.

Mathematical substance underlying the notation. There is a profound difference between most modern presentations and ours. Most authors try to fit everything that transforms tensorially into the mold of (p, q)−tensors (p times contravariant and q times covariant). Following Kähler in his generalization of Cartan's calculus, [46], [47], [48], we do not find that to be the right course of action. Here is why.

Faced with covariant tensor fields that are totally skew-symmetric, the modern approach that we criticize ignores that the natural derivative of a tensor field, whether skew-symmetric or not, is the covariant derivative. They resort to exterior derivatives, which belong to exterior algebra. That is unnatural and only creates confusion. Exterior differentiation should be applied only to exterior differential forms, and these are not skew-symmetric tensors. They only look that way.

Covariant tensor fields have subscripts, but so do exterior differential forms. For most of the authors that we criticize, the components of those two types of mathematical objects have subscripts, which they call q indices. But not all the q indices are born equal. There will be skew-symmetry and exterior differentiation in connection with some of them —"differential form" subscripts— but not in connection with the remaining ones, whether they are skew-symmetric with respect to those indices or not. They are tensor subscripts. Like superscripts, they are associated with covariant differentiation.

Correspondingly, the components of quantities in the Cartan and Kähler calculus have —in addition to a series of superscripts— two series of subscripts, one for integrands and another one for multilinear functions of vectors. This is explicitly exhibited in Kähler [46], [47], [48].

The paragon of quantities with three types of indices. Affine curvature is a $(1, 1)$−tensor-valued differential 2−form. The first "1" in the pair is for a superscript, and the other one is for a subscript. Torsions are $(1, 0)$−valued differential 2−forms and contorsions are $(1, 1)$−valued differential 1−forms.

Let $\mathbf{v}$ represent vector fields and let d be the operator that Cartan calls exterior differentiation. $d\mathbf{v}$ is a vector-valued differential $1-$form, and $dd\mathbf{v}$ is a vector-valued differential $2-$form. Experts not used to Cartan's notation need be informed that $dd\mathbf{v}$ is $(v^\mu R^\nu_{\mu\lambda_1\lambda_2})\omega^{\lambda_1} \wedge \omega^{\lambda_2}\mathbf{e}_\nu$. Relative to bases of $(p = 1,$ $q = 0)-$valued differential $2-$forms, the components of $dd\mathbf{v}$ are $(v^\mu R^\nu_{\mu\lambda_1\lambda_2})$. One can then define a $(1,1)-$valued differential $2-$form whose components are the $R^\nu_{\mu\lambda_1\lambda_2}$'s, and whose evaluation on $\mathbf{v}$ (responding to the $q = 1$ part of the valuedness) yields $dd\mathbf{v}$. Hence, the traditional $(p, q)-$characterization falls short of the need for a good understanding of issues concerned with the curvature differential form.

Bundles are of the essence. The perspective of valuedness that we have just mentioned is one which best fits sections of frame bundles, and transformations relating those sections. Lest be forgotten, the set of all inertial frames (they do not need to be inertial, but that is the way in which they appear in the physics literature) constitutes a frame bundle. Grossly speaking, a bundle is a set whose elements are organized like those inertial frames are. The ones at any given point constitute the fiber at that point. We have identical fibers at different points. There must be a group acting in the bundle (like Poincaré's is in our example), and a subgroup acting in the fibers (the homogenous Lorentz group in our example).

An interesting example of section of a bundle is found in cosmology. One is computing in a particular section when one refers quantities to the frame of reference of matter at rest in the large.

A section is built with one and only one frame from each fiber, the choice taking place in a continuous way. But, for foundational purposes, it is better to think in terms of the bundle than of the sections. At an advanced level, one speaks of Lie algebra valuedness of connections, the Lie algebra being a vector space of the same dimension as the bundle. All this is much simpler than it sounds when one really understands Euclidean space. We will.

It is unfortunate that books on the geometry of physics deal with connections valued in Lie algebras pertaining to auxiliary bundles (i.e. not directly related to the tangent vectors) and do not even bother with the Lie algebras of bundles of frames of tangent vectors. Which physicist ever mentions what is the Lie algebra where the Levi-Civita connection takes its values? Incidentally, the tangent vectors themselves constitute a so called fiber bundle, each fiber being constituted by all tangent vectors at any given point. It is the tangent bundle.

This author claims that the geometry of groups such as $SU(3)$ and $U(1) \times SU(2)$ fits in appropriately extended tangent bundle geometry, if one just knows where to look. One does not need auxiliary bundles. That will not be dealt with in this book, but in coming papers. This book will tell you whether I deserve your trust and should keep following me where I think that the ideas of Einstein, Cartan and Kähler take us.

Assume there were a viable option of relating $U(1) \times SU(2) \times SU(3)$ to bundles of tangent vectors, their frames, etc. It would be unreasonable to remain satisfied with auxiliary bundles (Yang-Mills theory). In any case, one should understand "main bundles geometry" (i.e. directly related to the tangent bundle) before studying and passing judgement on the merits and dangers of Yang-Mills theory.

Specific features distinguishing this book are as follows:

1. Differential geometry is presented from the perspective of integrability, using so called moving frames in frame bundles. The systems of differential equations in question emerge in the study of affine and Euclidean Klein geometries, those specific systems being integrable.

2. In this book, it does not suffice whether the equations of the general case (curved) have the appropriate flat limit. It is a matter of whether we use in the general case concepts which are the same or as close as possible to the intuitive concepts used in flat geometry. Thus, the all-pervasive definition of tangent vectors as differential operators in the modern literature is inimical to our treatment.

3. In the same spirit of facilitating understanding by non-mathematicians, differential forms are viewed as functions of curves, surfaces and hypersurfaces [65] (We shall use the term hypersurface to refer to manifolds of arbitrary dimension that are not Klein spaces). In other words, they are not skew-symmetric multilinear functions of vectors but cochains.

This book covers almost the same material as a previous book by this author [85] *except for the following*:

1. The contents of chapters 1, 3 and 12 has been changed or extended very significantly.

2. We have added the appendices. Appendix A presents the classical theory of curves and surfaces, but treated in a totally novel way through the introduction of the concept of canonical frame field of a surface (embedded in 3-D Euclidean space). We could have made it into one more chapter, but we have not since connections connect tangent vectors in the book except in that appendix; vectors in 3-D Euclidean space that are not tangent vectors to the specific curves and surfaces being considered are nevertheless part of the subject matter.

Appendix B speaks of the work of the mathematical geniuses Élie Cartan and Hermann Grassmann, in order to honor the enormous presence of their ideas in this book. Appendix C is the list of publications of this author for those who want to deal further into topics not fully addressed in this book but directly related to it. You can find there papers on Finsler geometry, unification with teleparallelism, the Kähler calculus, alternatives to the bundle of orthonormal frames, etc.

3. Several sections have been added at the end of several chapters, touching subjects such as diagonalization of metrics and orthonormalization of frames, Clifford and Lie algebras, etc.

Contents

Part I

INTRODUCTION

Chapter 1

ORIENTATIONS

1.1 Selective capitalization of section titles

While this book is primarily intended for non-experts on differential geometry, experts may enjoy the numerous advanced comments sprinkled through the text, which one does not find together in a single source. Some comments qualify as advanced simply because they are absent in other approaches to differential geometry, but are essential here. If those comments were omitted, this book would not have been worth writing. Their absence from most of the literature may be at the root of numerous errors found in it.

In order to accomodate the wide audience extending from neophytes to experts, we have included sections meant to achieve greater rigor in some key area, but too abstract for non-experts. Whenever suitable, those sections have been placed at the end of chapters.

We have used selective capitalization in titles of sections as guidance to degree of difficulty, though no universal agreement may be achievable in this regard. Thus, we have written in full capitals the titles of sections that should be skipped by non-expert readers. If we think that some comments but not a whole section are too advanced, we alert of that fact by writing the last word(s) of the title of the section in full capitals. In section 8.1, which is an exception, the middle rather that the end part should be skipped, if any.

This book should be read linearly for the most part. However, linear understanding may not be an achievable goal for many readers. That is no problem. Most of us do not understand everything the first time that we encounter some new theory. We keep learning the same subject at ever greater depths in successive approximations to a complete understanding.

We advocate that non-experts skip the sections with titles in full capitals, and that, when reading the ones where only part of the title is in capitals, they disregard any passage that they find obscure. They should try to pay attention to the bottom line of arguments. Such a skipping risks that a reader may overlook something important. In order to minimize this problem, concepts and remarks have often been repeated throughout the text.

Learning differential geometry should not be different from learning a foreign language. After learning differential geometry at the level of the command of his mother language by a five year old, one can learn the grammar of differential geometry by re-reading this book more thoroughly. Those who need a more thorough or more precise vocabulary —not every physicist does— can always read other books on the subject. This one is not a bible, their value being in any case religion dependent.

1.2   The meaning of classical in "classical differential geometry"

One might think that the term *classical differential geometry* refers to all those geometries that preceded (in physics) the formulation of Yang-Mills geometry. It certainly must include the theory of curves and surfaces in Euclidean space, which was virtually completed by the mid nineteenth century (Appendix A deals with that theory, except in language not known at the time). One would expect the italicized term to also include the Riemannian generalization of Euclidean geometry, and its further generalization by Cartan. But that is not how the term is used.

In Riemann's and Cartan's theories, a surface is a differentiable manifold of dimension two, meaning that we need two independent coordinates to label its points. But, as we shall see in an appendix, the theory of differentiable manifolds of dimension two and the theory of surfaces developed before Riemann do not coincide. Thus, whereas we may speak of the torsion of a differentiable 2-manifold, we may not speak of the torsion of a surface in the theory of curves and surfaces of that time. Let us look at this issue more in detail.

In general differentiable manifolds, there is no room for vectors. For instance, they do not fit on the sphere (Spanish readers not too converse with the English language should note that what we call *esfera* translates into *ball* in English, and what we call *spherical surface* translates into *sphere)*. Vectors stick out of the sphere or are like cords; if they are very small, we do not notice this, which is the reason why it was thought for ages that the earth was flat. We may and do, however, speak of *tangent vectors* precisely because of this property of small vectors lying on the surface as if it were flat. Of course, this is not a rigorous way of saying things; we will become more rigorous later on.

A flag pole, lying down on a flat floor, closely represents the concept of tangent vector for the purposes of this section. A connection is like a rule to compare flag poles lying down at different points of a surface. Differential geometry in the modern sense of the word is mainly the study of such rules. In that theory and in order to avoid confusion with classical differential geometry. We should then speak of 2-dimensional manifolds rather than surfaces, at least until we understand what is studied in each case.

In the classical differential theory of curves and surfaces (i.e. geometry), we care also about rules for upright poles, or sticking at an angle. Those rules cannot be assigned specifically to the surface, but to the latter being living in 3-D Euclidean space. Something similar occurs with curves, which are one-dimensional manifolds. When one refers to the (in general, not null) curvature and torsion of curves, one is referring to concepts different from what they are in differentiable manifold theory. The differences just stated between the theory of curves and surfaces, on the one hand, and the theory of differentiable manifolds of dimensions 1 and 2, on the other hand, has led us to relegate the first of those to Appendix A. Having made the foregoing distinctions, we shall use the term surface in both contexts, i.e. whether in classical differential geometry or in modern one.

Consider now the geometry associated with the electroweak and strong interactions. Here we are also dealing with modern theory of connections on differentiable manifolds, but one is not connecting tangent vectors but vectors of some auxiliary spaces, and frames of auxiliary bundles constructed upon those spaces. Auxiliary means precisely not being directly related to tangent vectors.

This book is about modern differential geometry as conceived and, equally important, as presented by the great master Élie Cartan (his son, Henri Cartan, also was a famous mathematician, cofounder of the Bourbaki group; in this book, we always refer to Élie, not Henri). The main concept is the *connection* on a *bundle of frames made of vectors tangent* to a differentiable manifold. Readers reaching all the way to the appendices should be well prepared to follow the argument that one does not need to invoke auxiliary bundles to do what Yang-Mills theory presently does. By the time that this book goes to press and falls in your hands, some paper making that point should have been posted by this author in the arXiv, or printed somewhere else.

1.3 Intended readers of this book

The backbone of this book is the theory of affine and Euclidean connections. There are other books on the subject, but they target a readership in the modern style of doing mathematics. We intend to serve readers unsatisfied by that style.

Contrary to most modern approaches, we use the same concepts of vector field and connection as in the first courses of calculus, i.e. as in the study of 3-D Euclidean space. In those elementary treatments, vector fields are not differentiable operators, nor will they be so here, but passive objects not acting on anything. They are actually acted upon, differentiated, to be specific. Connections are not explicitly mentioned in those treatments; they are not an absent concept, but their simplicity and the lack of thoroughness of standard treatments makes them be overlooked. Here connections are made explicit even in affine and Euclidean space.

The signature calculus of differential geometry is the exterior calculus. It deals with expressions such as $f(x,y,z)dx + g(x,y,z)dy$, or $dx \wedge dy$, etc., to which we shall refer as scalar valued differential forms or simply differential forms. Unlike $dx\mathbf{i}+dy\mathbf{j}+dz\mathbf{k}$, which we shall call vector valued, they contain no vectors or vector fields. In our approach, differential geometry is, in first approximation, the calculus of vector-valued differential forms, which includes the exterior calculus of scalar-valued ones. In order to differentiate vector fields or vector-valued differential forms, one needs to connect vectors at nearby points. A connection is thus involved. Other types of valuedness also are considered: covariant, mixed contravariant-covariant, Lie-algebra valuedness, Clifford algebra valuedness, etc. They will emerge in an organic way starting with the differentiation of contravariant (read ordinary) vector fields..

We deal *in extenso* with the concepts of differential forms and differentiable manifolds in chapter 2. The exterior calculus is introduced in chapter 4. Most of the algebra underlying this calculus is given in chapter 3. Affine and Euclidean spaces are respectively treated in chapters 5 and 6. For the uninitiated, affine space is just Euclidean space with the concept of distance removed from it. Their generalizations are the differentiable manifolds endowed with affine connections (chapter 8) and Euclidean connections (chapters 9 and 10). Chapter 7 is a transitional chapter constituted by simple examples of connected differentiable manifolds.

From the preceding considerations follows that our intended readers are mainly advanced physics and mathematics students, but also students of other sciences and engineering with unusual mathematics or physics interest. Because of the technique we use, they should be mature enough not to be discouraged by unfamiliarity with this calculus. Once again —for a variety of reasons (and following Rudin's baby book on analysis [65] and the ways of Cartan and Kähler)— our differential forms are integrands, i.e. functions of lines, surfaces, volumes, etc. (cochains in the main literature), not antisymmetric multilinear functions of vectors.

Although such advanced students constitute the main audience we target, we hope that professionals may find this approach to be refreshing because of the abundance of unusual advanced concepts. Finally, chapters 11-13, specially the last one, provides a glimpse of the many doors that this book opens.

1.4   The foundations of physics in this BOOK

This section and the next two reflect the fact that this book and other to follow are what they are because this author found that the mathematics already exists to carry out what Einstein called the logical homogeneity of differential geometry and the foundations of physics (Einstein used the term theoretical physics). In modern language, that would be the identification of the equations of structure of some geometry with the field equations of the physics.

Judging the *sources* (more accurately, lack thereof) used to train physicists in the foundations of physics, one can understand why that identification may be considered mission impossible *ab initio*. We shall present a critical panoramic view of theoretical and foundations physics (this section) and point out impediments to its progress (next two sections). May this book start to persuade its readers that, after all, Einstein's proposal for unification from teleparallelism might be the right way to go.

In the last decades physics has not solved one single "theory independent" problem, like the mass of the electron is. Nor has it solved "structural problems" of physical theory, like those we are about to mention. Much heralded new theories that promised to give us everything appear to be fizzling away. Some readers will context my negative perception. But there are authoritative opinions that are even more derogatory than mine. Thus, for instance, the very eminent physicist Carver Mead (emeritus professor at Caltech, winner of Lemelson-MIT 1999 Prize for invention and innovation, IEEE John von Neumann Medal, founder of several technology companies) opens the introduction of his wonderful book Collective Electrodynamics [55] with the following statement: "It is my firm belief that the last seven decades of the twentieth century will be characterized in history as the dark ages of theoretical physics". That was in 2000. There is no reason why he would not be writing today of more than eight decades of dark ages.

I would qualify Mead's statement by directing his criticism to the foundations of physics, not to theoretical physics. After all, there has been undeniable progress in the field which goes far beyond what is simply experimental, and for which we cannot think of a name if not theoretical physics, even if heavily loaded with phenomenology. I submit that the explosion in the development of technology over those eight decades has allowed theoretical physics to make progress —in spite of deficient foundations— by relying heavily on ever more sophisticated experiments. An anecdote will be illuminating.

When this author was an undergraduate student in the mid sixties, a more senior student (or it might have been an instructor) told him the following:

> "In Europe, we study theoretically how a stab breaks. One then breaks three stabs to check the result of the study. In the United States, one breaks three thousand stabs and then writes down the rule according to which they break."

Eventually, Europe became like the USA in matters of physics, very unlike what physics was there in the first three decades of the twentieth century. As for mathematics, money has also distorted its development. A differential geometer based in the United States of America once told me:

> "I do Riemannian geometry to obtain funding but, in my spare time, I do Finsler geometry, which is what I like."

In foundations of physics one questions some of its most basic assumptions. Virtually all but not all present assumptions will survive, but it is almost anathema to question any one of them, including certainly the one(s) which one day will be recognized as having being wrong. What does it have to do with this book?

If one knew better geometry and calculus, many basic physics results could be given better justification than at present. Since this is a book on differential geometry, let me start with something that pertains to general relativity (GR).

Nowadays, GR incorporates the Levi-Civita connection (LCC) [52], which is the connection canonically determined by the metric. It did not at one time, since GR and the LCC were born in 1915 and 1917 respectively. Under the LCC, there is no equality of vectors at a distance in GR. It is then not surprising that we do not yet know what is the expression for gravitational energy and, related to it, what form would the corresponding conservation law take. Suffice to say that it is meaningless to add the tiny energy-momentum vectors at small spacetime regions in order to get the energy-momentum for a whole region. All sort of nonsense has been spoken in order to sidestep this basic problem, as if it were an inescapable feature of any geometric theory of gravity. The problem lies with the adoption of the LCC by GR. Einstein understood it when he postulated the so called teleparallelism in his attempt at a unified theory in the late nineteen twenties [39], but he new very little geometry and calculus for what would be required to develop his postulate.

As serious is the problem with electromagnetic (EM) energy-momentum. The title of section 4 of chapter 27 of the second book in the Feynman Lectures on Physics [40] reads "The ambiguity of the field energy". In that section, speaking of the energy density, u, and the Poynting vector, $\mathbf{S}$, Richard Feynman stated: "There are, in fact, an infinite number of different possibilities for u and $\mathbf{S}$, and so far no one has thought of an experimental way to tell which one is right." Further down the text he insists on the subject as follows: "It is interesting that there seems to be no unique way to resolve the indefiniteness in the location of the field energy". Continuing with the argument, he repeats a similar statement with regards to experimental verification: "As yet, however, no one has done such a delicate experiment that the precise location of the gravitational influence of electromagnetic fields could be determined."

Consider on the other hand the field theoretic derivation of the EM energy-momentum tensor through variations of the Lagrangian pertaining to spacetime homogeneity. One obtains what is considered to be the "wrong" EM energy-momentum tensor. The "right" one is obtained in Landau and Lifshitz *Classical Field Theory* [51] (and in everybody's text who uses the same approach to EM energy-momentum) by adding to the wrong one a specific term whose integral over all of spacetime is zero. But, in doing so, one is changing the gravitational field and thus, in principle, the distribution of gravitational energy. Viewed in the context of the argument by Feynman, having to "cheat" to correct the result of a canonical argument implies that the Lagrangian approach cannot be totally trusted, or that the electromagnetic Lagrangian is wrong, or that the variational method cannot be relied upon, or some other equally troublesome consequence.

We will not waste our time with refutations of hypothetical alternative derivations of what is considered to be the right energy-momentum tensor, since we believe that one can do much better. We think that the problem of gravitational and electromagnetic energies are inextricably tied, and so are the problems related to gravitational physics mentioned earlier in this section. The solution that has not been tried is teleparallelism, which entails connection with zero affine curvature (better called Euclidean or Lorentzian curvature) on the standard "metric".

We now enter the most egregious example of insisting on the wrong mathematics for physics, though ignorance may largely excuse it. Young minds who aspire to one day change the paradigm put their sights in Yang-Mills theory, or strings, or something with "super" at the front, but not in tangent bundle related geometry. How would you achieve unification with the latter?

Assume that a common language for both, quantum mechanics and gravitation, did exist. One should start by using the same language in both cases. Here is the great tragedy: that language exists, namely Kähler calculus of differential forms and concomitant quantum theory. In spite of the many interpretational problems with the Dirac theory, it keeps being used even though Kähler's theory is easier to interpret and does not have problems like negative energy solutions, has spin at par with orbital angular momentum, etc. [46], [47], [48]. See also [83]. A good understanding of Kähler theory together with an approach to GR where one adopts Riemannian geometry for metric but not for affine relations may go a long way towards making GR and quantum physics come together. See chapter 13.

Let us mention an avenue that we have not explored and that, therefore, will not yet be considered in this book. The Higgs field is the difference between a covariant derivative and a Lie derivative [54]. In spite of its brevity, our brief treatment of Lie differentiation in this book will allow us to understand that Lie derivatives are disguised forms of partial derivatives. These, as well as exterior derivatives, differ from covariant derivatives in connection terms, which suggests the possibility that one may be unduly replacing, in the arguments where the Higss field is concerned, one type of quantity for another (Different derivatives result from application of a common operator to quantities of different valuedness). If we misrepresent one valuedness with another in describing some physical magnitude, we shall be getting the wrong derivatives. The detection of a Higgs particle might be in principle the experimental isolation of a term introduced to correct an otherwise defficient description. In other words, an experiment might perhaps be isolating a dominant energy term that represents that correction.

1.5 Mathematical VIRUSES

The physicist David Hestenes has devoted his life to the commendable tasks of preaching the virtues and advantages of Clifford algebra over vector and tensor algebra and of applying it to physics. He has had the courage of (and deserves credit for) advancing the concept of viruses in mathematics, and for suggesting what, in his opinion, are some of them [44]. His definition of mathematical virus is: "a preconception about the structure, function or method of mathematics which impairs one's ability to do mathematics". Needless to say that if practitioner A sees a virus in practitioner B, the latter may retort that it is A who has a virus. We enumerate instances of what this author considers to be viruses. We give them names. The best antidote against viruses is to follow the greatest masters, who climbed the greatest peaks because they presumably lived in the healthiest mathematical environments.

Bachelor Algebra Virus. It consists in trying to do everything in just one algebra when two or more of them are required. For instance, $f(x, y, z)dx + g(x, y, z)dy$ belongs to one algebra, and $f(x, y, z)\mathbf{i} + g(x, y, z)\mathbf{j}$ belongs to another algebra. The tensor calculus is one where the fundamentally different nature of those two algebras is not taken into account. It is a bachelor algebra. There are two algebras involved in $dx\mathbf{i} + dy\mathbf{j} + dz\mathbf{k}$. They come intertwined, but the result is not equivalent to a single one of them. This virus is totally absent in the work if E. Kähler and É. Cartan.

Unisex Virus. This is a mild virus consisting in that a symbol is used for two different concepts. For instance, it is common in the literature to define differential forms as antisymmetric multilinear functions of vector fields. Fine. But one cannot then say at the same time, as often happens, that they are integrands, since r-integrands are functions of r-surfaces. Either one or the other, but not both at the same time. It is correct (actually it is customary) to use the term differential forms for the first concept, and then use the term cochain for the second one. There is not a unisex virus there. We do prefer, however, to follow W. Rudin, E. Kähler and É. Cartan in referring to cochains as differential forms.

Transmutation Viruses (mutations of the unisex virus). Recall what we said in the preface about two types of subscripts in the Kähler calculus. One of them is for covariant tensor valuedness, the other one being for differential forms, abstraction made of their valuedness. Applying to quantities with one type of subscript the differentiation that pertains to the other type is an example of transmutation virus: a quantity is made to play a role that is not in its true nature. Let us be specific. We can evaluate (meaning integrate) differential 1-forms on curves of a manifold independently of what the rule to compare vectors on the manifold is. But we cannot compare vectors (whether contravariant or covariant) at different points of the curve without a connection.

Another example of the transmutation virus consists in thinking of the electromagnetic field —which is a function of 2-dimensional submanifolds of spacetime— as a (tangent) tensor. As Cartan pointed out, Maxwell's equations do not depend on the connection of spacetime [12] If Maxwell's equations were viewed as pertaining to an antisymmetric tensor field, Maxwell's equations would depend on the connection of spacetime.

Monocurvature Virus. It consists in ignoring (or writing or speaking) as if manifolds endowed with a Euclidean connection (i.e. metric-compatible affine connection) had just one curvature. Books dealing with Euclidean connections usually fail to make the point that the metric defines a metric curvature, which plays metric roles independently of what the affine connection of the manifold is. If the latter's affine connection is the LCC, the same curvature symbols then play two roles, metric and affine. This, however, has the effect of leaving readers unaware of this double representation. That is a pernicious virus. It affected Einstein when he attempted unification through teleparallelism [39].

Riemannitis Virus. This virus consists in viewing Riemannian geometry as the theory of the invariants of a quadratic differential form in n variables with respect to the infinite group of analytic transformations of those variables. Élie Cartan referred to the Riemannian spaces viewed from that perspective as the "false spaces of Riemann" (see page 4 of [13] and misconception (b) in next section). This virus has its root in the failure to realize that, as he pointed out in his note of 1922 *On the equations of structure of generalized spaces and the analytic expression of Einstein's tensor* [10], "the *ds* does not contain all the geometric reality of the space..." The Riemannitis virus has given rise to a plague that infects virtually all physicists and mathematicians who deal with this geometry. It causes a mental block impeding a frame bundle view of Finsler geometry and potentially successful efforts to unify the gravitational interaction with the other ones.

Curveism Virus. This virus is a mutant of the Riemannitis virus. It consists in viewing Finsler geometry as the geometry based on an element of arc, i.e. on curves. The Finsler spaces so defined are false spaces in the same sense as when Cartan referred to the original Riemannian spaces as false spaces. This virus does not make sick those who work on global geometry or variational problems since they are interested in connection independent results. But it causes blindness in those who need to (but fail to) see other connections which also are compatible with the same expression for the element of arc. In the spirit of Cartan, Finsler geometry should be defined in terms of a bundle of frames, or something equivalent to it, independently of distances. The metric, and thus the distance, should then be viewed as derived invariants caused by the restriction of the bundle. The key difference with pre-Finsler geometry lies not in the form of the metric, but in the type of fibration. Of course, geometry is what geometers do and, if they want to go that path, it is their privilege. But their frames do not then fit the nature of Einstein elevators, of great importance for physicists.

1.6 FREQUENT MISCONCEPTIONS

Readers should ignore the comments in this section that they do not understand, which may be all of them for newcomers to differential geometry. I regret if some experts may find some of these comments trivial. Judging by what goes in the literature, one never knows.

This book is not what readers likely expect just because of our use of differential forms and the moving frame method. Nor is it because, in addition, affine space is treated separately. It is first and foremost an attempt to mitigate the effects of the haphazard way in which the mathematical education of a theoretical physicist takes place, which is at the root of misunderstandings propagating in literature on differential geometry for physicists. Examples abound:

(a) Does a group determine a differential geometry in the classical, i.e. (non-Yang-Mills) sense of the word? (no, it does not; a pair of group and subgroup satisfying a certain property is needed [27], [67]). See also discussion in section 4 of chapter 12 in connection with Klein.

(b) Is the infinite group of coordinate transformations of the essence —i.e. a defining property— of classical differential geometry? (no, it is not; as Cartan put it, such a way of thinking in Riemannian geometry does not make evident and actually masks its geometric contents, in the intuitive sense of the term [21]). See also section 10.1.

(c) "Coordinates are numbers" (equivocal statement though not too dangerous in actual practice if one uses the same term and symbols for coordinate functions as for their values; if the term coordinates denoted only numbers, their differentials would vanish). See section 2.1.

(d) "When using the method of the moving frame using differential forms, there are two types of indices in the affine/Euclidean curvature" (wrong, because in addition to the superscripts there are two different types of subscripts). See preface and section 2.1.

(e) "$d\mathbf{r}$ ($=dx\mathbf{i}+dy\mathbf{j}+dz\mathbf{k}$) is a $(1,1)-$tensor" (this is an incorrect statement from the perspective of the Cartan calculus and its generalization known as the Kähler calculus; it is rather a $(1,0)-$tensor-valued differential $1-$form). See sections 2.1, 5.2, 5.7 and 8.2.

(f) "The ω^i's, which are components of $d\mathbf{r}$ ($=\omega^1\mathbf{e}_1 + \omega^2\mathbf{e}_2 + \omega^3\mathbf{e}_3$) more general than the dx^i's are not differential $1-$forms because they are not invariant under coordinate transformations but transform like the components of a vector" (they are $1-$forms in the bundle of frames of Euclidean 3-space, but their pullbacks to sections of that bundle are not, since they then transform like the components of a tangent vector field). See sections 5.8 and 6.1.

(g) "An affine space is a vector space where we replace points with vectors" (not until we choose a point in the former to play the role of the zero vector, since affine space does not have a special point that we could call the zero). See section 5.1.

(h) "Skew-symmetric covariant tensors constitute a subalgebra of the tensor algebra" (wrong since they do not constitute a closed set under the tensor product; they rather constitute a quotient algebra). See sections 1.3 and 3.6.

(i) "3-dimensional Euclidean space does not have a connection defined on it" (wrong, it has the trivial one where a section of constant bases ($\mathbf{i}$, $\mathbf{j}$, $\mathbf{k}$) exists, i.e. equal to themselves everywhere; if one had made a different choice for the comparison of vectors, the space would no longer be Euclidean). See sections 6.1 and 7.2.1.

(j) "The canonical connection of a space endowed with a metric is its Levi-Civita connection" (not necessarily true, as the torus has the natural connection where the circles resulting from intersecting it with horizontal and vertical planes are lines of constant direction; this is not its Levi-Civita connection, which is canonical of the metric, but not of the manifold itself). See section 7.4.

(k) Is a metric compatible affine connection actually an affine connection? (no, it is rather a Euclidean connection, which lives in the smaller bundle defined by the Euclidean group). See sections 7.2, 9.1 and 9.2.

(l) Is the so called affine extension of an Euclidean connection by the linear group an actual affine connection? (no, it only looks that way since the frame bundle has been extended; but the number of independent components of the connection remains what it was, not the same as for affine connections). See sections 6.2 and 6.6.

(m) "There is such a thing as the teleparallel equivalent of general relativity" (wrong, the general theory of relativity that had not yet acquired the Levi-Civita connection did not have any connection and, for lack of greater geometric contents, was not equivalent to a theory containing an affine connection; and the general theory of relativity with Levi-Civita connection cannot be equivalent to one with teleparallelism, since the first one does not allow for equality of vectors at a distance and the second one does; as an example of the difference, suffice to say that one cannot say at the same time that some space is and is not flat). See remark (v) and sections 7.2 to 7.4. See also section 8.10 for the impact on the type of conservation law that we may (or may not!) have depending on the type of connection.

(n) Is the exterior derivative of a scalar-valued differential form a covariant derivative? (it certainly has the right covariant transformation properties, though the concept does not respond to what goes by covariant derivative in the tensor calculus). See section 6.3.

(o) Is the conservation law of vector-valued quantities, like energy-momentum, dependent on the behavior of partial derivatives, or of covariant derivatives in the sense of the tensor calculus? (none of the two, as it depends on the behavior of what Cartan and Kähler refer to as exterior derivatives of vector-valued differential forms, more often called exterior covariant derivatives; and, yet, not even this refinement is quite correct as we shall show). See section 8.10.

(p) Does the annulment of the exterior covariant derivative imply a conservation law? (only if the affine/Euclidean curvature is zero, for only then can one add –read integrate– the little pieces of "conserved" vector or tensor at different points; notice that the conservation law of the torsion is given by the first Bianchi identity, which, in addition to the exterior covariant derivative of the torsion, contains the affine/Euclidean curvature). See section 8.10.

(q) Is it possible to speak of a $(1,3)$−tensor curvature? (yes, a $(1,3)$−tensor with geometric meaning that has the same components as the affine curvature exists, but as a concept different from the structural concept of curvature, which is a $(1,1)$−valued differential 2−form). See section 8.6.

(r) Does the (1,3) nature of the tensor in (q) make it a vector-valued differential 3−form? (no; in addition to the reason already mentioned in (h), there is not total skew-symmetry with respect to the three subscripts). Combine contents of sections 5.7 and 8.6.

(s) Is the Riemannian curvature an affine (or, said better, Euclidean) curvature? (not necessarily, as it need not play —and did not play for half a century— any affine or Euclidean role; it was conceived by Riemann for another role). See sections 7.5, 9.8 and 10.1.

(t) The quantities that appear in the equations of structure under the names of affine and Euclidean curvatures, are they tensors of rank four or vector-valued 3−forms? (none of the two; from a section's perspective, they are $(1,1)$−tensor-valued differential 2−forms, and they are Lie algebra valued from the perspective of bundles). See sections 5.7, 8.13 and 9.2.

(u) In Riemannian geometry, can we integrate the energy-momentum "tensor"? (no, as implied in (p); however, at a time when Riemannian geometry did not have a concept of affine curvature but only of metric curvature, one would be entitled to do many of the integrations that general relativists presently do since, without knowing it, the unstated but implicit assumption that the affine curvature of spacetime is zero would justify those computations). See sections 8.10 and 9.8.

(v) In Einstein's general theory of relativity as of 1916, was the Riemannian curvature the one determining the change of a vector when transported along a closed curve? (wrong statement, as the concept of vector transport, which is an affine concept, did not exist in the mathematical market until one year later; the Riemannian "symbols" were then seen as also playing an affine role, but this was not a necessary course of action for general relativity, since one could have extended the old Riemannian theory with an affine connection other than Levi-Civita's). Go to remark (m) and see also sections 10.1 to 10.3.

(w) Is the torsion the antisymmetric part of the connection? (no, not in general; the statement, which actually refers to the components of the torsion, fails to be correct in non-coordinate basis fields). See section 8.2.

(x) "Torsion is related to spin" (incorrect statement, as the torsion has to do with translations and spin has to do with rotations [59]). See section 8.1 and, in more elaborate form, see [88].

(y) "The cyclic property of Riemann's curvature, when considered as the curvature of an Euclidean connection, is simply one more of its properties": (wrong, it is the disguised form that the first Bianchi identity takes when the exterior covariant derivative of the torsion is zero, and, in particular, when the torsion itself is zero). See sections 8.11 and 10.3.

(z) "Using connections, one can develop on flat spaces curves of non-flat spaces. Do closed curves on non-flat spaces develop onto closed curves on flat spaces if the torsion is zero?" (no!, those developments fail to close in general, regardless of whether the torsion is zero or not; the difference that the torsion makes is that, as closed curves are made smaller and smaller, their developments tend towards closed curves much faster when the torsion vanishes than when it does not). See section 7.3.3 and 7.4.1; on the smallness of contours, see 8.9.

1.7 Prerequisite and anticipated mathematical CONCEPTS

For non-expert readers, we proceed to briefly mention or explain what structures are involved in this book. If the brief remarks given here are not enough, nothing is lost; readers will at least have a starting point for their consultations in books on algebra, analysis or geometry.

Loosely speaking, a ring is a set of objects endowed with two operations, called addition and product. The ring is a group under addition and has, therefore, an "inverse under addition", called opposite. Among the rings there are the fields, like the field of reals and the field of complex numbers. A field has more structure than a general ring, namely that each element except the zero (i.e. the neutral element under addition) has an inverse. Functions constitute rings, but not fields, since a function does not need to be zero everywhere (the zero function) in order not to have a multiplicative inverse.

And there are structures which, like a vector space, involve two set of objects, say the vectors on the one hand, and the scalars by which those vectors are multiplied, on the other hand. The other most common structures involving two sets of objects are modules and algebras. Although one first defines modules and then vector spaces as modules of a special kind, general familiarity with vector spaces recommends a different perspective for non-expert readers. A module is like a vector space, except for the following. In a vector space we have the vectors and the scalars, the latter constituting a field. In a module we again have the "vectors", but the scalars constitute only a ring, like a ring of functions. We may, inadvertently or not, abuse the language and speak of vector spaces when we should speak of modules. Finally, an algebra is a module that is also a ring.

In anticipation of material to be considered in chapter two, let us advance that the differential forms of a given grade (say differential 2−forms, i.e. integrands of surface intgrals, which are of grade 2) constitute a module. And the set of all differential forms of all grades (together with those of mixed grade) built upon a given space or upon a module of differential 1−forms constitute an algebra, exterior and Clifford algebras in particular.

Another important algebraic concept in this book, though it will make only brief appearance, is the concept of ideal. Normally one speaks of an ideal within an algebra, but the concept is extendable to any set where there is a multiplication product. A left ideal is a subset I of a set S such that multiplications of elements of I by elements of S on the left always returns elements in I. It must be obvious now what a right ideal is. As simple examples, let us mention that integers do not make an ideal within the real numbers, neither on the right nor on the left, since the product of an integer and a real number is not an integer number. The even numbers constitute right and left ideals within the integers since the product of even and integer is even (counting zero).

Ideals are important in quantum mechanics, not so much, if at all, in standard differential geometry. But, since this author aspires to bring differential geometry and quantum mechanics closer to each other, readers will find an inkling about the proximity to geometry of physically relevant ideals in the last section of chapter 13.

Let us now deal with an issue closer to analysis. We are going to be rather casual with our use of the similar concepts of map, function and functional. For the record and so that anybody who wants may make the text more precise in this regard, we define those three concepts. A *map* is an application from an arbitrary set to another arbitrary set. It can thus be used in all cases where the more specific terms of function and functional also apply. The domain of maps called *functions* also is arbitrary, but the range or codomain must be at least a ring. Finally, a *functional* maps an algebra of functions to another algebra, or to itself in particular.

What follows will be presented formally in later chapters. A little incursion at this point may be helpful. Where the terms affine and Euclidean are concerned, we may be dealing with algebra or with geometry. A Euclidean vector space is a vector space endowed with a dot product. That is algebra.

An affine space is a space of objects called points such that, after arbitrarily taking a point to play the role of zero, we can establish a one-to-one correspondence between its points and the vectors of a vector space, called its associated vector space. If the vector space is Euclidean, the affine space is called Euclidean space (i.e. without "vector" between the words "Euclidean" and "space"). The study of affine and Euclidean space is geometry.

According to the above, a point separates algebra from geometry. In algebra there is a zero; in geometry, there is not any but we introduce it arbitrarily in order to bring algebra into it. There are additional subtle differences. They are related to what we have just said. They have to do with what is it in algebra that algebraists and geometers consider most relevant. Grossly speaking, it is matrices and linear equations for algebraists. And it is groups for geometers. More on this in section 9 of chapter 6.

Rather than viewing affine geometry as a generalization of Euclidean geometry, which it is in some sense, it is better to think in terms of the latter being a restriction of the former. Here is why. Anything affine is related to the comparison of vectors and their frames. The restriction enriches the contents with an invariant, the distance. It is invariant under the Euclidean group, not under the affine group.

The main topic of this book is a different type of generalization, attributable to Cartan. Affine and Euclidean geometries are Klein geometries (think flat geometries for the time being). The "generalized spaces" of the type to which Riemannian geometry belongs are continuous spaces endowed with Euclidean connections, improperly known as metric-compatible affine connections. For the first six decades of its life (since mid nineteenth century), a Riemannian space was viewed as a generalization of Euclidean space in the sense of having a more general expression for distance. But there are generalizations which –rather than the distance or in addition to it— generalize the affine contents. The affine contents of the original Riemannian geometry started to be considered only six decades after its birth.

Clifford algebra is an algebraic concept largely overlooked by geometers, though not by certain algebraists and mathematical physicists. The algebra of introductory college courses cannot be extended to Euclidean spaces of higher dimensions and cannot, therefore, be properly called Euclidean algebra. So called Clifford algebra is Euclidean algebra proper. A brief incursion into it is given at the end of chapter three. There are several books that deal with that algebra. We shall not take sides and mention any since we would also speak of its limitations.

Because of tradition, one studies Euclidean geometries and their Cartan generalizations with tools that pertain to affine geometry, i.e. linear algebra and matrices. But one could study Euclidean geometry with technique specific to it, i.e. Clifford algebra. It could rightly be called Euclidean algebra, as it pertains to anything Euclidean (For experts, the fact that we can put aside part of its structural contents and do with it projective geometry does not negate this statement precisely because "we are putting aside structural contents").

The term Cartan generalization evokes (or certainly should evoke) the study of structure of manifolds from the perspective of their frame bundles, not of sections thereof (also known as frame fields). When you do that, Lie algebras jump to the fore through the Lie groups involved in the definition of those bundles since a concept of differentiation of a Lie group is involved in defining a Lie algebra. We shall thus make the first acquaintance with this concept at the end of chapter 4. Key examples of them will emerge naturally in later chapters.

The following remark on structure is required because of how classical differential geometry is dealt with in books on the geometry of physics. As pointed out in the preface, bundles and Lie algebras are rarely used for introducing the concept of connection in classical differential geometry, unlike in Yang-Mills theory. This "lack of usage" virtually impedes the realization of the following that, as pointed out by (Fields Medalist) Donaldson and Kronheimer in the opening paragraph of chapter 2 of their book [35]

"The key feature of Yang-Mills theory is that it deals with connections on auxiliary bundles, not directly tied to the geometry of the base manifold.'

Thus it is not a matter of whether there are gauge transformations or not in classical differential geometry. There are, since those transformations are simply changes of frame field.

Here is a question that almost nobody raises nowadays: could it be that those auxiliary bundles that are presently considered to be of the essence of Yang-Mills theory are disguised forms of some sophisticated bundles directly related to the tangent bundle? Physicists have not seen a problem with viewing $SU(2)$ in that way. But they cannot see $SU(3)$ in the same way. That is the problem. Apparently you cannot mix an $SU(2)$ from the tangent bundle with $SU(3)$ from an auxiliary bundle. Well, the answer is simple. One may view $SU(3)$ also from a tangent bundle perspective. One needs for that a still better knowledge of differential forms than what you are going to get in this book. So start with this book, since you will first need to know its contents. Later publications by this author will largely answer the opening question of this paragraph.

Part II

TOOLS

Chapter 2

DIFFERENTIAL FORMS

2.1 Acquaintance with differential forms

The most general objects that we have to deal with in this book are tensor-valued differential forms. The simplest ones are the scalar-valued ones, or simply differential forms. The literature on this subject is fraught with dangers in the definitions and their implications. For instance, one may encounter the statement that a differential form is a covariant skew-symmetric tensor. If this is the case, what is a tensor-valued differential form? Is it perhaps a contravariant-tensor-valued covariant skew-symmetric tensor? I shall not pursue this line of argument further to avoid confusing readers.

We here make some acquaintances. A (scalar-valued) differential $1-$form is the differential of a function, which can take forms as simple as dx, or as

$$2dx + 3dy - 7dz, \tag{1.1}$$

and, more generally, as

$$f dx^1 + g dx^2 + h dx^3 + ..., \tag{1.2}$$

where the x^i are arbitrary curvilinear coordinate functions in some generalized space (in differential geometry, the generalized spaces of interest receive the name of differentiable manifolds). The coordinates should not be necessarily Cartesian, and the space not necessarily 3-dimensional. The $f, g, ...$are functions of those coordinates, so they are functions of functions, i.e. composite functions. The components of these differential $1-$forms are said to be $(2, 3, -7)$ and $(f, g, h, ...)$ relative to bases of differential $1-$forms in respective spaces.

We have spoken of coordinates as functions. To be specific, the x of the set of (x, y, z) is a function which assigns to a point of (say) a Euclidean space a number. The symbol x also is used to denote the value taken by that function at that point. We shall use the term coordinates to refer to both, such functions and the values that they take, the appropriate meaning in each instance being obvious from the context.

The reason for sometimes using the qualification scalar-valued is to make clear that, when saying it, we are excluding from our statements differential forms like

$$dx\mathbf{i} + dy\mathbf{j} + dz\mathbf{k}. \tag{1.3}$$

This is a vector-valued differential $1-$form. If we use the symbols x^l ($l = 1, 2, 3$) to refer to (x, y, z), and $\mathbf{a}_l$ to refer to $(\mathbf{i}, \mathbf{j}, \mathbf{k})$, we may rewrite this vector-valued differential $1-$form as

$$dx^l \mathbf{a}_l \quad \left(= \sum_{l=1}^{l=3} dx^l \mathbf{a}_l \right), \tag{1.4}$$

where we have used Einstein summation convention (This convention consists in implying a summation sign in front of expressions with repeated indices, one of the indices being a superscript and the other being a subscript). If, in addition, δ_l^m is the so called Kronecker delta (i.e. $\delta_l^m = 1$, if $l = m$, and $\delta_l^m = 0$, if $l \neq m$), we may rewrite (1.3) as

$$\delta_l^m dx^l \mathbf{a}_m, \tag{1.5}$$

which exhibits δ_l^m as components in terms of the basis $(dx^l \mathbf{a}_m)$ of vector-valued differential $1-$forms. We must emphasize that (1.3) is not a (1,1) tensor, since differential forms are not tensors.

A function which, like

$$F(x, y, ...), \tag{1.6}$$

exhibits no differentials is a scalar-valued differential $0-$form. Similarly,

$$x\mathbf{i} + y\mathbf{j} + z\mathbf{k} \tag{1.7}$$

is a vector-valued $0-$form, equivalently a vector field. Readers who have problems seeing this as a vector field because all the vectors (1.7) "start" at the same point need only think of it as the negative of the vector field $-(x\mathbf{i} + y\mathbf{j} + z\mathbf{k})$, with may be viewed as extending from all over the space and ending at the origin.

We shall later learn why the $dxdy$ under the integral sign of surface integrals should be written as

$$dx \wedge dy, \tag{1.8}$$

where the symbol $\wedge$ denotes exterior product (soon to be defined), which is the same reason why the $dxdydz$ that we find under the integral sign of volume integrals should be written as

$$dx \wedge dy \wedge dz. \tag{1.9}$$

The last two displayed expressions are referred to as scalar-valued differential $2-$form and $3-$form respectively. Another example of differential $2-$form is

$$ady \wedge dz + bdz \wedge dx + cdx \wedge dy. \tag{1.10}$$

The likes of

$$a\mathbf{i} \wedge \mathbf{j} \tag{1.11}$$

are called bivectors. They are to vectors what differential 2−forms are to differential 1−forms. Something like

$$a(\mathbf{i} \wedge \mathbf{j})dx \qquad (1.12)$$

would be called a bivector-valued differential 1−form, which, by an abuse of language, is (often) called 2−tensor-valued differential 1−form. And one could also have vector-valued differential 2−forms, etc.

Except for indication to the contrary, the term differential form is reserved here as in most of the literature for the skew-symmetric ones with respect to any pair of indices within a given monomial (single terms, as opposed to sums thereof). The terms in the sums will be all of the same grade in this book. We need to distinguish such differential forms from those which, like the metric, $g_{ij}dx^i dx^j$, are quadratic symmetric differentials.

Of great importance is the fact that the tensor valuedness comes in two types, contravariant and covariant. The usual one is the contravariant one. The covariant one is usually induced by and accompanies the former. Suffice to know at this point that those vectors that even the most inexperienced readers are familiar with are called contravariant. A covariant vector is a linear function of contravariant vectors. The next paragraph can be skipped by readers bothered by statements containing terms with which they are not yet familiar.

Tensor-valuedness (1,1) means once contravariant and once covariant. As already mentioned in the previous chapter, the affine curvature is a $(1, 1)$−valued differential 2−form. One can associate with the affine curvature a $(1, 3)$−valued differential 0−form, or simply (1,3) tensor. It is not a differential 3−form, among other reasons because it is not skew-symmetric with respect to its three subscripts. For lack of a better name, we shall call it curvature tensor. It plays a relatively minor role, unlike affine curvature and torsion, which belong to the core of geometry. Warning: the notation (p, q) has to be viewed in each author's context, since it may refer in some of them to a differential q−form that has valuedness p. Of course, two indices cannot play the role that the three indices in (s, t)−valued r−forms play.

2.2 Differentiable manifolds, pedestrianly

There is a great virtue in calculus of differential forms. It is appropriate for differentiation in general spaces known as differentiable manifolds, which, in general, lack a metric and a connection or rule to compare vectors at different points. We must, however, have a concept of continuity in the space in question, which eliminates automatically discrete sets of points. The set must be such that regions of the same can be represented unequivocally by open sets of n-tuples of real numbers (it could also be complex numbers, but not in this book). The openness of the set has to do with the behavior at the borders of the region.

We said "unequivocally" in order to deal with situations like the following. Consider the Euclidean plane punctured at one point, which we take as the origin of polar coordinates. We can put all its points in a one-to-one correspondence with all the pairs of numbers satisfying the following properties. One element of the pair, ρ, takes all positive real values. The other element of the pair, ϕ, takes all non-negative real values less than 2π. But the full plane is not covered by a system of polar coordinates since the origin of the system is covered by specifying just $\rho = 0$. We would need another system of coordinates that would cover the origin of the first in order to achieve that every point of the plane will be covered by at least one coordinate system. It is for this reason that one speaks of systems of local coordinates.

The term region will be used in this book to denote open subsets with the same dimension as the manifold. The reason to often consider a region and not the whole manifold is that one may need to avoid some special point, like the origin of the polar coordinates.

A coordinate system then is a map from a region of the differentiable manifold to the set of n-tuples. If we choose a point, the map determines an n-tuple of coordinates of the point. Hence, we have the coordinate map and the coordinates as components, or values taken by the set of coordinate functions that constitute the map.

When two coordinate systems overlap, it is required that the functions expressing the coordinate transformation are continuous and have continuous derivatives up to some order, appropriate to achieve some specific purpose.

A pair of region and coordinate assignment is called a chart. The region in the chart of Cartesian coordinates in the plane is the whole plane. And the region in the chart for the system of polar coordinates is the plane punctured at the chosen origin of the system.

The inverse of a coordinate function will certainly take us from the coordinates to the points on the manifold represented by those coordinates. Let us call them inverse coordinate functions. Let m and n be the respective dimensions of manifolds M and N, i.e the number of components needed to represent their points in the manner prescribed. In terms of coordinates, a function from a manifold M to another manifold N is a composite of three functions in succession. The first one takes one from m-tuples to the manifold M. The second one is the given function from M to N and the third one takes us from N to n-tuples. We say that the function from M to N is differentiable if the forgoing composite function is differentiable. A change of coordinates on a (region of) a manifold M is the particular case where $M = N$.

The identity function on the real line assigns a coordinate to each of its points. In other words, a point in the real line is its own coordinate. Thus the expression in terms of coordinates of a real function on M (i.e. a real valued function or simply real function) reduces to the composition of two functions, namely from the coordinates to the point on the manifold and from the latter to $\mathbb{R}^1$.

Let f denote the set of the f^i that take the values x^i. The inverse function f^{-1} takes us from the coordinates to the manifold. One defines the differential dg of a real function g on the manifold as (recall Einstein's summation convention)

$$dg = \frac{\partial g}{\partial x^i} dx^i,\qquad(2.1)$$

where

$$\frac{\partial g}{\partial x^i} = \frac{\partial(g \circ f^{-1})}{\partial x^i}.\qquad(2.2)$$

It will be shown in the next section that dg is independent of the coordinate system chosen for the definition. It is convenient to introduce the notation $g(x^i)$ for $g \circ f^{-1}$. The expression for dg given above then becomes the dg of the calculus in $\mathbb{R}^n$. The notation $g(x^i)$ reflects the fact that one normally gives points P of a manifold by giving their coordinates in some region. With this notation, equations look as relations on $\mathbb{R}^n$ rather than as relations on the manifold.

2.3   Differential 1−forms

The integrands of the integrals on a differential manifold (on its curves, surfaces, etc.; all of them oriented) constitute a structure called an algebra. To be more specific, it is a graded algebra. As a first approach to the concept of grade in a graded algebra, let us say the following. The integrands in integrals on curves, surfaces, volumes ... are said to be of grade $1, 2, 3, ...$ The real functions are said to be of grade zero (We shall justify the last statement when dealing with Stokes theorem in chapter 4).

The subspaces constituted by all the elements of a graded algebra that have the same grade constitute modules (generalization of the concept of vector space). For the present expository purposes, let us say that the algebra may be considered as built from the module of elements of grade one by some kind of product (exterior, tensor, Clifford). In algebras of differential forms, the module of grade one is the module of differential 1−forms. In this book, the product of scalar-valued differential forms is mainly the exterior product. In Kähler (barely considered in this book), it will be the Clifford product.

The meaning of the elements in the algebra will be induced by the meaning of the differential 1−forms. These are linear combinations of the differentials of the coordinates. So, what is the meaning of the differentials of the coordinates? Non-expert readers are invited to check this in the mathematical books of their libraries. The comparison of definitions is a very interesting exercise. Authors sometimes state in the introduction of their books that, say, a differential 1−form is a function of oriented curves (whose evaluation is their integration), and in the main text it is a linear function of vector fields. This is unsatisfactory since these are different concepts.

One can speak of a exterior (i.e. Cartan's) calculus for both concepts of differential forms. In the opinion of this author, it is better to think of them as integrands if one wants to understand Cartan's approach to differential geometry well. This is no longer a matter of opinion in the Kähler calculus, where components can have two series of subscripts (and one of superscripts). The two types of subscripts correspond to two different types of objects and associated

differentiations. One type fits naturally (in the case of just one subscript of each
type) the functions of curves and the other one the linear functions of vector
fields. In Rudin's book [65] in analysis, (scalar-valued) differential forms are
introduced as integrands in the main text, rather than just paying lip service to
them in the introduction of the book and then being ignored.

Regardless of how differential forms are defined, it is the case that

$$\int_{x=a}^{x=b} dx = b - a,$$ (3.1)

or

$$\int_c dx = x_B - x_A,$$ (3.2)

where c is an oriented curve with A and B as origin and end points. That
is for us the definition of dx. It is simply a consequence of the definition of a
differential $1-$form as a map (or function) that assigns to each open curve c on
a manifold the integral

$$\int_c a_i(x)dx^i.$$ (3.3)

The map may be expressed in coordinate independent form as a formal linear combination of differentials of smooth functions with smooth functions as
coefficients

$$f_1 dg^1 + f_2 dg^2 + \ldots + f_m dg^m.$$ (3.4)

Here m is any positive integer. It can be equal to, greater than, or smaller than
n. At this point, the position of the indices in (3.4) is not of great importance.

When one develops the dg^i as linear combinations of differentials of coordinates, one obtains a more practical expression. Since

$$dg^i = \frac{dg^i}{dx^j}dx^j,$$ (3.5)

$(i = 1, ..., m, \, j = 1, ..., n)$, (3.4) becomes

$$\alpha = f_i \frac{\partial g^i}{\partial x^j}dx^j = a_j(x)dx^j,$$ (3.6)

where

$$a_j(x) = f_i \frac{\partial g^i}{dx^j}.$$ (3.7)

Two arbitrary differential $1-$forms α and β,

$$\alpha \equiv a_i(x)dx^i, \qquad \beta \equiv b_i(x)dx^i,$$ (3.8)

are said to be equal on some subset U of the manifold (curve, surface, ..., region)
if

$$a_i = b_i \qquad i = 1, \ldots, n$$ (3.9)

at all points of the subset. We define an associative sum of differential 1—forms by

$$\alpha + \beta + \gamma... = (a_i + b_i + c_i...)dx^i, \tag{3.10}$$

and the product by a function f as

$$f\alpha = f a_i dx^i. \tag{3.11}$$

If $(x^1 ... x^n)$ constitutes a set of n independent variables, the set $(dx^1 ... dx^n)$ constitutes a basis for differential 1—forms. A change of coordinates induces a change of such bases at the intersection of the corresponding sets U_1 and U_2 of $\boldsymbol{M}$. Let us refer the same differential form α to two different coordinate systems. We have

$$\alpha \equiv a_j dx^j = a'_i dx'^i. \tag{3.12}$$

Since

$$dx^j = (\partial x^j / \partial x'^i)_{x'} dx'^i, \tag{3.13}$$

it follows by substitution of (3.13) in (3.12) and comparison of components that

$$a'_j = (\partial x^i / \partial x'^j)_{x'} a_i. \tag{3.14}$$

We proceed to provide important examples of differential 1—forms. The most pervasive one other than the differential of the coordinates is the differential of scalar-valued functions,

$$df = (\partial f / \partial x^i) dx^i = f_{,i} dx^i. \tag{3.15}$$

Notice the new notation for partial derivatives in (3.15). Several important 1—forms in physics are specific cases of this equation. But which are the manifolds in such cases? There are many different ones in physics. One of them is, for instance, the manifold of states of a thermodynamical system. Typical coordinate patches on this manifold are constituted by the energy E, volume V, and numbers of molecules $N_1, N_2, ...$ of the different substances that constitute, say, a mixture of gases (if such is the system). A function on the manifold is the entropy S of the system, $S = S(U, V, N_1, N_2, ...)$. The differential of the entropy function is a 1—form with coefficients that contain the temperature T, pressure P, and chemical potentials μ_1, μ_2, namely:

$$dS = T^{-1}dU - PT^{-1}dV + \mu_1 T^{-1}dN_1 + \mu_2 T^{-1}dN_2 ... \tag{3.16}$$

The most familiar manifold, however, is the space-time manifold, which has 4 dimensions. It is described by a time coordinate and three spatial coordinates. In addition to being a differentiable manifold, space-time has much more additional structure. It is at the level of this additional structure, which will be the subject of later chapters, that the space-time of Newtonian physics differs from the space-time of special relativity and from the space-times of Einstein's theory of gravity (also called general relativity). The contents of this chapter applies equally well to all these different space-time manifolds and we shall proceed to discuss some specific physical examples of 1—forms in space-time.

The phase of a wave is usually given in Cartesian coordinates, $\phi(t, x, y, z)$. Its differential is

$$d\phi = (\partial\phi/\partial t)dt + (\partial\phi/\partial x)dx + (\partial\phi/\partial y)dy + (\partial\phi/\partial z)dz$$
$$= \omega dt - k_x dx - k_y dy - k_z dz, \tag{3.17}$$

where $\omega, -k_x, -k_y$, and $-k_z$ are nothing but those partial derivatives up to sign. As we shall eventually see, there is no room for Cartesian coordinates in the space-times of Einstein's gravity theory. However, the concept of phase of a wave is of such a nature that it does not require the space-time to be flat, the wave to be a plane wave and the coordinates to be Cartesian. We always have

$$d\phi = (\partial\phi/\partial x^0)dx^0 + (\partial\phi/\partial x^i)dx^i, \tag{3.18}$$

with summation over i's. Generalized frequency and wave-numbers can be defined as $k_0 \equiv \partial\phi/\partial x^0$ and $k_i \equiv -\partial\phi/\partial x^i$ so that

$$d\phi = k_0 dx^0 - k_1 dx^1 - k_2 dx^2 - k_3 dx^3, \tag{3.19}$$

in terms of arbitrary coordinates of whatever space-time differentiable manifold is concerned.

Another important physical example is given by the action S_0 of a free particle when considered as a function of the space-time coordinates. The action is the integrand in the variational principle $\delta \int dS = 0$ that gives rise to the equations of motion of a physical system in classical physics (We use the subscript zero when the system is just a free particle). In this case, we have:

$$-dS_0 = -(\partial S_0/\partial t)dt - (\partial S_0/\partial x)dx - (\partial S_0/\partial y)dy - (\partial S_0/\partial z)/dz$$
$$= Edt - p_x dx - p_y dy - p_z dz, \tag{3.20}$$

where E is the energy and the p_i's are the components of the momentum of the particle. As in the case of the phase of a wave, we could generalize this concept to arbitrary coordinates in the space-times of Newtonian physics, special relativity, and gravitational theory.

Finally, if $-S_1$ is the interaction part of the action of a unit-charge particle in an electromagnetic field, we have:

$$-dS_1 = -(\partial S_1/\partial t)dt - (\partial S_1/\partial x)dx - (\partial S_1/\partial y)dy - (\partial S_1/\partial z)dz$$
$$= \phi dt - A_x dx - A_y dy - A_z dz, \tag{3.21}$$

where ϕ, A_x, A_x, A_y, and A_z constitute the so called scalar and vector potentials, actually components of the four-potential. Said better, they are the components of a differential form, and we could express it in terms of arbitrary coordinates.

It is worth reflecting on the change of perspective that the language of forms represents relative to other languages. Take for instance, Eq. (3.15). Readers are probably familiar with the idea of considering the set of partial derivatives $(\partial f/\partial x^i)$ as the components of the gradient vector field, and the set (dx^i) as components of $dx^i \mathbf{a}_i$. It is, however, much simpler to consider $\partial f / \partial x^i$ as the components of the form df, with the dx^i's constituting a basis of 1−forms. Not to mention the fact that the gradient requires a metric structure, and df does not. From this new perspective, df can be referred to as the gradient form.

Similarly, the differential of the phase of a wave could be referred to as the frequency-wave-number form. The negative differential of the actions S_0, S_1, and $S_0 + S_1$ could be referred to as, respectively, the free particle energy-momentum form, the four-potential form, and the generalized energy-momentum form for a particle in an electromagnetic field. All these concepts are meaningful *ab initio* in any coordinate system. Finally, readers should notice that $(dx^2 + dy^2 + dz^2)^{1/2}$, i.e. the line element, is neither a differential 1−form nor a differential 2−form (these are considered in the next section).

We used equations (3.12) to effect a change of components under a change of basis of 1−forms. We shall also be interested in bases of a more general type. Two independent differentials dx^1 and dx^2 constitute a basis of 1−forms in the plane, where they constitute a coordinate (also called holonomic) basis. It is obvious that, say,

$$\omega^1 = dx^1, \qquad \omega^2 = x^1 dx^2, \tag{3.22}$$

also constitute a basis of differential 1−forms, since the transformation is invertible. However, there is no coordinate system (y^1, y^2) such that $\omega^1 = dy^1$ and $\omega^2 = dy^2$, which is the reason why it is called an anholonomic or non-holonomic basis.

We shall represent changes of bases as

$$\omega'^j = A_i^{j'}(x)\omega^i, \qquad \omega^j = A_{i'}^j(x)\omega'^i, \tag{3.23}$$

where x represents as many as all the coordinates in the system, four of them in spacetime. Of course, if $(\omega'^j) = (dx'^j)$ and $(\omega^i) = (dx^i)$, then $A_i^{j'} = \partial x'^j / \partial x^i$ and $A_{i'}^j = \partial x^j / \partial x'^i$.

We may wish to express the differential form df in terms of the ω^i's. We shall use the symbols $f_{,i}$ and $f_{/i}$ to denote the coefficients of df in the respective bases dx^i and ω^i, i.e.,

$$df = f_{,i} dx^i = f_{/j} \omega^j. \tag{3.24}$$

We want to have the explicit relation between $f_{/i}$ and $f_{,i}$. Equations (3.23) apply to any pair of bases of 1−forms and, in particular, they apply to the pair of bases $\{dx^i\}$ and $\{\omega^i\}$. Thus, defining $\{\omega^{i'}\} \equiv \{\omega^i\}$, we have

$$\omega^j \equiv \omega^{j'} = A_i^{j'} dx^i, \tag{3.25}$$

$$dx^i = A_{j'}^i \omega^{j'} = A_{j'}^i \omega^j. \tag{3.26}$$

Substitution of (3.26) in (3.24) yields:

$$f_{,i} A_{j'}^i \omega^j = f_{/j} \omega^j. \tag{3.27}$$

Hence, since (ω^j) constitutes a basis, we have,

$$f_{/j} = A_{j'}^i f_{,i}, \tag{3.28}$$

which is the sought relation.

Expert readers will have noticed that what we have called and shall be calling differential forms are called cochains in some fields of mathematics. Since the terms exterior calculus and differential forms almost always come together, and since said calculus certainly applies to cochains even more directly than to skew-symmetric multilinear functions of vectors (think of what the generalized Stokes theorem states), it makes sense to use the term differential form to refer to what goes by the name of cochains, at least when doing calculus.

2.4 Differential $r-$forms

We proceed to introduce differential forms of "higher grade". You have met them in the vector calculus, where they are dealt with in an unfortunate way. Indeed, consider the integral

$$\iint p^2 dp dq \tag{4.1}$$

and perform the change of variables $t = (p+q)/2$, $x = (p-q)/2$ or, equivalently, $p = t + x$, $q = t - x$. We readily get that $dpdq = (dt + dx)(dt - dx) = dt^2 - dx^2$. So the expression $p^2 dpdq$ is incontrovertibly,

$$(t + x)^2(dt^2 - dx^2). \tag{4.2}$$

However, the expression

$$\iint (t + x)^2(dt^2 - dx^2) \tag{4.3}$$

does not make sense and, in any case, it is not the way in which we change variables in a multiple integral. One may wonder what the problem is with what we have done. The problem is the inadequate notation $dpdq$, as we now explain.

When we write $dpdq$ under the integral sign, we really mean a "new" type of product, whose properties we proceed to discuss. We can give two orientations to surface elements by giving two different orientations to the boundary. These two oriented elements of surface integral can be represented respectively by $dp \wedge dq$ (read "dp exterior product dq") and $dq \wedge dp$ (read "dq exterior product dp", and also "dq wedge dp"). The skew-symmetry reminds us of the cross product, which actually is the composition of the exterior product, which we are about to define, with the assignment of a set of quantities with one index to a set of quantities with two indices.

The interpretation of the exterior product as an element of surface integral requires that, as part of the definition of that product, we must have

$$dp \wedge dq = -dq \wedge dp. \tag{4.4}$$

Consequently

$$dp \wedge dp = dq \wedge dq = 0, \tag{4.5}$$

which corresponds to the lack of integrands proportional to dp^2 and dq^2.

Suppose now that the two elements of line integral constituting the surface element were given by the $1-$forms $\alpha = a_1 dp + a_2 dq$ and $\beta = b_1 dp + b_2 dq$. The corresponding element of surface integral would be written as

$$\alpha \wedge \beta = (a_1 dp + a_2 dq) \wedge (b_1 dp + b_2 dq). \tag{4.6a}$$

We should like to reduce this to the basic object $dp \wedge dq$. One readily infers that linearity with respect to both factors should be assumed (as suggested for instance by the case $(a_2 = b_1 = 0)$). Therefore:

$$\alpha \wedge \beta = a_1 b_2 dp \wedge dq + a_2 b_1 dq \wedge dp = (a_1 b_2 - a_2 b_1) dp \wedge dq. \tag{4.6b}$$

If α and β are the differential forms dt and dx in terms of dp and dq, then $(a_1 b_2 - a_2 b_1)$ is simply the Jacobian

$$\frac{\partial(t, x)}{\partial(p, q)}. \tag{4.7}$$

Thus:
$$dt \wedge dx = \frac{\partial(t, x)}{\partial(p, q)} dp \wedge dq. \tag{4.8a}$$

If we interchange the role of the variables in this expression, we get

$$dp \wedge dq = \frac{\partial(p, q)}{\partial(t, x)} dt \wedge dx. \tag{4.8b}$$

Substitution of (4.8b) in (4.8a) yields the known fact that these two Jacobians are the inverse of each other. This is precisely the way in which we change variables in a double integral. In general and not only in two dimensions, we call differential $2-$forms the skew-symmetric products of two differential $1-$forms, and sums thereof.

Let us return to (4.1). Assume that instead of $\iint p^2 dp dq$ we considered $\iint p^2 dp \wedge dq$. Then

$$dp \wedge dq = (dt + dx) \wedge (dt - dx) = -dt \wedge dx + dx \wedge dt = -2dt \wedge dx. \tag{4.9}$$

The surface integral would become

$$-2 \iint (t + x)^2 dt \wedge dx. \tag{4.10}$$

The coefficient -2 is the (constant) Jacobian of the transformation. The minus sign expresses the fact that the orientation of the "sets of axes" (t, x) and (p, q) is the reverse of each other. We conclude that we obtained the inappropriate expression $dt^2 - dx^2$ in (4.3) because inappropriate notation led us to change variables in $dp dq$ rather than in $dp \wedge dq$.

Let us forget differential forms for the moment and recall the cross product, $\mathbf{a} \times \mathbf{b}$. It is designed as a vector that has the same direction as one of the two oriented normals to the parallelogram with sides $\mathbf{a}$ and $\mathbf{b}$. The magnitude of $\mathbf{a} \times \mathbf{b}$ is the area of the parallelogram. We could think of $\mathbf{a} \wedge \mathbf{b}$ as representing the parallelogram itself. Hence, in a fist step we assign to the pair $(\mathbf{a}, \mathbf{b})$ the parallelogram $\mathbf{a} \wedge \mathbf{b}$ and, in a second step, the normal to the parallelogram $\mathbf{a} \times \mathbf{b}$. With differential forms, one ignores taking the normal and stops at the wedge product.

We shall reach the concept of differential $r-$form by means of the so-called exterior product $\wedge$ of a number r of differential 1$-$forms. This is defined as an algebraic operation that satisfies the properties

(a) it is linear with respect to each factor and

(b) it is anticommutative (i.e. skew-symmetric) with respect to the interchange of any two factors.

The result of the exterior product of a number r of differential 1$-$forms $(\omega^1, \ldots, \omega^r)$ is called a simple $r-$ form and is denoted as $\omega^1 \wedge \ldots \wedge \omega^r$. Any linear combination of $r-$forms also is called an $r-$ form, which in general, need not be simple. For instance, if (t, x, y, z) are four independent coordinates, the 2$-$form $dt \wedge dx + dy \wedge dz$ is not simple. If (x, y, z) are three independent coordinates, the form $dx \wedge dy + dx \wedge dz + dy \wedge dz$ can be written as a simple 2$-$form in many different ways.

An exchange of two 1$-$form factors in a product changes the sign of the product. To see this move the last of the two factors to be exchanged to just behind the other one, and then this one to where the last one previously was. If s is the number of places involved in the first move, then the number involved in the second move is $s + 1$ for a total of $2s + 1$, which is an odd number and thus implies a change of sign.

Because of skew-symmetry, a simple $r-$form becomes zero if two of the factors in the product coincide. But the interchange of a pair of indices in a non-simple $r-$form does not necessarily change its sign. So, the interchange of the superscripts 3 and 4 in $\omega^1 \wedge \omega^2 + \omega^3 \wedge \omega^4$ yields $\omega^1 \wedge \omega^2 + \omega^4 \wedge \omega^3$ which is equal to $\omega^1 \wedge \omega^2 - \omega^3 \wedge \omega^4$.

Consider all possible forms that we can construct by exterior product of the differentials (dx^1, dx^2, dx^3) of three independent coordinates (x^1, x^2, x^3). There are only six non-null products of these three one-forms. They can be written as

$$dx^{i_1} \wedge dx^{i_2} \wedge dx^{i_3}, \tag{4.11}$$

where (i_1, i_2, i_3) is $(1, 2, 3)$ or a permutation thereof. Obviously

$$dx^{i_1} \wedge dx^{i_2} \wedge dx^{i_3} = \varepsilon_{i_1 i_2 i_3} dx^1 \wedge dx^2 \wedge dx^3, \tag{4.12}$$

where $\varepsilon_{i_1 i_2 i_3}$ is the sign of the permutation (i_1, i_2, i_3). It follows that all 3$-$forms in 3-dimensional manifolds are a scalar function times $dx^1 \wedge dx^2 \wedge dx^3$. This applies to any coordinate system. To be specific, let (x, y, z) and (r, θ, ϕ) respectively denote the Cartesian and spherical coordinates of our usual 3-space. It follows from the above considerations that the 3−form $dx \wedge dy \wedge dz$ is a multiple of the 3−form $dr \wedge d\theta \wedge d\phi$. Thus

$$dx \wedge dy \wedge dz = f(r, \theta, \phi)dr \wedge d\theta \wedge d\phi. \tag{4.13}$$

We shall soon find the function $f(r, \theta, \phi)$.

Let us now comment on the postulate of linearity. It implies that, if one of the factors is a linear combination of one-forms, the given simple r−form can be written as a sum of other simple r−forms. As an example, if $\omega^1 = \alpha + \beta$ then

$$\omega^1 \wedge \ldots \wedge \omega^r = \alpha \wedge \omega^2 \wedge \ldots \wedge \omega^r + \beta \wedge \omega^2 \wedge \ldots \wedge \omega^r. \tag{4.14}$$

We are now ready to relate $dr \wedge d\theta \wedge d\phi$ to $dx \wedge dy \wedge dz$.

The Cartesian coordinates are given in terms of the spherical coordinates by the relations

$$x = r \sin \theta \cos \phi, \qquad y = r \sin \theta \sin \phi, \qquad z = r \cos \theta. \tag{4.15}$$

Differentiation of these equations and substitution in $dx \wedge dy \wedge dz$ yields

$$\begin{aligned} dx \wedge dy \wedge dz = {} & (\sin \theta \cos \phi dr + r \cos \theta \cos \phi d\theta - r \sin \theta \sin \phi d\phi) \\ & \wedge (\sin \theta \sin \phi dr + r \cos \theta \sin \phi d\theta + r \sin \theta \cos \phi d\phi) \\ & \wedge (\cos \theta dr - r \sin d\theta). \end{aligned} \tag{4.16}$$

If we denote the eight different one-forms in the right hand side of this equation by the symbols A_1 to A_8 in the same order in which they have appeared, we get

$$\begin{aligned} dx \wedge dy \wedge dz = {} & A_1 \wedge A_6 \wedge A_8 + A_2 \wedge A_6 \wedge A_7 \\ & + A_3 \wedge A_4 \wedge A_8 + A_3 \wedge A_5 \wedge A_7 \\ = {} & -r^2 \sin^3 \theta \cos^2 \phi \, dr \wedge d\phi \wedge d\theta + r^2 \sin \theta \cos^2 \theta \cos^2 \phi \, d\theta \wedge d\phi \wedge dr \\ & + r^2 \sin^3 \theta \sin^2 \phi \, d\phi \wedge dr \wedge d\theta - r^2 \cos^2 \theta \sin \theta \sin^2 \phi \, d\phi \wedge d\theta \wedge dr \\ = {} & r^2 \sin \theta \, dr \wedge d\theta \wedge d\phi. \end{aligned} \tag{4.17}$$

An efficient way of getting the terms involved is to start by considering that there is no $d\phi$ factor in the last line of (4.16). The preceding exercise embodies several manipulations with differential r−forms, and we recommend using it for practice.

Given n linearly independent 1−forms, there are as many linearly independent r−forms as there are r-combinations of n elements, i.e. as there are subsets with r objects taken from a given set of n different objects. If the space of the 1−forms is n dimensional, the r−forms constitute a space of dimension $n!/r!(n-r)!$. In particular, the space of the n−forms is of dimension 1. All the forms of grade $r > n$ are null.

As an example, consider the space of 2−forms in a 3-dimensional manifold. A basis in this space is given by $(dx^1 \wedge dx^2, dx^2 \wedge dx^3, dx^3 \wedge dx^1)$. However, we might as well have taken $(\omega^1 \wedge \omega^2, \omega^2 \wedge \omega^3, \omega^3 \wedge \omega^1)$ as a basis of 2−forms if $(\omega^1, \omega^2, \omega^3)$ represents a basis of 1−forms.

In Einstein's theory of gravity, one often comes across changes of bases of r -forms, especially $2-$forms. A couple of examples will help us to further become familiar with these ideas. A useful basis of $1-$forms in central body problems in Einstein's theory of gravity is given by:

$$\omega^0 = e^{\lambda(r)}dt, \quad \omega^1 = e^{\mu(r)}dr, \quad \omega^2 = rd\theta, \quad \omega^3 = r\sin\theta d\phi, \qquad (4.18a)$$

where ν and μ are functions of the variable r. Let us denote $(dt, dr, d\theta, d\phi)$ as (dx^0, dx^1, dx^2, dx^3). We are interested in writing the $2-$forms $dx^i \wedge dx^j$ as linear combinations of the $2-$forms $\omega^k \wedge \omega^i$. In this way the expression in terms of $dx^i \wedge dx^j$ can be translated into an expression in terms of $\omega^k \wedge \omega^i$. From the given equations, we obtain:

$$dt = e^{-\lambda}\omega^0, \quad dr = e^{-\mu}\omega^1, \quad d\theta = r^{-1}\omega^2, \quad d\phi = r^{-1}(\sin\theta)^{-1}\omega^3. \qquad (4.18b)$$

Thus, for example,

$$dr \wedge dt = e^{-(\lambda+\mu)}\omega^0 \wedge \omega^1, \qquad dr \wedge d\theta = r^{-1}e^{-\mu}\omega^1 \wedge \omega^2 \qquad (4.19)$$
$$dr \wedge d\phi = r^{-1}(\sin\theta)^{-1}e^{-\mu}\omega^1 \wedge \omega^3, \qquad d\theta \wedge d\phi = r^{-2}(\sin\theta)^{-1}\omega^2 \wedge \omega^3.$$

Suppose now that we are given the following, slightly more complicated problem, where the forms (ω^i) are not simply proportional to the forms (dx^i) but are rather given by

$$\omega^0 = dt - ar^2 d\phi, \quad \omega^1 = e^{-\alpha^2 r^2/2}dr, \quad \omega^2 = e^{-\alpha^2 r^2/2}dz, \quad \omega^3 = rd\phi, \qquad (4.20a)$$

where α and a are constants. We shall consider these forms again in Chapter 10. We get

$$dt = \omega^0 + ar\omega^3, \quad dr = e^{\alpha^2 r^2/2}\omega^1, \quad dz = e^{\alpha^2 r^2/2}\omega^2, \quad d\phi = r^{-1}\omega^3. \qquad (4.20b)$$

It then follows that

$$\begin{aligned}
dt \wedge dr &= e^{\alpha^2 r^2/2}\omega^0 \wedge \omega^1 - are^{\alpha^2 r^2/2}\omega^1 \wedge \omega^3, \\
dt \wedge dz &= e^{\alpha^2 r^2/2}\omega^0 \wedge \omega^2 - are^{\alpha^2 r^2/2}\omega^2 \wedge \omega^3, \\
dt \wedge d\phi &= r^{-1}\omega^0 \wedge \omega^3, \qquad dr \wedge dz = e^{\alpha^2 r^2}\omega^1 \wedge \omega^2, \\
dr \wedge d\phi &= r^{-1}e^{\alpha^2 r^2/2}\omega^1 \wedge \omega^3, \qquad dz \wedge d\phi = r^{-1}e^{\alpha^2 r^2/2}\omega^2 \wedge \omega^3.
\end{aligned} \qquad (4.21)$$

As can be seen, there is nothing complicated about exterior products of $1-$forms. Let us now consider the case when the grades of the factors are greater than one.

2.5   Exterior products of differential forms

The concept of exterior products of $1-$forms can be extended to $r-$forms, the "grades" of the factors not being necessarily the same, much less one. Consider the product of two forms, ρ and σ, of respective grades r and s,

$$\rho = \omega^{i_1} \wedge \ldots \wedge \omega^{i_r}, \qquad \sigma = \omega^{j_1} \wedge \ldots \wedge \omega^{j_s}. \qquad (5.1)$$

Their exterior product is defined as:

$$\rho \wedge \sigma = \omega^{i_1} \wedge \ldots \wedge \omega^{i_r} \wedge \omega^{j_1} \wedge \ldots \wedge \omega^{j_s}. \tag{5.2}$$

Notice that we did not need to consider a coefficient in ρ (and in σ), say $\rho = R\omega^{i_1} \wedge \ldots \wedge \omega^{i_r}$, since the coefficient can always be absorbed in one of the factors, e.g. $\rho = (R\omega^{i_1}) \wedge \ldots \wedge \omega^{i_r}$. The exterior product is linear with respect to each factor, and is associative with respect to the multiplication by a scalar f (0-grade form), i.e.

$$(f\rho) \wedge \sigma = \rho \wedge (f\sigma) = f(\rho \wedge \sigma). \tag{5.3}$$

The following property is very useful:

$$(\rho \wedge \sigma) = (-1)^{rs}\omega^{j_1} \wedge \ldots \wedge \omega^{j_s} \wedge \omega^{i_1} \wedge \ldots \wedge \omega^{i_r} = (-1)^{rs}\sigma \wedge \rho. \tag{5.4}$$

This can be seen as follows. In (5.2), we move ω^{j_1} to the front by exchanging places with ω^{i_r}, then with $\omega^{i_{r-1}}$ and so on until it exchanges places with ω^{i_1}. In doing so, there have been r changes of sign. We repeat the process with all the ω^j's. The number of changes of sign is rs, and the above expression follows.

We further demand the distributive property,

$$(\rho_1 + \rho_2) \wedge \sigma = \rho_1 \wedge \sigma + \rho_2 \wedge \sigma, \tag{5.5}$$

where ρ_1 and ρ_2 are usually of the same grade in the exterior calculus, but need not be so. Similarly

$$\rho \wedge (\sigma_1 + \sigma_2) = \rho \wedge \sigma_1 + \rho \wedge \sigma_2. \tag{5.6}$$

Exterior products of differential forms will be used extensively in the exterior calculus, also called Cartan calculus, which will be presented in chapter four.

2.6 Change of basis of differential forms

Differential forms are invariants: they do not transform, their components do. A change of basis of differential 1−forms, like from the basis (dp, dq) to (dx, dy) in 2-D, induces the change of basis in the one-dimensional module of differential 2−forms, from $dp \wedge dq$ to $dx \wedge dy$.

Consider 3−forms on a 3-D manifold. They also constitute a 1-dimensional space. Thus $dx^1 \wedge dx^2 \wedge dx^3$ constitutes a basis. If the dx^i ($i = 1, 2, 3$) are the differentials of a set of three independent coordinates, a change of coordinates from x's to y's induces a change to the basis $dy^1 \wedge dy^2 \wedge dy^3$. If

$$dy^i = A^{i'}_j \, dx^j, \tag{6.1}$$

the corresponding change of basis of 3−forms is given by

$$dy^1 \wedge dy^2 \wedge dy^3 = A^1_i A^2_j A^3_k dx^i \wedge dx^j \wedge dx^k. \tag{6.2}$$

We end up on the right side with the full panoply of all possible exterior products of the three 1−forms dx^i, but all of them are just proportional to each other.

We shall distinguish a sum such as

$$A_{ijk}dx^i \wedge dx^j \wedge dx^k, \tag{6.3}$$

which is extended to all possible values of the indices, from a sum like

$$A'_{ijk}(dx^i \wedge dx^j \wedge dx^k), \tag{6.4}$$

where the parenthesis is used to denote summation only over all possible combinations of indices with $i < j < k$. The forms of (6.3) do not constitute a basis (in general). Those of (6.4) do. It is clear that, if $n = 3$, the right hand side of Eqs. (6.3) and (6.4) contain 27 (of which only 6 are not zero) and one term respectively. The notation (6.4) is needed so that we can automatically equate the coefficients on both sides of equations that state the equality of two $r-$ forms. We illustrate this with a simple example.

Suppose that the exterior product of the two 1−forms $\alpha = A_i\omega^i$ and $\beta = B_j\omega^j$ is zero. The sum $A_iB_j\omega^i \wedge \omega^j = 0$ is extended to all pairs of indices. It can alternatively be written as $(A_iB_j - A_jB_i)(\omega^i \wedge \omega^j) = 0$. It then follows that $A_iB_j - A_jB_i = 0$, not that $A_iB_j = 0$.

Needless to say that the coefficients of a differential 3−form written as in (6.4) will not be the same as when written as in (6.3). The expression (6.4) picks up the skew-symmetric part of the coefficients in the expression (6.3), in the sense of the example given for $r = 2$. For $r = 3$, however, six (in general non-zero and potentially different) A_{ijk}'s go into the obtaining of each single A'_{ijk}, which is the sum of six A_{ijk} for given triple of values of (i, j, k). Once these primed quantites have been obtained (with $i < j < k$), one can rewrite the same differential form as

$$\frac{1}{3!}A'_{ijk}dx^i \wedge dx^j \wedge dx^k, \tag{6.5}$$

with summation over all permutations of the indices (i, j, k) without the restriction $i < j < k$. Two A'_{ijk} such that the indices of one are a permutation of the indices of the other are equal if the permutation is even, and opposite if the permutation is odd. We then have:

$$\begin{aligned} A'_{ijk} &= 0 &&\text{unless all three indices are different} \\ A'_{ijk} &= \pm A'_{123} &&\text{in the other cases.} \end{aligned} \tag{6.6}$$

Of course, for $n > r$ the three indices may be any. It should be clear also that, for $r-$forms, we have

$$A'_{i_1 i_2 \ldots i_r}(dx^{i_1} \wedge dx^{i_2} \wedge \ldots \wedge dx^{i_r}) = \frac{1}{r!}A'_{i_1 i_2 \ldots i_r}dx^{i_1} \wedge dx^{i_2} \wedge \ldots \wedge dx^{i_r}, \tag{6.7}$$

with the $A'_{i_1 i_2 \ldots i_r}$ totally skew-symmetric.

Let us practice with the simple example of the electromagnetic 2−form. Its version of the (6.4) type is

$$F = E_x dt \wedge dx + E_y dt \wedge dy + E_z dt \wedge dz - B_z dx \wedge dy - B_y dz \wedge dx - B_x dy \wedge dz. \quad (6.8a)$$

If we wish to write it in terms of the 2−forms generated by dx', dy', dz' and dt' where

$$x' = \gamma(x - vt), \quad y' = y, \quad z' = z, \quad t' = \gamma(t - vx), \quad (6.9a)$$

we use the transformations inverse to these, which are given by

$$x = \gamma(x' + vt'), \quad y = y', \quad z = z', \quad t = \gamma(t' + vx'). \quad (6.9b)$$

We differentiate (6.9b) at constant v, substitute in (6.8a) and obtain

$$\begin{aligned} F = &\, E_x dt' \wedge dx' + \gamma(E_y - vB_z)dt' \wedge dy' + \gamma(E_z + vB_y)dt' \wedge dz' \\ &- \gamma(B_z - vE_y)dx' \wedge dy' - \gamma(B_y + vE_z)dz' \wedge dx' - B_x dy' \wedge dz'. \end{aligned} \quad (6.8b)$$

If we denote as E'_x, E'_y, E'_z, $-B'_x$, $-B'_y$ and $-B'_z$ the coefficients of F with respect to the primed basis of 2−forms, we obtain (E', B') components in terms of (E, B) components.

On the other hand, we might be given F in the form

$$\frac{1}{2!} F_{ij} dx^i \wedge dx^j, \quad (6.10)$$

with F_{ij} skew-symmetric and with components which are nothing but the coefficients in (6.8a). We then have

$$F'_{km} = A^i_k A^j_m F_{ij}. \quad (6.11)$$

where the A^i_k are read from the equations (6.9b), with $x^i = (t, x, y, z)$.

2.7 Differential forms and measurement

In this section, we discuss lengths, areas and volumes. For simplicity, let us start with the 3-dimensional manifold of daily life (Euclidean 3-space, to be defined later). We ask ourselves whether volumes are given by the integral

$$\int dx \wedge dy \wedge dz = \int r^2 \sin\theta \, dr \wedge d\theta \wedge d\phi \quad (7.1)$$

over corresponding domains or by the integral

$$\int dr \wedge d\theta \wedge d\phi. \quad (7.2)$$

The solution is well-known, more because of one's familiarity with Euclidean space than because of familiarity with the theory of integration. But, what happens in the case when the $n-$dimensional manifold (say, the n-sphere, not to confuse with the n-ball), unlike Euclidean space, does not admit n Cartesian coordinates?

Inexperienced readers should not overlook the fact that, when we use Cartesian coordinates (x, y, z) on the 2-sphere, the number of such coordinates is $n + 1$, with $n = 2$; they are not independent in the sphere since, given the x and y of a point on it, its z coordinate is determined. Let us consider a n-manifold given in such a way that we do not know whether it is contained (although it is) in some Euclidean space of higher dimension, and which thus admits Cartesian coordinates. We would not know whether to read the n-volume differential form as the coefficient of the $n-$form $dx^1 \wedge \ldots \wedge dx^n$ associated with a given coordinate system or with the form $dy^1 \wedge \ldots \wedge dy^n$ associated with some other coordinate system. Looking for guidance, we can refer to the transformation of $n-$forms. A change of bases of $1-$forms induces a change of bases of $n-$forms given by

$$\omega^1 \wedge \ldots \wedge \omega^n = A^1_{i_{1'}} A^2_{i_{2'}} \ldots A^n_{i_{n'}} \omega^{i_{1'}} \wedge \omega^{i_{2'}} \wedge \ldots \wedge \omega^{i_{n'}}$$
$$= |A^j_{i'}| \omega^1 \wedge \omega^2 \wedge \ldots \wedge \omega^n, \tag{7.3}$$

where $|A^j_{i'}|$ is the determinant of the matrix whose elements are the $A^j_{i'}$'s. If $(\omega'^i) = (dy^i)$ and $(\omega^i) = (dx^i)$, Eq. (7.3) becomes

$$dy^1 \wedge dy^2 \wedge \ldots \wedge dy^n = J(\partial y / \partial x) dx^1 \wedge dx^2 \wedge \ldots \wedge dx^n, \tag{7.4}$$

where $J(\partial y / \partial x)$ is the Jacobian. But we have as well

$$dx^1 \wedge dx^2 \wedge \ldots \wedge dx^n = J(\partial x / \partial y) dy^1 \wedge dy^2 \wedge \ldots \wedge dy^n. \tag{7.5}$$

From these expressions, there is no way to tell which $3-$form, whether $dx^1 \wedge dx^2 \wedge \ldots \wedge dx^n$ or $dy^1 \wedge dy^2 \wedge \ldots \wedge dy^n$ or none of the two represents the volume element. Actually, one such element may not be defined.

What we are missing is a yardstick for every direction so that, even along a coordinate line, we shall be able to tell how long something is. Surfaces and volumes make clear that we also need to know how much a coordinate line x^1 is tilted with respect to another coordinate line x^j (i.e., the "angle of two directions"). All this is achieved by the introduction of a metric, which will be done in a later chapter (angle, like length, is a metric concept). We conclude that, unless we introduce more structure on our manifold, we cannot say what the area of a surface is. We can, however, calculate fluxes through surfaces and, in general, we can integrate. As an example, the flux of the magnetic field through a surface at constant time is given by the integral of the negative of the electromagnetic 2$-$form after removing the terms containing dt as a factor.

2.8   Differentiable manifolds DEFINED

We now give a formal definition of differentiable manifold $\boldsymbol{M}$, previously treated informally. A differentiable manifold of dimension n and class C^r consists of a set M of "points" and a set $\boldsymbol{U}_M$ (called atlas) of coordinate patches or charts. A chart consists of a pair (U, f_U) where U is a subset of M and f_U is a one-to-one correspondence from U onto an open subset $f_U(U)$ of $\mathbb{R}^n$. One refers to n as the dimension of the manifold. Let the symbol "$\circ$" denote composition of functions. The manifold has to satisfy the following two conditions:

1) The sets $U \subset \boldsymbol{U}_M$ cover the whole set M.

2) If U_1 and U_2 overlap, $f_{U_2} \circ f_{U_1}^{-1}$ is a function from $f_{U_1}(U_1 \cap U_2)$ into $\mathbb{R}^n$ that is continuous together with all its derivatives up to order r, i.e. it is C^r.

Let us now describe in less technical words the object that we have just defined. Open sets are required so as to avoid boundaries, since points in boundaries do not have neighborhoods that look like $\mathbb{R}^n$. The one-to-one and onto requirements for the charts together make the functions f_U invertible. Notice that $f_{U_i}(U_i)$ is nothing but a set of coordinate functions. If $P \in U$, $f_U(P)$ represents a set $(x^i(P))$ of coordinates of the point P. We have guaranteed that we can introduce coordinates all over the set of points (first condition above).

A set of coordinates need not cover the whole manifold. For instance, there is no one chart that covers the whole spherical surface (i.e. with just two but never less than two coordinates). Thus the values $\theta = 0$ and $\theta = \pi$ describe the poles of the spherical coordinate system regardless of the value of ϕ. We require that, piece by piece, we cover the whole set of points M with charts, also called coordinate patches.

Condition 2 achieves that when two charts overlap (the overlap being $U_1 \cap U_2$), the transformation from one system of coordinates to the other is C^r. Two charts (U_1, f_{U_1}) and (U_2, f_{U_2}) that satisfy condition 2 above are said to be C^r compatible. Once an atlas has been given or found, we shall include in the atlas all the charts that are C^r compatible with those already there. The atlas so completed is called a maximal atlas. The atlas in the definition of manifold will be understood to be maximal. Let I be the identity map on $\mathbb{R}^n$; $\mathbb{R}^n$ then becomes a manifold if we add to it the atlas constituted by the one chart $(\mathbb{R}^n, I)$ together with all the charts that are C^r compatible with it.

As a summary, a C^r differentiable manifold is the pair made of a set of points and a maximal atlas that covers the set of points of the manifold. Changes of chart within the atlas (i.e. coordinate transformations) are required to be C^r functions. In the following, we shall speak indistinctly of $\boldsymbol{M}$ and M.

Let us next consider differentiable functions on manifolds. Let g be a function from manifold $\boldsymbol{M}_1$ to manifold $\boldsymbol{M}_2$ and let $U_1 \subset \boldsymbol{U}_{M_1}$ and $U_2 \subset \boldsymbol{U}_{M_2}$. We say that g is differentiable to order r at a point P of M_1 if $f_{U_2} \circ g \circ f_{U_1}^{-1}$ is differentiable to order r at $f_U(P)$. Because of condition 2 on manifolds, this definition does not depend on which patches we choose around P. The function g may in particular be a real valued function, i.e. $\boldsymbol{M}_2$ may be $\mathbb{R}$ itself. In this case, f_{U_2} is simply the identity map.

Since the composite map $g \circ f_U^{-1}$ takes us from $\mathbb{R}^n$ to $\mathbb{R}^1$ (see Figure 1), we define the differential dg of g as $dg = (\partial g/\partial x^i)dx^i$, where

$$\partial g/\partial x^i = \partial(g \circ f_U^{-1})/\partial x^i, \tag{8.1}$$

with x^i defined by $(x^i) = f_U(P)$. It is convenient to introduce the notation

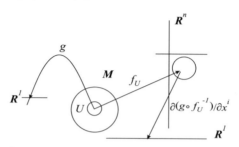

Figure 1: Manifold related concepts

$g(x^i)$ for $g \circ f_U^{-1}$. The expression for dg given above then becomes the dg of the calculus in $\mathbb{R}^n$. Most authors with a utilitarian approach to mathematics usually eliminate any reference to the class of the manifold. Explicitly or implicitly, they assume, like we shall do, that the class is as high as needed for our purposes at any given time.

To those who do not think like experts on analysis (which includes this author), let us say that there are 1-D figures which are not 1-D manifolds in the strict sense just defined. A continuous curve is a 1-D differentiable manifold if it does not intersect itself. Points of intersection or points of tangency (think of the letter q) are not differentiable manifolds. They fail to be so by virtue of the intersection and/or tangency points. As for polygons, they are not differentiable manifolds because of not existence of a derivative at the vertices (one has to first smooth them in order to apply any specific results). We shall assume the validity in the only case where it is needed in this book (in a lemma to the Gauss-Bonnet theorem in Appendix A).

2.9   Another definition of differentiable MANIFOLD

The concept of differentiable manifold just given is the usual one in the modern literature, but it is not the only accepted one. We now present an equivalent one [92] (not completely, since it only has to do with Euclidean structure) in which one does not make explicit reference to the charts.

Let start by saying in plain language what we got in the previous section. Charts play the role of providing the arena that supports differentiation in a continuous set of points. The purpose of having different charts and not just one is to be able to adequately cover all points in the set, one small piece at the time. We then endow the set with additional structure, materialized through a system of connection equations and a system of equations of structure, the latter speaking of whether the former is integrable or not.

For simplicity, consider 2-D surfaces and try to ignore the 3-D Euclidean space in which we visualize them. You may think that you are ignoring the 3-D space, but you are not when you deal with the normal to the surface: it is not defined except through the embedding of the surface in a higher Euclidean space. Thus, without the (in the present case three dimensional) Euclidean space, one has to think of even a 2-D surface in just some abstract, mathematical way. But the equations of structure deny in general that the set is a flat Euclidean structure, unless we are speaking of planes, cones and cylinders (i.e. surfaces developable on a plane). The most structure we then get on those manifolds is that small neighborhoods of each point are almost like small pieces of 2-D Euclidean space. Hence, it helps to think of an alternative concept of differentiable manifold.

Consider the practical aspects of defining a set of which we say it is a differentiable manifold. A 2-sphere is a differentiable manifold, but so is the punctured sphere, and also the idealized part of surface of the earth covered by land, and also the smoothed part covered by sea, and the also idealized surface of any contiguous country, etc. What is behind the possibility of distinguishing between all these cases? They have something in common but are still given by different subsets of the 3-D Euclidean space. This brings us to the following alternative definition.

Let E_n denote Euclidean space of dimension n. A differentiable manifold is a subset of E_n defined near each point by expressing some of the coordinates in terms of the others by differentiable functions" [92]. This replaces the charts, but introduces the differentiable functions to determine the subset of interest. When one starts by defining differentiable manifolds as in the previous section, this new definition is the result of a theorem by Whitney [92]. The theorem assigns to each Euclidean differentiable manifold defined as in the previous section a Euclidean space of higher dimension in which it fits. Notice, in order to avoid wrong inferences, that common spirals (for example) do not fit in two dimensions, but fit in three. They are born ab initio in E_3. Just $n + 1$ may not suffice.

The new definition is very interesting, but that is not the way in which, for practical reasons, one does differential geometry. It involves, in addition to the differentiable manifolds being defined, Euclidean spaces of higher dimensions, not to speak of the fact that it is limited to Riemannian structures, which are of just a very specific type. (Affine is more general than Euclidean; Riemannian is Euclidean at the level of the very small). Nor does it comprise projective, conformal, etc. manifolds.

We thus keep the definition of the previous section, though one barely uses that definition in actual practice, not in this book, not in other books. We assume that such a structure justifies the development of the theory in part four of this book. The present section provides the vision that the tangible thing in such a theory are the flat spaces (projective, conformal, affine, Euclidean, etc.) and (in general not totally integrable) associated systems of differential equations. One can hardly say that the non-flat spaces are very tangible, very imaginable. The flat ones are predominantly algebraic concepts, and thus far closer to what we can imagine.

Chapter 3

VECTOR SPACES AND TENSOR PRODUCTS

3.1 INTRODUCTION

In this book, the tensor calculus is absent, but the concept of tensor product is needed. The familiar expression (1.3) of chapter 2 lives in a space which is the tensor product of a space of vectors and a space of differential 1−forms. One does not need to know the formal definition of such products in order to be able to operate with those expressions.

More subtle is the tensor product of different copies of one and the same vector space. We devote the last section of this chapter to this concept. We call attention to the fact that what one actually defines is the tensor product of spaces, tensors being the elements of the resulting product structure. It is the tensorial composition law in the structure that makes tensors be tensors. Thus, elements of a structure may look like tensors but not be so because its tensor product by a similar object does not yield another similar one. Example: the tensor product of two so called skew-symmetric (also called antisymmetric) tensors is not skew-symmetric in general. The skew-symmetric product in this case is the exterior product. Their algebra is exterior algebra and its elements should be called multivectors.

In the tensor algebra constructed upon some vector space (or over some module), the vectors (or the elements of the module) are considered as tensors of grade or rank one. The vectors that one finds in the vector calculus are called tangent vectors. They constitute so called tangent vector spaces (one at each point of the manifold), though it is difficult to realize in Euclidean spaces why should one call them tangent. This will later become evident. Elements of the space that results from tensor product of copies of a tangent vector space are called tangent tensors.

We shall also have another type of vector space, called the dual vector space, to be introduced in section 3. In differential geometry, its members are called cotangent vectors. One can also construct tensor products of copies of this vector space. The members of these product spaces are called cotangent tensors. For the purpose of clarity, we shall use the term *tensor* in this section only to refer to those of grade two or higher.

Our tensor-valued differential forms are going to be tensor products of three "spaces": (a) a space of tangent vectors (rarely if ever tangent tensors); (b) a space of cotangent vectors (in a few exceptional cases, cotangent tensors) and (c) a space of differential forms. Very seldom shall we have tensors that are a product of copies of the same vector space. The term (p, q)-valued differential form will be used for a (sum of) members of the product of a tangent tensor of grade p, a cotangent tensor of grade q (whether skew-symmetric or not) and a differential form. We shall say that the valuedness is (p, q). In view of what we have just said, our valuedness will be almost exclusively of the types $(0, 0)$, $(1, 0)$, $(0, 1)$ and $(1, 1)$.

We said, *almost* exclusively. There are exceptions. On the one hand, there is the so called metric tensor and the curvature tensor. The latter —of minor importance relative to the curvature differential form— is a $(1, 3)$ tensor-valued differential $0-$form. It is derived from the affine curvature $(1, 1)$-valued differential $2-$form and is skew-symmetric with respect to only two of its three subscripts. As a concept, it is an end in itself and not the gate to other concepts. As for the metric, one can adopt a more advanced perspective of geometry than in this book, a Finslerian perspective [73]. We say perspective since the geometry need not be properly Finslerian but, say, Riemannian, though viewed as if it were Finslerian. In this case, the role of providing the length of curves is taken over from the metric by a differential $1-$form, which is after all what differential $1-$forms are: line integrands. The Finslerian structure corresponding to a spacetime manifold is a space-time-velocity manifold.

The contents of this chapter reflects those considerations. The cotangent valuedness is induced by the vector valuedness (see the argument on $d d\mathbf{v}$ and curvature in the preface, and later in section 5.7). It thus becomes fair to say that modern differential geometry is the calculus of vector-valued differential forms.

We shall start with a formal though not exceedingly abstract presentation of vector spaces in section 2 (A more compact definition of a vector space exists, but it resorts more heavily to concepts in algebra). The points of importance here are that vectors do not need to have magnitude, and that what makes a vector a vector is its belonging to a particular structure. Section 3 deals with the dual tangent space. The vector spaces of section 2 may be called plain. If they have the additional structure conferred by a dot product, they are called Euclidean vector spaces. They are considered in section 4. In section 5, we deal in greater detail with the entities that constitute the tensor products of different structures that appear in this presentation of differential geometry. Section 6, addresses the issue of tensor products of copies of the same vector space. Readers with access to the excellent book by Lichnerowicz on tensor calculus [53] will find there deep algebraic involvement with tensors, necessary for a good understanding of that calculus, but unnecessary in this book.

3.2 Vector spaces (over the reals)

The structure "vector space" appears repeatedly in differential geometry and is at the root of almost any formal development. We commence with a definition of this structure. A vector space (over the reals) is a set of elements $\mathbf{u}, \mathbf{v}, \ldots$ with two operations obeying the following properties:

(1) Addition

    a. $\mathbf{u} + \mathbf{v} = \mathbf{v} + \mathbf{u}$ Commutative
    b. $\mathbf{u} + (\mathbf{v} + \mathbf{w}) = (\mathbf{u} + \mathbf{v}) + \mathbf{w}$ Associative
    c. $\mathbf{0}, \ \mathbf{u} + \mathbf{0} = \mathbf{u}$ Existence of Zero (Vector)
    d. $\mathbf{u} + (-\mathbf{u}) = 0$ Opposite Vector defined

(2) Multiplication by a scalar (real number: $a, b, c \ldots$a)

    a'. $1\mathbf{u} = \mathbf{u}$
    b'. $a(b\mathbf{u}) = (ab)\mathbf{u}$ Associative
    c'. $(a + b)\mathbf{u} = a\mathbf{u} + b\mathbf{u}$ Distributive for addition of scalars
    d'. $a(\mathbf{u} + \mathbf{v}) = a\mathbf{u} + a\mathbf{v}$ Distributive for addition of vectors

The concept of basis permits us to define the dimensionality of a vector space. A basis of a vector space is defined as a set of linearly independent vectors that is not contained in a larger set of linearly independent vectors. A vector space is finite-dimensional if it has finite basis. One can show that, in this case, all bases have the same number of elements, say n, where n is called the dimension of the vector space V^n. A set of r independent vectors that is not maximal, $r < n$, spans an r-dimensional vector space which is said to be a subspace of V^n. The polynomials in one variable of degree smaller or equal to n constitute a vector space of dimension $n + 1$. A basis for this vector space is $(1, x, x^2 \ldots x^n)$. The polynomials of degree smaller or equal to r (with integer $r < n$), constitute a subspace of dimension $r + 1$ of that $n + 1$ dimensional space. In particular, the real numbers can be viewed as a one-dimensional subspace.

The $n \times n$ square matrices constitute a vector space of dimension n^2. For $n = 2$, a simple basis is the set

$$\mathbf{a}_1 = \begin{pmatrix} 1 & 0 \\ 0 & 0 \end{pmatrix} \quad \mathbf{a}_2 = \begin{pmatrix} 0 & 1 \\ 0 & 0 \end{pmatrix} \quad \mathbf{a}_3 = \begin{pmatrix} 0 & 0 \\ 1 & 0 \end{pmatrix} \quad \mathbf{a}_4 = \begin{pmatrix} 0 & 0 \\ 0 & 1 \end{pmatrix}.$$

In this case, the arbitrary element $\begin{pmatrix} a & b \\ c & d \end{pmatrix}$ of this vector space can be written as:

$$\begin{pmatrix} a & b \\ c & d \end{pmatrix} = a\mathbf{a}_1 + b\mathbf{a}_2 + c\mathbf{a}_3 + d\mathbf{a}_4.$$

A final example: The solutions to the differential equation $\ddot{x} + \omega_0^2 x = 0$ (each overdot indicating one derivation with respect to t) constitute a vector space of dimension 2. A basis for this vector space is $(\cos \omega_0 t, \ \sin \omega_0 t)$.

Given a vector $\mathbf{v} \in V^n$, its components relative to a basis $(\mathbf{a}_1, \ldots, \mathbf{a}_n)$ will be denoted as $(v^1, \ldots, v^n)$ and we shall, therefore, have

$$\mathbf{v} = v^1 \mathbf{a}_1 + v^2 \mathbf{a}_2 + \ldots + v^n \mathbf{a}_n = v^i \mathbf{a}_i. \tag{2.1}$$

This decomposition is unique. If there were two (or more), by subtracting them we would obtain that a nontrivial linear combination of basis vectors is zero which is a contradiction with the assumption that $(\mathbf{a}_1, \ldots, \mathbf{a}_n)$ constitutes a basis. In case you missed it previously, the right hand side of (2.1) is an abbreviation known as Einstein's summation convention: whenever there is a repeated index (once as a subscript and once as a superscript), summation from one to the largest possible value of the index, n, is understood.

It is important to notice that the left hand side, $\mathbf{v}$, does not have indices and that the i of v^i has been written as a superscript (in contradistinction to the subscript i of $\mathbf{a}_i$). We shall soon learn that the difference in position of i in v^i and $\mathbf{a}_i$ conveys much information.

We proceed to consider changes of basis. Let $(\mathbf{a}_i)$ and $(\mathbf{a}'_i)$ be two different bases, which allows us to write

$$\mathbf{v} = v^i \mathbf{a}_i = v'^i \mathbf{a}'_i. \tag{2.2}$$

We express the basis $(\mathbf{a}'_i)$ in terms of the basis $(\mathbf{a}_j)$ as

$$\mathbf{a}'_i = A^j_{i'} \mathbf{a}_j, \tag{2.3a}$$

where we have used Einstein's convention. We can think of the apostrophe in $\mathbf{a}'_i$ as if belonging to the subscript, i.e. as if $\mathbf{a}'_i$ were $\mathbf{a}_{i'}$. The use of such primed subscripts (and, later on, also primed superscripts) has the following purpose. We want to distinguish the coefficients $A^j_{i'}$ from the coefficients $A^{j'}_i$ that give $\mathbf{a}_i$ in terms of $\mathbf{a}'_j$,

$$\mathbf{a}_i = A^{j'}_i \mathbf{a}'_j. \tag{2.3b}$$

Notice that the coefficient of, say, $\mathbf{a}_1$ in terms of, say, $\mathbf{a}'_2$ should be written as $A^{2'}_1$, in order to distinguish it from the coefficient $A^2_{1'}$ of $\mathbf{a}'_{1'}$ in terms of $\mathbf{a}_2$. So, $A^{2'}_1$ can be viewed as $A^{j'}_i$ with $i = 1$ and $j' = 2'$. Substitution of (2.3b) in (2.3a) yields, after appropriate change of indices,

$$\mathbf{a}'_i = A^j_{i'} A^{k'}_j \mathbf{a}'_k \tag{2.4}$$

and, therefore,

$$A^j_{i'} A^{k'}_j = \delta^{k'}_{i'} = \delta^k_i, \tag{2.5}$$

where the apostrophes to the indices of the Kronecker delta are no longer needed. We now use Eqs. (2.3a) and (2.3b) in (2.2) and get

$$v'^i = A^{i'}_j v^j, \tag{2.6a}$$

and

$$v^i = A^i_{j'} v'^j. \tag{2.6b}$$

It is easy to remember these equations by observing that the superscript on the left is also a superscript on the right and that Einstein's summation convention has to be reflected in the notation. Notice that Eqs. (2.3) express the elements of one basis in terms of the elements in the other basis of vectors. Thus, whereas v^i and v'^i represent the ith component of the same vector $\mathbf{v}$ in two different bases, $\mathbf{a}_i$ and $\mathbf{a}'_i$ are two different vectors. The expressions

$$\mathbf{v} = v^i \mathbf{a}_i = v'^i \mathbf{a}'_i \tag{2.6c}$$

admit the following parallel expressions for explicitly chosen $\mathbf{a}_i$:

$$\mathbf{a}_i = \delta_i^j \mathbf{a}_j = A_i^{j'} \mathbf{a}'_j. \tag{2.6d}$$

Let us finally consider the relation of vector bases to the general linear group. It is a trivial matter to show that the matrices $[A_{i'}^j]$ and $[A_i^{j'}]$ that relate two bases have determinants different from zero. Conversely, if $(\mathbf{a}_j)$ is a basis and $[A_{i'}^j]$ is any matrix such that $\det [A_{i'}^j] \neq 0$, then $(\mathbf{a}'_i)$ given by

$$\{\mathbf{a}'_i\} = [A_{i'}^j]\{\mathbf{a}_j\} \tag{2.7}$$

also is a basis. The set of all such $n \times n$ matrices constitutes the (general) linear group for dimension n, group denoted as $GL(n)$. It has a preferred member, i.e. the unit matrix. The set of all bases does not have a preferred member. By explicitly choosing a particular basis in $\boldsymbol{V}^n$ and expressing all other ones in terms of it, we generate the set of all matrices in the linear group $GL(n)$. We are thus able to put all the bases in a one-to-one correspondence with the elements of the linear group. The set of elements $(A_{i'}^j)$ of each particular matrix $[A_{i'}^j]$ in the group constitutes a system of coordinates in the set of all bases. The coordinatization just introduced is of great importance in the Cartan calculus.

All vector spaces of the same dimension are isomorphic.

3.3 Dual vector spaces

Given a vector space $\boldsymbol{V}^n$, the set of linear functions $\phi : \boldsymbol{V}^n \to \mathbb{R}^1$ constitute a vector space $\boldsymbol{V}^{*n}$ of the same number n of dimensions if we introduce in said set a law of addition of those functions and a law of multiplication by scalars (the same set of scalars as for $\boldsymbol{V}^n$). In this book, unlike in most modern presentations of differential geometry, we make very minor use of this concept. Hence, our treatment of dual spaces will be schematic.

Bases (ϕ^i) in $\boldsymbol{V}^{*n}$ can be chosen such that

$$\phi^i(\mathbf{a}_j) = \delta_j^i. \tag{3.1}$$

They are called dual bases, meaning dual of corresponding basis of $\boldsymbol{V}^n$. Any linear transformation can be written in terms of a basis as

$$\phi = \lambda_i \phi^i, \tag{3.2}$$

where the coefficients λ_i's are its components. A change of basis $\mathbf{a}_i \to \mathbf{a}'_i$ in V^n induces a change of basis $\phi^i \to \phi'^i$ in V^{*n} so that Eq. 3.1 continues to hold, i.e., so that $\phi'^i(\mathbf{a}'_j) = \delta^i_j$. If A is the matrix for the change of basis $\{\mathbf{a}_i\} \to \{\mathbf{a}'_i\}$, then $[A^T]^{-1}$ is the matrix for the change of basis $\{\phi^i\} \to \{\phi'^i\}$, where the superscript T stands for the transpose matrix. In modern notation, Eq. (3.1) is also written as

$$\phi^i \, \llcorner \, \mathbf{a}_j = \delta^i_j. \tag{3.3}$$

It readily follows that

$$\phi^i \, \llcorner \, (v^j \mathbf{a}_j) = v^i. \tag{3.4}$$

In later chapters, we shall refer to spaces V^n as tangent vector spaces, and to spaces V^{*n} as cotangent vector spaces. The reason for these denominations will then become clear.

3.4 Euclidean vector spaces

3.4.1 Definition

The concept of Euclidean vector space is at the root of the often mentioned but not always well defined concept of Euclidean space. In this book, the term Euclidean will comprise also what is often called pseudo-Euclidean (soon to be defined), except for specification to the contrary. A Euclidean vector space is a vector space V^n together with a map $V^n \times V^n \to \mathbb{R}$ called dot or scalar product such that:

(a) $\mathbf{u} \cdot \mathbf{v} = \mathbf{v} \cdot \mathbf{u}$

(b) $(a\mathbf{u}) \cdot \mathbf{v} = \mathbf{u} \cdot (a\mathbf{v}) = a(\mathbf{u} \cdot \mathbf{v})$

(c) $\mathbf{u} \cdot (\mathbf{v} + \mathbf{w}) = \mathbf{u} \cdot \mathbf{v} + \mathbf{u} \cdot \mathbf{w}$

(d) $\mathbf{u} \cdot \mathbf{v} = 0 \quad \forall \mathbf{u} \Rightarrow \mathbf{v} = 0.$

Given a vector basis $(\mathbf{a}_i)$, let g be the matrix whose elements are given as

$$g_{ij} = \mathbf{a}_i \cdot \mathbf{a}_j = g_{ji}. \tag{4.1}$$

Thus

$$\mathbf{u} \cdot \mathbf{v} = u^i v^i g_{ij}. \tag{4.2}$$

Property (d) implies that $det\, g \neq 0$. Indeed if $\mathbf{u} \cdot \mathbf{v}$ equals zero for all $\mathbf{u}$, i.e. if

$$g_{ij} u^i v^i = 0, \quad \forall \{u^i\},$$

then

$$g_{ij} v^i = 0, \quad \forall i.$$

This is a homogeneous linear system in the v^j's. Its only solution is $v^i = 0$, for all j. Then $det\ g \neq 0$, which we wanted to prove. As a particular case of (4.2), we have that, for any vector $\mathbf{v}$,

$$\mathbf{v} \cdot \mathbf{v} = g_{ij}v^i v^j. \tag{4.3}$$

If the expression $g_{ij}v^i v_j$ is definite positive, the space is said to be properly Euclidean. If it is definite negative, one can redefine the dot product as the negative of the previous one so that it becomes definite positive and the vector space is again properly Euclidean. If the expression (4.3) is not definite, one speaks of a pseudo-Euclidean vector space.

The symbols used may remind readers of those used in differential geometry. But we are dealing here with algebra. The g_{ij} do not depend on coordinates here, as there are not yet coordinates to speak of. Eventually, in later chapters, we shall have fields of vector spaces. g_{ij} will then be point dependent and we shall start to refer to g_{ij} as the components of the metric.

3.4.2 Orthonormal bases

In properly Euclidean vector spaces, a basis $(\mathbf{a}_i)$ is said to be orthonormal if

$$\mathbf{a}_i \cdot \mathbf{a}_j = \delta_{ij}. \tag{4.4}$$

"Ortho" stands for

$$\mathbf{a}_i \cdot \mathbf{a}_j = 0 \quad i \neq j, \tag{4.5}$$

and "normal" stands for

$$\mathbf{a}_i \cdot \mathbf{a}_i = 1 \tag{4.6}$$

(without summation over repeated indices). In terms of the components v^i of a vector $\mathbf{v}$ relative to an orthonormal basis, we have,

$$\mathbf{v} \cdot \mathbf{v} = \sum_i (v^i)^2. \tag{4.7}$$

The quantity $\left| \sum(v^i)^2 \right|^{1/2}$ is denoted as the magnitude or norm of $\mathbf{v}$. If a vector space is pseudo-Euclidean, no basis can be found in it such that Eq. (4.4) is satisfied for all values of the indices. Indeed, Eq. (4.4) implies Eq. (4.7) and, therefore, positive definiteness, contrary to the hypothesis. Instead of Eq. (4.4), one now has

$$\mathbf{a}_i \cdot \mathbf{a}_i = \pm 1, \tag{4.8}$$

where the $+$ sign applies to at least one but not all the values of the subscript. In the following, we shall use the term orthonormal bases to refer to both properly orthonormal and pseudo-orthonormal bases.

One says that the orthogonal group $O(n)$ acts transitively on the set of orthonormal bases in n dimensions, meaning that any orthonormal basis can be reached from any other orthonormal basis by some element in the group. The subgroups $O(m)$ for $m < n$ act on the same set, but not transitively.

Unless required by the specificity of a topic, we shall often use terminology that does not make explicit whether the orthogonal matrices we shall be dealing with are special (determinant +1), or general (also -1).

Notice that the map $V^n \times V^n \to \mathbb{R}$ is not necessarily preserved by the isomorphism of all vector spaces of the same number of dimensions. For instance, consider an Euclidean and a pseudo-Euclidean vector spaces, both of dimension 2. We can choose one basis, respectively orthonormal and pseudo-orthonormal, in each of these two spaces. We establish a one-to-one correspondence between the vectors of both spaces by identifying each vector in one space with the vector that has the same components in the other space. The scalar product of the two vectors is not in general the same number as the scalar product of their homologue vectors in the other space. In particular, homologue vectors need not have the same norm.

3.4.3   Reciprocal bases

For simplicity, let us consider properly Euclidean spaces. Given a basis $(\mathbf{a}_i)$, consider another basis $(\mathbf{b}_j)$ such that

$$\mathbf{a}_i \cdot \mathbf{b}_j = \delta_{ij}, \quad \forall (i, j). \tag{4.9}$$

We say that the two bases are the reciprocal of each other. We denote this relation by representing $(\mathbf{b}_j)$ as $(\mathbf{a}^j)$. Thus:

$$\mathbf{a}_i \cdot \mathbf{a}^j = \delta_i^j. \tag{4.10}$$

Eventually, as we consider specific structures the position of the indices will acquire meaning. The $(\mathbf{a}^j)$ is then called the reciprocal basis of the "original" basis $(\mathbf{a}_i)$.

Given a basis $(\mathbf{a}_i)$, Eqs. (4.10) can be used to obtain $(\mathbf{a}^j)$ in terms of $(\mathbf{a}_i)$. It is suggested that junior readers explicitly solve Eqs. (4.10) for $\mathbf{a}^j$ when the basis $(\mathbf{a}_i)$ is given as

$$\mathbf{a}_1 = \mathbf{i}, \quad \mathbf{a}_2 = \frac{\mathbf{i} + \mathbf{j}}{\sqrt{2}}. \tag{4.11}$$

Notice that $\mathbf{a}_1$ and $\mathbf{a}_2$ are normal (magnitude one) but not orthogonal.

We now show that Eqs. (4.10) uniquely determine $(\mathbf{a}^i)$ in terms of $(\mathbf{a}_i)$. If $(\mathbf{a}^i)$ exists, we can expand it in terms of the original basis:

$$\mathbf{a}^i = D^{ij} \mathbf{a}_k. \tag{4.12}$$

If we substitute this in (4.10), we obtain

$$D^{jk} g_{ki} = \delta_i^j, \tag{4.13}$$

which shows that the matrix with entries D^{jk} is the inverse of the matrix g, which always has an inverse. Hence

$$\{\mathbf{a}^i\} = g^{-1} \{\mathbf{a}_j\}, \tag{4.14}$$

where curly bracket denotes column matrix. This equation can be used for obtaining the reciprocal of the basis $\{\mathbf{a}_i\}$ in (4.11) by first obtaining $g_{ij} = \mathbf{a}_i \cdot \mathbf{a}_j$ and then inverting this matrix. This is an expedient way of obtaining reciprocal bases if one is efficient at inverting matrices. We have used and will use curly brackets to emphasize that the elements of the bases are written as columns when a square matrix acts on them from the left.

In parallel to the definition of g_{ij} in Eq. (4.1), we define

$$g^{ij} \equiv \mathbf{a}^i \cdot \mathbf{a}^j = g^{ji}. \tag{4.15}$$

Substitution of (4.12) in (4.15) yields:

$$g^{ij} = D^{il}D^{jk}\mathbf{a}_l \cdot \mathbf{a}_k = D^{il}g_{lk}D^{kj}, \tag{4.16}$$

where we have made use of the symmetry of D^{ij}, i.e. of the matrix g^{-1}. Equations (4.13) and (4.16) together imply

$$g^{ij} = \delta^i_k D^{kj} = D^{ij} \tag{4.17}$$

and, therefore

$$[g^{ij}] = g^{-1}, \tag{4.18}$$

where $[g^{ij}]$ denotes the matrix of the g^{ij}. Hence, we rewrite Eq. (4.14) as

$$\mathbf{a}^i = g^{ij}\mathbf{a}_j \tag{4.19}$$

which in turn implies, premultiplying by g_{ki} and contracting:

$$g_{ki}\mathbf{a}^i = g_{ki}g^{ij}\mathbf{a}_j = \delta^j_k\mathbf{a}_j = \mathbf{a}_k. \tag{4.20}$$

Equations (4.19)-(4.20) state that we may raise and lower indices of basis vectors with g^{ij} and g_{ij} respectively. It must be clear from the definition (4.17) how the g^{ij} transform under a change of basis.

In the same way as we have introduced components v^i of a vector with respect to the basis $(\mathbf{a}_i)$, we now introduce its components v_i with respect to the basis $(\mathbf{a}^i)$. Thus

$$\mathbf{v} = v^i\mathbf{a}_i = v_i\mathbf{a}^i. \tag{4.21}$$

The basis with subscripts will be called contravariant and so will the accompanying components, i.e. the v^i's. The bases $(\mathbf{a}^i)$ and the components (v_i) will be called covariant.

The dot product of $\mathbf{v}$ and $\mathbf{a}_j$ yields, using (4.21),

$$\mathbf{v} \cdot \mathbf{a}_j = (v_i\mathbf{a}^i) \cdot \mathbf{a}_j = v_i\delta^i_j = v_j. \tag{4.22}$$

Hence

$$v_j = \mathbf{v} \cdot \mathbf{a}_j = (v^i\mathbf{a}_i) \cdot \mathbf{a}_j = v^ig_{ij}. \tag{4.23}$$

One equally shows that we raise indices with g^{ij}. Hence the rules for raising and lowering the indices of the components are the same rules as for the basis elements with indices of the same type.

Let A denote the matrix that transforms $(\mathbf{a}_i)$ into $(\mathbf{a}'_i)$. One readily shows that the matrices g' and g obtained from the respective bases $(\mathbf{a}'_i)$ and $(\mathbf{a}_i)$ are related by

$$g' = AgA^T. \tag{4.24}$$

We invert (4.24) and obtain

$$g'^{-1} = (A^T)^{-1}g^{-1}A^{-1} = (A^T)^{-1}g^{-1}[(A^T)^{-1}]^T. \tag{4.25}$$

the symmetry relation existing between a basis and its dual allows us to immediately infer from Eqs. (24)-(25):

$$\{\mathbf{a}'^i\} = (A^T)^{-1}\{\mathbf{a}^i\}, \tag{4.26}$$

a result which is obvious in any case in view of previous considerations.

Exercise. Consider a (pseudo)-Euclidean two dimensional vector space such that $g_{11} = 1$, $g_{22} = -1$ and $g_{12} = g_{21} = 0$ in certain basis $\{\mathbf{a}^i\}$, $i = 1, 2$. Find g'_{ij} in another basis $\{\mathbf{a}'_i\}$ such that an arbitrary vector $\mathbf{v} = a\mathbf{a}_1 + b\mathbf{a}_2$ becomes:

$$\mathbf{v} = a(1 - \rho^2)^{1/2}\mathbf{a}'_1 + (b - \rho a)(1 - \rho)^{1/2}\mathbf{a}'_2, \tag{4.27}$$

where ρ is a parameter entering the change of basis. (Your result must contain only numbers and the parameter ρ.)

The transformation of the g_{ij}'s under a change of bases is obvious:

$$g'_{ij} = \mathbf{a}'_i \cdot \mathbf{a}'_j = A^k_{i'} A^l_{j'} \mathbf{a}_k \cdot \mathbf{a}_l = A^k_{i'} A^l_{j'} g_{kl}, \tag{4.28}$$

and similarly, but not identically, for the transformation of the g'^{ij}.

We return to the opening statement of this section, with a remark to keep going until we address Euclidean spaces in chapter 6. The surface of a table extended to infinity and the space of ordinary life as conceived before the advent of the theories of relativity are said to be Euclidean spaces of dimensions 2 and 3 respectively. They are not Euclidean vector spaces, or even simply vector spaces for that matter since they do not have a special point, a zero. If we choose some arbitrary point to play the role of the zero, we may then use in Euclidean spaces, the tools of Euclidean vector spaces.

3.4.4   Orthogonalization

Two related processes of orthogonalization

There are two related main problems of orthogonalization. One of them is: given a vector basis that is not orthonormal, find in terms of it another one that is. A direct way to solve it is through the so called Schmidt's method.

The second problem is: given a quadratic form, reduce it to a sum of squares. We shall solve it through a direct method that we found in Cartan [23]. It could also be solved indirectly through its relation to the first problem since the two problems are in essence the same, as we show in the third subsection.

The issue is essentially about orthogonality, since normalizing is trivial once we have orthogonality. For this reason, we have so far spoken only of orthogonalization in the titles of the section and of this subsection.

Schmidt's orthogonalization procedure

Given any basis of a space E^n, one can always find n independent linear combinations of its elements such that these linear combinations constitute an orthonormal basis. Of course, the solution is not unique, since, once we have generated one orthonormal basis, we can generate all the others by action of orthogonal matrices. The basic problem is thus the generating of just one orthonormal basis, which Schmidt's method achieves.

In order to minimize clutter in the computation with symbols that follows, it is best to first orthogonalize the basis and only then normalize it (as opposed to normalizing after each step of diagonalization). We assume the dot products $\mathbf{e}_\mu \cdot \mathbf{e}_\mu$ to be known, including when μ equals ν (if not so, what would it mean to know the basis?). We build an orthogonal basis $(\mathbf{a}_\mu)$ as follows:

$$\mathbf{a}_1 \equiv \mathbf{e}_1, \tag{4.29}$$

$$\mathbf{a}_2 \equiv \mathbf{e}_2 + \lambda_2^1 \mathbf{a}_1, \tag{4.30}$$

$$\mathbf{a}_3 \equiv \mathbf{e}_3 + \lambda_3^1 \mathbf{a}_1 + \lambda_3^2 \mathbf{a}_2, \tag{4.31}$$

$$\mathbf{a}_4 \equiv \mathbf{e}_4 + \lambda_4^1 \mathbf{a}_1 + \lambda_4^2 \mathbf{a}_2 + \lambda_4^3 \mathbf{a}_3, \tag{4.32}$$

and so on. Notice that, except for the first term on the right hand of these equations, all the others involve the $\mathbf{a}_\mu$'s, not the $\mathbf{e}_\mu$'s. In this way, we make maximum use of orthogonality.

The λ's will be determined so that $(\mathbf{a}_\mu)$ be orthogonal. Dot-multiplying (4.30) by $\mathbf{a}_1$, we get

$$0 = \mathbf{e}_2 \cdot \mathbf{a}_1 + \lambda_2^1, \tag{4.33}$$

which allows us to obtain λ_2^1.

Next we dot-multiply (4.31) by $\mathbf{a}_1$. We also multiply (4.31) by $\mathbf{a}_2$. We thus get

$$0 \equiv \mathbf{e}_3 \cdot \mathbf{a}_1 + \lambda_3^1, \tag{4.34}$$

$$0 \equiv \mathbf{e}_3 \cdot \mathbf{a}_2 + \lambda_3^2, \tag{4.35}$$

which determine λ_3^1 and λ_3^2. Notice that we knew at this point what $\mathbf{a}_1$ and $\mathbf{a}_2$ are in terms of $\mathbf{e}_1$ and $\mathbf{e}_2$, so that we can compute the dot products. We now multiply (4.32) by $\mathbf{a}_1$, $\mathbf{a}_2$ and $\mathbf{a}_3$ to obtain three equations that yield λ_4^1, λ_4^2 and λ_4^3. And so on. Finally, as stated, we orthonormalize the elements $\mathbf{a}_\mu$ proceeding in the same order in which they were obtained.

Orthonormalization of bases and reduction of quadratic forms

Let us refer a vector $\mathbf{v}$ to the bases $(\mathbf{e}_\mu)$ and $(\hat{\mathbf{a}}_\mu)$, where $\hat{\mathbf{a}}_\mu$ results from the normalization of $\mathbf{a}_\mu$:

$$\mathbf{v} = v^\mu \mathbf{e}_\mu = V^\nu \hat{\mathbf{a}}_\nu. \tag{4.36}$$

We could also have used on the right hand side the index μ, but inexperienced readers may inadvertently think that this is an equality term by term. We then have:

$$v^2 = v^\mu v^\nu g_{\mu\nu} = (V^1)^2 + (V^2)^2 + ... + (V^n)^2. \tag{4.37}$$

Hence the problem of orthonormalizing a basis is equivalent to the problem of reducing a symmetric quadratic form to the Cartesian form

$$\sum_\mu (V^\mu)^2. \tag{4.38}$$

This reduction is equivalent to finding linear combinations V^μ's of the v^ν's so that, when we substitute them in (4.38), we recover $v^\mu v^\nu g_{\mu\nu}$.

Remark concerning bases of spaces that are not properly Euclidean

Assume a 2-D Lorentzian space. Consider a basis $(\mathbf{a}_0, \mathbf{a}_1)$ such that $\hat{\mathbf{a}}_0^2 = 1$, $\hat{\mathbf{a}}_1^2 = -1$ and $\hat{\mathbf{a}}_0 \cdot \hat{\mathbf{a}}_1 = 0$. Define another basis, $(\mathbf{e}_+, \mathbf{e}_-)$, as

$$\mathbf{e}_+ \equiv \frac{\hat{\mathbf{a}}_0 + \hat{\mathbf{a}}_1}{\sqrt{2}}, \quad \mathbf{e}_- \equiv \frac{\hat{\mathbf{a}}_0 - \hat{\mathbf{a}}_1}{\sqrt{2}}. \tag{4.39}$$

We then have

$$\mathbf{e}_+^2 = 0, \quad \mathbf{e}_-^2 = 0, \quad \mathbf{e}_+ \cdot \mathbf{e}_- = 1. \tag{4.40}$$

The equation in

$$\mathbf{P} \equiv t\hat{\mathbf{a}}_0 + x\hat{\mathbf{a}}_1 = y^+ \mathbf{e}_+ + y^- \mathbf{e}_-, \tag{4.41}$$

allows us to relate (t, x) to (y^+, y^-) by using (4.39) in (4.41),

$$t\hat{\mathbf{a}}_0 + x\hat{\mathbf{a}}_1 = y^+ \frac{\hat{\mathbf{a}}_0 + \hat{\mathbf{a}}_1}{\sqrt{2}} + y^- \frac{\hat{\mathbf{a}}_0 - \hat{\mathbf{a}}_1}{\sqrt{2}}, \tag{4.42}$$

thus obtaining

$$t = \frac{y^+ + y^-}{\sqrt{2}}, \quad x = \frac{y^+ - y^-}{\sqrt{2}}. \tag{4.43}$$

Hence,

$$\mathbf{P} \cdot \mathbf{P} = t^2 - x^2 = 2y^+ y^-. \tag{4.44}$$

In other words

$$\mathbf{P} \cdot \mathbf{P} = g_{\mu\nu} y^\mu y^\nu, \tag{4.45}$$

with

$$y^1 \equiv y^+, \quad y^2 \equiv y^- \tag{4.46}$$

and

$$g_{11} = g_{22} = 0, \quad g_{12} = g_{21} = 1. \tag{4.47}$$

We have thus seen that, when a quadratic expression is not positive definite — $(t^2 - x^2)$ is not—, we can have bases where all the $g_{\mu\mu}$ (diagonal terms) are zero.

Cartan's reduction of a quadratic symmetric form to a sum of squares

We have seen that diagonalization and orthonormalization are two aspects of the same problem. We shall now be interested in directly diagonalizing a given symmetric quadratic form. The process we are about to consider applies when at least one of the $g_{\mu\mu}$ is not zero. That will be the only case we are here interested in, since there is not significant motivation that we know of to consider the alternative case. Readers interested in the case $g_{11} = g_{22} = ... = g_{nn} = 0$ should refer to Cartan [23].

Let g_{11} denote a $g_{\mu\mu}$ different from zero. Let $a, b = (2, 3, ...n), ...$ The quadratic form

$$g'_{ab}x^a x^b \equiv g_{\mu\nu}x^\mu x^\nu - \frac{1}{g_{11}}(g_{11}x^1 + g_{12}x^2 + ... + g_{1n}x^n)^2 \qquad (4.48)$$

does not contain the variable x^1. Define

$$y^1 \equiv g_{1\mu}x^\mu, \qquad (4.49)$$

and rewrite (4.48) as

$$g_{\mu\nu}x^\mu x^\nu = \frac{1}{g_{11}}(y^1)^2 + g'_{ab}x^a x^b. \qquad (4.50)$$

We would now proceed in the same way with $g'_{ab}x^a x^b$. End of proof.

The metric on a surface depends on only two independent coordinates, usually called parameters to distinguish them from the coordinates of the 3-D Euclidean space in which we consider the surface to be embedded. Let x^λ refer to these two coordinates. Equation (4.50) then reduces to a sum of squares:

$$g_{\mu\nu}x^\mu x^\nu = \left(\frac{y^1}{\sqrt{g_{11}}}\right)^2 + \left(g'_{22}x^2\right)^2. \qquad (4.51)$$

This will be used in appendix A.

3.5 Not quite right concept of VECTOR FIELD

Many readers will have come across the concept of tensor field as a set of functions that under a change of coordinates transform in such a such a way. The transformation involves partial derivatives of one system of coordinates with respect to another. This is a horrible approach to the concept of tensor fields, and in particular to vector fields (which are tensor fields of "rank or grade or degree one"). One can certainly act in reverse if circumstances so advice, by endowing some set of quantities with rules that make it become a vector or tensor space. One must then understand the structure that, in so doing, one would be creating. To avoid clutter and make things simple, we only discuss vector fields.

Notice how we defined vectors in section 2. We first defined a set with appropriate properties. Vectors are simply the members of those sets, now structures. The definition of the previous paragraph does not make reference to the structure to which tensor and tensor fields belong. It has to be inferred from the specific application that one makes of those tensor fields, if possible.

One should first define the tensor product of vector spaces. Tensors are the members of those product spaces, like vectors are the members of vector spaces. In this section, we are interested in vector fields. A vector field on some manifold, region, surface, etc. is a vector at each point of the manifold, region, surface etc. The vectors at different points will in general belong to different vector spaces. That is where differential geometry becomes interesting.

When specialized to contravariant vector fields, the definition of the opening statement of this section reads: "a contravariant vector field is a set of quantities v^i that under a change of coordinates transforms like

$$v'^j = \frac{\partial x'^j}{\partial x^i} v^i."$$
(5.1)

This expression represents how the components of a tangent vector field in one vector basis field are related to the components of the same vector in another vector basis field. At each point, the $\partial x'^j / \partial x^i$ are values of coordinates $A^{i'}_j$ in structures that we shall introduce later under the names of frame bundles and bundles of bases. In those very important structures, the $A^{i'}_j$ are additional to and independent of the x coordinates.

The vector fields components in (5.1) are not so relative to general basis fields, but to "coordinate basis fields" of tangent vectors. The concept of tangent vector will be introduced in due time. From the definition (5.1), it is not clear what the field is a field of.

The preceding considerations will be unintelligible to many readers. We have not yet developed the machinery to understand with examples why (5.1) is a wrong concept of contravariant vector field, since there are also other entities with components that transform in the same way. It is much simpler —because we have clearer examples at this point— to discuss the definition of components of covariant vector fields given in the same publications where (5.1) is introduced. They are sets of quantities λ_i that under a change of coordinates transform like

$$\lambda'_i = \frac{\partial x^j}{\partial x'^i} \lambda_j.$$
(5.2)

This correctly states how the components of a cotangent vector field (also called covariant vector field) transform under a change of coordinate field of bases. But (5.2) also apply to the components of our differential 1−forms, which are functions of curves, not functions of vectors. In addition, if the vector space V^n is Euclidean, we might be referring to a tangent vector field (also called contravariant vector field) in terms of reciprocal bases. Hence, how something transforms is not an indicator of the nature of an object, unless, of course, one is dealing with the blunt instrument of the tensor calculus, where there is a poorer palette of concepts. Notice that we criticized tensor calculus, not tensor algebra. The former is contra natura. The second is present almost anywhere in differential geometry, indirectly through quotient algebras (see a little bit more on this in the next section)..

To summarize: (5.1) does not refer only to contravariant vector fields. When it does, it characterizes only sets of components of vector fields relative to arbitrary bases of these.

3.6 Tensor products: theoretical minimum

In this book, we shall restrict ourselves to a minimum of concepts related to tensors, the next section being an exception. As we are about to see, readers have already been dealing with tensor products when multiplying different structures rather than different copies of one structure. These products are so natural that they go unnoticed. That familiarity is almost sufficient for our purposes.

Consider the expression $dx\mathbf{i} + dy\mathbf{j} + dz\mathbf{k}$, also written as

$$\delta_i^j dx^i \mathbf{a}_j. \tag{6.1}$$

It is a member of the tensor product of a module of differential $1-$forms with a (tangent) vector space. The module is a submodule of the algebra of differential forms, itself a module as any algebra is in the first place. The $(\mathbf{a}_j)$ retrospectively represent, as we shall see, a field of vector bases which, in the particular case of the $(\mathbf{i}, \mathbf{j}, \mathbf{k})$, simply happens to be a constant basis field.

Inasmuch as we know how to handle these quantities, we know how to deal with tensor products of different structures. The more common version of the tensor calculus does not make explicit use of bases, only of their components; thus (6.1) would be given as δ_i^j, in that tensor calculus. One does not need to know how components transform if the new components (in terms of the old ones) can be read from the expression that results from substitution of the old basis in terms of the new one in expressions such as (6.1). Notice also that we can change just the basis of differential forms, or just the basis of tangent vectors. This is not the case in the most common version of the tensor calculus, where both changes go together. The freedom just mentioned is an advantage of making the bases explicit.

Since the expression $\delta_i^j dx^i \mathbf{a}_j$ is a vector-valued differential form, it is clear why an in-depth understanding of what vector-valued (and, in general, tensorvalued) differential forms are involves tensor products. But a lack of major immersion in tensor algebra in this book will not produce a large loss in understanding of the topics of practical interest. For this reason, the next section can be ignored by readers to which these concepts are totally foreign, unless they are truly curious and/or challenged.

Another way in which tensor algebra is involved in this book is that exterior algebra, as in the previous chapter, is a so called quotient algebra of the general tensor algebra constructed upon the module of differential $1-$forms. The general tensor algebra for a given vector space is the algebra of all tensors of all ranks. Quotient algebra is a powerful concept but too subtle for those not very knowledgeable in algebra. It need not be dealt with, provided we know how to work in it. We must understand, however, that quotient algebras are not subalgebras, since the product in subalgebras remain the same as in the respective algebras. In the quotient algebras (exterior, Clifford), on the other hand, the products are something different, even if its objects look like those in the mother algebra. In order to connect with the contents of the previous paragraph, let us state that tensor-valued differential forms are members of a structure which is the tensor product of some tensor space by the algebra of scalar-valued differential forms. But this statement needs to be revisited as we go deeper and deeper into differential geometry.

In the tensor calculus, the concept of connection, which is central in Yang-Mills differential geometry, is undervalued. A pernicious effect of this is the lack of a good concept of curvature in modern differential geometry for physicists. Both connection and curvature are as legitimate Lie algebra valued differential forms in "non-Yang-Mills" as in Yang-Mills differential geometry. With this perspective, $SU(2)$ Yang-Mills theory is identical to the classical Euclidean geometry of algebraic spinors in three dimensions (see chapter 6). The issue is whether the Lie algebra $su(2)$ that appears in non-Yang-Mills differential geometry has anything to do with the $su(2)$ algebra of Yang-Mills theory. One has implicitly assumed in the literature that the answer is in the negative, but this might one day be viewed as an incorrect inference done at the time when the Lie algebras emerging in non-Yang-Mills context are largely being overlooked or ignored because of the tensor calculus.

3.7 Formal approach to TENSORS

3.7.1 Definition of tensor space

This section can be ignored without major consequence. On the other hand, readers who are interested in knowing even more about this subject may refer to a beautiful book by Lichnerowicz [53].

Consider the abstract vector spaces U^m, V^n, and $W^{m \times n}$. Let $\mathbf{u} \in U^m$, $\mathbf{v} \in V^n$ and $\mathbf{w} \in W^{m \times n}$. Consider a map $(\mathbf{u}, \mathbf{v}) \to \mathbf{w}$, which we shall designate as $(\mathbf{u}, \mathbf{v}) \to \mathbf{u} \otimes \mathbf{v}$, with the following properties:

(a)

$$\mathbf{u} \otimes (\mathbf{v}_1 + \mathbf{v}_2) = \mathbf{u} \otimes \mathbf{v}_1 + \mathbf{u} \otimes \mathbf{v}_2, \quad \forall (\mathbf{u}, \mathbf{v}_1, \mathbf{v}_2),$$
$$(\mathbf{u}_1 + \mathbf{u}_2) \otimes \mathbf{v} = \mathbf{u}_1 \otimes \mathbf{v} + \mathbf{u}_2 \otimes \mathbf{v}, \quad \forall (\mathbf{u}_1, \mathbf{u}_2, \mathbf{v}),$$

(b) if α is an arbitrary scalar: $(\alpha \mathbf{u}) \otimes \mathbf{v} = \mathbf{u} \otimes (\alpha \mathbf{v}) = \alpha(\mathbf{u} \otimes \mathbf{v})$,

(c) if $(\mathbf{a}_1, \ldots, \mathbf{a}_m)$ and $(\mathbf{b}_1, \ldots, \mathbf{b}_n)$ designate arbitrary bases, respectively of U^m and V^n, then the $\mathbf{a}_i \otimes \mathbf{b}_\alpha$, $(i = 1, \ldots, m, \alpha = 1, \ldots, n)$ constitute a basis $\mathbf{c}_{i\alpha}$ of the vector space $W^{m \times n}$.

We then say that the vector space $W^{m \times n}$ is (can be viewed, becomes) the tensor product $U^m \otimes V^n$ of the vector spaces U^m and V^n. The vectors of

$W^{m \times n}$, when considered as elements of $U^m \otimes V^n$, are called tensors. Pedestrian approaches to the tensor calculus based on behavior under coordinate transformations just miss the algebraic nature of these objects.

A linear combination of tensor products is a tensor, but the converse is not true: sums of tensors may or may not be written as a tensor products. A sum of tensors is a tensor and multiplication of a tensor by a scalar is another tensor.

Properties (a) and (b) are familiar properties. Less familiar and more subtle is (c). How could it not be satisfied? The dot, exterior and vector products do not satisfy (c). Suppose for simplicity that U^m and V^n were just two copies of a two dimensional vector space, V^2, and let $(\mathbf{a}_i)$ be a basis in V^2. We have

$$\mathbf{a}_1 \wedge \mathbf{a}_1 = \mathbf{a}_2 \wedge \mathbf{a}_2 = 0, \quad \mathbf{a}_1 \wedge \mathbf{a}_2 = -\mathbf{a}_2 \wedge \mathbf{a}_1. \tag{7.1}$$

Notice that the subspace of those products for this particular example is now 1-dimensional, rather than 4-dimensional. This product, the exterior product (which we used for differential forms), is not a tensor product. For another example, we could have

$$\mathbf{a}_1 \vee \mathbf{a}_1 = \mathbf{a}_2 \vee \mathbf{a}_2 = 1, \quad \mathbf{a}_1 \vee \mathbf{a}_2 = -\mathbf{a}_2 \vee \mathbf{a}_1. \tag{7.2}$$

The algebra of $\vee$ is one where the product of two vectors is constituted by the sum of their exterior and dot products. In the $\vee-$algebra, there are subspaces of scalars, vectors, bivectors, stopping at multivectors of grade n. The "1" in (7.2) is the unit scalar. We shall not enter to discuss this product further, as it involves new subtleties.

From the axioms one readily obtains that

$$\mathbf{u} \otimes \mathbf{v} = u^i v^\alpha \mathbf{c}_{i\alpha}, \tag{7.3}$$

where $\mathbf{u} = u^i \mathbf{a}_i$ and $\mathbf{v} = v^\alpha \mathbf{b}_\alpha$. Vice versa, one can show that this composition law is the only one that satisfies all three axioms.

3.7.2 Transformation of components of tensors

In differential geometry, we shall encounter tensor products $\mathbf{u} \otimes \mathbf{v}$ where $\mathbf{u}$ and $\mathbf{v}$ belong to two different vector spaces. At this point, let us concentrate on tensors which result from tensor product $V^n \otimes V^n$ of a vector space by itself. We shall use the term second rank tensors to refer to the elements of $V^n \otimes V^n$. A second rank tensor, alternatively called "of rank 2", can be written as

$$\mathbf{T} = T^{ij} \mathbf{a}_i \otimes \mathbf{a}_j. \tag{7.4}$$

If we choose a new basis in V^n, the same tensor can be written as

$$\mathbf{T} = T'^{ij} \mathbf{a}'_i \otimes \mathbf{a}'_j = T'^{ij} \mathbf{a}_{i'} \otimes \mathbf{a}_{j'}, \tag{7.5}$$

with $T'^{i'j'} \equiv T'^{ij}$ and $\mathbf{a}'_{i'} \equiv \mathbf{a}'_i$ in order to facilitate operations where the $A^{i'}_k$ or the $A^i_{k'}$ are involved. Substitution of (2.3b) in (7.4) and comparison with (7.5) yields

$$T'^{kl} = A^{k'}_i A^{l'}_j T^{ij}, \tag{7.6a}$$

where we have removed on the left the redundant use of primed indices. Similarly, substitution of (2.3a) in (7.5) and comparison with (7.5) yields:

$$T^{ij} = A^i_{k'} A^j_{l'} T'^{kl} \tag{7.6b}$$

In obtaining Eqs. (7.6), we have used that the $(\mathbf{a}_i \otimes \mathbf{a}_j, \forall i, j \leq n)$ and the $(\mathbf{a}'_i \otimes \mathbf{a}'_j, \forall i, j \leq n)$ constitute bases of $\boldsymbol{V}^n \otimes \boldsymbol{V}^n$.

Notice that, in Eq. (7.6a), n^2 quantities $A^{i'}_k$ suffice to describe the transformation of the tensors belonging to $\boldsymbol{V}^n \otimes \boldsymbol{V}^n$. Whereas the linear transformations of $\boldsymbol{V}^{n \times n}$ are given by all the invertible $n^2 \times n^2$ matrices which constitute the group $GL(n^2)$ of n^4 parameters, in $\boldsymbol{V}^n \otimes \boldsymbol{V}^n$ one only considers a subgroup of all these possible transformations. It only has n^2 parameters $A^i_{k'}$ $(i, k' = 1 \ldots n)$, the transformations now being quadratic in these parameters.

A second rank *tensor field* is a tensor at each point in some domain. In this case, we have

$$T^{ij}(x) = A^i_{k'}(x) A^j_{l'}(x) T'^{kl}(x) \tag{7.7}$$

where x denotes any number of coordinates in a given space. We shall see in future chapters that the $A^i_{k'}(x)$ could take the form $\partial x^i / \partial x'^k$. When that is the case, the last equation becomes

$$T^{ij}(x) = \frac{\partial x^i}{\partial x'^k} \frac{\partial x^j}{\partial x'^l} T'^{kl}(x). \tag{7.8}$$

This formula is used for the type of definition of tensor field to which we referred in section 5. It applies only to certain types of bases, the only bases that are used in pedestrian approaches to the tensor calculus.

By successive application of the concept of tensor product of two spaces, one can generate the tensor product of several vector spaces. Since $\boldsymbol{U}^m \otimes \boldsymbol{V}^n$ is a vector space itself, we define $\boldsymbol{U}^m \otimes \boldsymbol{V}^n \otimes \boldsymbol{X}^p$ as the tensor product $(\boldsymbol{U}^m \otimes \boldsymbol{V}^n) \otimes \boldsymbol{X}^p$. This will in turn be a vector space of dimension $m \times n \times p$. We could also have defined the tensor product $\boldsymbol{U}^m \otimes (\boldsymbol{V}^n \otimes \boldsymbol{X}^p)$, which is also a vector space of dimension $m \otimes n \otimes p$. Let $(\mathbf{a}_i)$, $(\mathbf{b}_\alpha)$ constitute respective bases of $\boldsymbol{U}^m$, $\boldsymbol{V}^n$ and $\boldsymbol{X}^p$. We shall identify $(\mathbf{a}_i \otimes \mathbf{b}_\alpha) \otimes \mathbf{c}_A$ with $\mathbf{a}_i \otimes (\mathbf{b}_\alpha \otimes \mathbf{c}_A)$, and shall refer to these products as simply $\mathbf{a}_i \otimes \mathbf{b}_\alpha \otimes \mathbf{c}_A$. In the same way, we do not need the parentheses when referring to the tensor product of $\boldsymbol{U}^m$, $\boldsymbol{V}^n$ and $\boldsymbol{X}^p$. Hence the tensor product $\boldsymbol{U}^m \otimes \boldsymbol{V}^n \otimes \boldsymbol{X}^p$ is the vector space $\boldsymbol{V}^{m \times n \times p}$ in which we only consider those changes of bases introduced by the changes of bases in $\boldsymbol{U}^m$, $\boldsymbol{V}^n$ and $\boldsymbol{X}^p$ through the mapping $(\mathbf{u}, \mathbf{v}, \mathbf{x}) \to (\mathbf{u} \otimes \mathbf{v} \otimes \mathbf{x})$.

Tensors of rank r that are members of the tensor product of a vector space by itself (r factors) can be written as

$$\mathbf{T} = T^{i_1 \ldots i_r} \mathbf{a}_{i_1} \otimes \ldots \otimes \mathbf{a}_{i_r}. \tag{7.9}$$

If we choose a new basis for $\boldsymbol{V}^n$, we induce a choice of a new basis for $\boldsymbol{V}^n \otimes \boldsymbol{V}^n \otimes \ldots$, and the same tensor can now be expressed as

$$\mathbf{T} = T'^{i_1 \ldots i_r} \mathbf{a}'_{i_1} \otimes \ldots \otimes \mathbf{a}'_{i_r}. \tag{7.10}$$

Substitution of (2.3a) in (7.10) and comparison with (7.9) yields

$$T^{i_1 \cdots i_r} = A^{i_1}_{j'_1} \cdots A^{i_r}_{j'_r} T'^{j_1 j_2 \cdots j_r}. \tag{7.11}$$

Similarly, substitution of (2.3b) in (7.9) and comparison with (7.10) yields

$$T'^{i_1 \cdots i_r} = A^{i'_1}_{j_1} \cdots A^{i'_r}_{j_r} T^{j_1 j_2 \cdots j_r}. \tag{7.12}$$

Notice again that the transformations in V^n yield the transformations in $V^n \otimes V^n \otimes \ldots \otimes V^n$. The n^2 independent quantities $A^{i'}_j$ describe the transformations of tensor of any rank.

Once more, scalars (which usually are real or complex numbers) and vectors are said to be of ranks zero and one respectively. We speak of rank. It would be more technical to speak of grade, since the set of all tensors of all grades and their sums (generally inhomogeneous, meaning that the grades are mixed) is a so called graded algebra. We do as we do in order to emphasize that differential forms are not tensors. Using the terms rank for tensor algebra and grade for quotient algebras will help us to remember that.

In all of the above, V^n could have been replaced with V^{*n}. And both in turn could be replaced with fields of such vector spaces, one of each at each point of some manifold. And these fields (same notation as the vector spaces themselves) can be tensor multiplied by a module of differential forms. Thus, one way of looking at curvatures is as elements of the tensor product of $V^{*n} \otimes V^n$ by a module of differential 2−forms (There is also a concept of curvature that is an element of $V^{*n} \otimes V^n \otimes V^{*n} \otimes V^{*n}$). We shall later see what the concept of Lie algebra valuedness has to do with this.

Algebras and ideals also could be factors in tensor product of structures, which is everywhere. And we do not need to be aware of them because their defining properties (a) and (b) are all too natural.

3.8 Clifford algebra

3.8.1 Introduction

Some readers might consider reading this section just before section 6.4. By then, they will have acquired practice with exterior algebra, which brings them closer to Clifford algebra.

We here give a slight idea of an associative algebra, called Clifford algebra, that goes beyond the exterior algebra on which the exterior calculus is based. It will be used in section 6.4. When it makes occasional appearance in differential geometry, its presence is disguised, taking the form of ad hoc concepts. In the tensor calculus, the Levi-Civita tensor is one such concept. One contracts it with other tensors in order to obtain what, in Clifford algebra, are called their Hodge duals. One useful purpose of Hodge duality is that it permits one to obtain interior derivatives (following Kähler, the term derivative is not restricted only to operations satisfying the standard Leibniz rule) by combining Hodge duality with exterior differentiation in order to obtain "interior differentiation". This conceptual replacement is only possible in Euclidean, pseudo–Euclidean, Riemannian and pseudo-Riemannian spaces.

The vector product and, therefore, vector algebra are peculiarities of 3-D Euclidean vector space (in the rest of this book, Euclidean and Riemannian will also mean pseudo-Euclidean and pseudo-Riemannian, except when indicated otherwise). They do not exist in arbitrary dimension. All those spaces, however, have a Clifford algebra associated with them, different from one signature to another.

A vector product is the combination of a exterior product and the duality operation. Except in the most general cases (meaning when both factors in a product are of grade two or greater), Clifford product is the combination of the exterior and interior product (read dot product if you prefer). The vector product does not combine with the dot product. That is one of its handicaps.

Many elements of the algebra have inverses. Among those that do not, most interesting are the elements that are equal to their squares (and, therefore, to all their integer powers). They are called idempotents. Other important concepts are spinors, which are members of ideals in the algebra.

In view of what has been said in the last two paragraphs, we should consider Clifford algebra as the true, natural, canonical, proprietary, defining algebra of spaces endowed with a metric. Examples follow. Complex algebra is the Clifford algebra of one-dimensional pseudo-Euclidean vector space. Quaternions constitute an algebra isomorphic to the Clifford algebra of 2-D Euclidean space with signature (-1,-1), although it is formulated as if it were associated with three dimensions. The algebras of the Pauli and Dirac matrices are the algebra of 3-D Euclidean and 4-D Lorentzian vector spaces respectively.

Corresponding to a Euclidean vector space of dimension n, there is a Clifford algebra which, as a vector space, is of dimension 2^n. Thus, the Pauli algebra is a space of dimension 8. A basis of the algebra is constituted by the matrices

$$\mathbf{I},\ \sigma_1,\ \sigma_2,\ \sigma_3,\ \sigma_1\sigma_2,\ \sigma_2\sigma_3,\ \sigma_3\sigma_1,\ \sigma_1\sigma_2\sigma_3. \qquad (8.1)$$

It happens that the square of $\sigma_1\sigma_2\sigma_3$ is the unit matrix times the unit imaginary. But it is better to think of the unit imaginary times the unit matrix as the unit element of grade three in the algebra, like σ_1, σ_2 and σ_3 are units of grade one and $\sigma_1\sigma_2$, $\sigma_2\sigma_3$ and $\sigma_3\sigma_1$ are units of grade two. Warning, we may speak of $\sigma_1\sigma_2$, $\sigma_2\sigma_3$, $\sigma_3\sigma_1$ as units of grade two only because they are orthogonal and, therefore, they are equal to $\sigma_1 \wedge \sigma_2$, $\sigma_2 \wedge \sigma_3$, $\sigma_3 \wedge \sigma_1$. Exterior products are of grade two, whether the factors σ_i are orthogonal or not.

3.8.2   Basic Clifford algebra

Assume the existence of a product which is associative and distributive with respect to addition, but not necessarily commutative or anticommutative. We form the following identity

$$ab \equiv (1/2)(ab + ba) + (1/2)(ab - ba). \qquad (8.2)$$

We introduce symbols to name these two parts of the product as individual products themselves:

$$a \cdot b \equiv (1/2)(ab + ba), \qquad a \wedge b \equiv (1/2)(ab - ba). \qquad (8.3)$$

It is clear that $a \wedge a = 0$ and $a \wedge b = -b \wedge a$. In anticipation of what we are about to say, we have introduced the symbol for dot product to represent the symmetric part of the Clifford product.

We have not yet characterized Clifford algebra. This must be clear when we realize that tensor products satisfy the properties of being associative and distributive with respect to addition, since $(1/2)(ab + ba)$ would be a two tensor if a and b were vectors. We associate a scalar with it, specifically what we know as the dot product of the two vectors. ab then is their Clifford product. If the foregoing sounds too abstract, think of the products of the (four gamma) matrices of relativistic quantum mechanics. They are Clifford products. $(1/2)(ab + ba)$ then is a scalar multiple of the unit matrix. Such multiples play the roles of scalars in matrix algebras.

It is obvious from equations (8.2) and (8.3) that

$$ab = a \cdot b + a \wedge b, \qquad (8.4)$$

which brings the exterior and dot products closer to each other than the vector and dot products are.

We are interested in understanding how basic elements of Euclidean vector calculus (we said Euclidean in order to have the dot product) are present in the exterior calculus. Consider any two non-colinear vectors in a 3-dimensional subspace of an n-dimensional vector space. We shall denote as $(\mathbf{i}, \mathbf{j}, \mathbf{k})$ an orthonormal basis of this subspace. We then have

$$\mathbf{a} \wedge \mathbf{b} = (a_1\mathbf{i} + a_2\mathbf{j} + a_3\mathbf{k}) \wedge (b_1\mathbf{i} + b_2\mathbf{j} + b_3\mathbf{k}) =$$

$$= (a_1 b_2 - a_2 b_1)\mathbf{i} \wedge \mathbf{j} + (a_2 b_3 - a_3 b_2)\mathbf{j} \wedge \mathbf{k} + (a_3 b_1 - a_1 b_3)\mathbf{k} \wedge \mathbf{i}, \qquad (8.5)$$

where we have used antisymmetry. In terms of $\mathbf{i} \wedge \mathbf{j}$, $\mathbf{j} \wedge \mathbf{k}$ and $\mathbf{k} \wedge \mathbf{i}$, the components of the expression on the right hand side of (8.5) are the same as for the vector product, but one no longer associates $\mathbf{i} \wedge \mathbf{j}$, $\mathbf{j} \wedge \mathbf{k}$ and $\mathbf{k} \wedge \mathbf{i}$ with $\mathbf{k}$, $\mathbf{i}$ and $\mathbf{j}$. We shall say that $\mathbf{a} \wedge \mathbf{b}$ represents oriented 2-dimensional figures in the plane determined by $\mathbf{a}$ and $\mathbf{b}$, and of size $ab\sin\theta$, where θ is the angle between $\mathbf{a}$ and $\mathbf{b}$. $\lambda(\mathbf{a} \wedge \mathbf{b})$ and $\lambda(\mathbf{b} \wedge \mathbf{a})$ represent the same figures but with opposite orientation and magnitude, $\lambda ab\sin\theta$.

The contents of the parentheses on the right hand side of (8.5) are said to represent the components of the "bivector" $\mathbf{a} \wedge \mathbf{b}$ in terms of the bivectors $\mathbf{i} \wedge \mathbf{j}$, $\mathbf{j} \wedge \mathbf{k}$ and $\mathbf{k} \wedge \mathbf{i}$. The exterior product itself represents two dimensional figures of the same size as the parallelogram built with the two vectors as sides and in the same plane.

Let us denote as $\mathbf{a}_i$ the vectors of a basis in arbitrary dimension. From the first of equations (8.3), we have

$$\mathbf{a}_i \mathbf{a}_j + \mathbf{a}_j \mathbf{a}_i - 2g_{ij} = 0, \qquad (8.6)$$

where g_{ij} is the dot product of $\mathbf{a}_i$ and $\mathbf{a}_j$. Equation (8.6) is taken as the equation that defines the Clifford algebra, of which many definitions exist.

The Clifford product of a vector and a multivector (bivector, trivector, etc.) also satisfy a relation like (8.4), namely

$$aA = a \cdot A + a \wedge A. \tag{8.7}$$

If A is of homogeneous grade r, aA will have in general a part of grade $r - 1$ and another part of grade $r + 1$.

A Clifford algebra may be constructed not only upon a vector space but also upon a module of differential 1-forms. In that case, the algebra receives the specific name of Kähler algebra. I shall sometimes use the symbol $\vee$ for Clifford product, instead of juxtaposition; otherwise $dxdy$, meaning $dx \vee dy$, could be confused with $dxdy$ under the integral sign. In parallel with (8.6), the defining relation of the Kähler algebra of "clifforms" is

$$dx^\mu \vee dx^\nu + dx^\nu \vee dx^\mu = 2g^{\mu\nu}. \tag{8.8}$$

3.8.3   The tangent Clifford algebra of 3-D Euclidean vector space

A little bit of practice with in 3-D Euclidean space without the use of matrices will be helpful to get the gist of this algebra. In order to emphasize that matrices do not have to do with its essence, we do not use the symbols $\boldsymbol{\sigma}_i$ but rather notation typical of geometry.

Because of orthogonality, we have

$$\mathbf{i} \cdot \mathbf{j} = 0, \qquad and \qquad \mathbf{i} \vee \mathbf{j} = \mathbf{i} \wedge \mathbf{j}. \tag{8.9}$$

On the other hand,

$$\mathbf{i} \vee \mathbf{i} = \mathbf{i} \cdot \mathbf{i}. \tag{8.10}$$

Members of the Clifford algebra of 3-D Euclidean three space can be expanded in terms of the following basis of unit elements:

$$1, \ \mathbf{i}, \ \mathbf{j}, \ \mathbf{k}, \ \mathbf{j} \wedge \mathbf{k}, \ \mathbf{k} \wedge \mathbf{i}, \ \mathbf{i} \wedge \mathbf{j}, \ \mathbf{i} \wedge \mathbf{j} \wedge \mathbf{k}, \tag{8.11}$$

which is nothing but (8.1). A generic name for the elements in (8.11) is multivectors.

More important than even literally reproducing standard vector identities in this language is the knowledge of what expressions play the same role in the Clifford algebra. Thus, for instance, the volume scalar $\mathbf{u} \cdot (\mathbf{v} \times \mathbf{w})$ is replaced with the trivector

$$\mathbf{u} \wedge \mathbf{v} \wedge \mathbf{w}, \tag{8.12}$$

which is not a scalar. To be a volume is to be a trivector, like to be an area is to be a bivector.

As another example consider the vector identity

$$\mathbf{u} \times (\mathbf{v} \times \mathbf{w}) = \mathbf{v}(\mathbf{u} \cdot \mathbf{w}) - \mathbf{w}(\mathbf{u} \cdot \mathbf{v}). \tag{8.13}$$

What matters is not its translation to Clifford algebra, but rather that its role is played by

$$\mathbf{u} \cdot (\mathbf{v} \wedge \mathbf{w}) = (\mathbf{u} \cdot \mathbf{v})\mathbf{w} - (\mathbf{u} \cdot \mathbf{w})\mathbf{v}, \tag{8.14}$$

which is a particular case of the rule to dot-multiply a vector by a multivector

$$\mathbf{u} \cdot (\mathbf{v} \wedge \mathbf{w} \wedge \mathbf{r}...) = (\mathbf{u} \cdot \mathbf{v})\mathbf{w} \wedge \mathbf{r}.. - (\mathbf{u} \cdot \mathbf{w})\mathbf{v} \wedge \mathbf{r}... + (\mathbf{u} \cdot \mathbf{r})\mathbf{v} \wedge \mathbf{w}... - ..., \tag{8.15}$$

with alternating signs.

An important contribution of Clifford algebra is its treatment of rotations, which is similar for vectors, general members of the algebra and spinors. A multiplication on the left by a certain exponential rotates a left spinor. The same multiplication followed by multiplication by the inverse exponential on the right rotates elements in the Clifford algebra. Two theorems of 3-D Euclidean geometry (namely that reflections of a vector with respect to two perpendicular directions in the same plane are opposite, and that the product of two reflections is a rotation) allow one to readily get the generators of SU(2) as the bivectors of the tangent Clifford algebra of 3-space.

3.8.4 The tangent Clifford algebra of spacetime

This subsection is meant for readers familiar with gamma matrices but not with Clifford algebra. The Clifford algebra of spacetime is defined by

$$\gamma_\mu \vee \gamma_\nu + \gamma_\nu \vee \gamma_\mu = 2\eta_{\mu\nu}, \tag{8.16}$$

where $\eta_{\mu\nu} = (1, -1, -1, -1)$. We proceed to state what remains to be done with the symbols for those algebras in order to work with them as we do in Clifford algebra. We further define

$$\gamma_\mu \wedge \gamma_\nu \equiv \frac{1}{2}(\gamma_\mu \vee \gamma_\nu - \gamma_\nu \vee \gamma_\mu), \tag{8.17}$$

and

$$\gamma_\mu \cdot \gamma_\nu \equiv \frac{1}{2}(\gamma_\mu \vee \gamma_\nu + \gamma_\nu \vee \gamma_\mu), \tag{8,18}$$

where we have used the symbol $\vee$ in order to emphasize that we are dealing with unit vectors and not matrices. Obviously

$$\gamma_1 \gamma_2 = \gamma_1 \wedge \gamma_2 + \gamma_1 \cdot \gamma_2. \tag{8.19}$$

Because of orthonormality, we clearly have

$$\gamma_\mu \gamma_\nu = \gamma_\mu \wedge \gamma_\nu = -\gamma_\nu \gamma_\mu \quad \text{for} \quad \mu \neq \nu = 0; \qquad \gamma_\mu \gamma_\mu = \gamma_\mu \cdot \gamma_\mu = \eta_{\mu\mu}. \tag{8.20}$$

These rules for dealing with gamma matrices will be too obvious for practitioners.

3.8.5   Concluding remarks

The Clifford product supersedes the exterior one because it provides the concept of inverse of a vector, which the exterior product does not. Let $\mathbf{a}$ be a vector, or a vector field. The dot product of $\mathbf{a}$ by itself permits one to define its inverse, $\mathbf{a}^{-1}$, namely $\mathbf{a}/a^2$, where $\mathbf{aa} \equiv \mathbf{a} \cdot \mathbf{a} = a^2$.

Clifford products of vectors also have inverses: $(\mathbf{abc}\ldots)^{-1}$ is $\ldots\mathbf{c}^{-1}\mathbf{b}^{-1}\mathbf{a}^{-1}$. If we had exterior and other products, we would first express them in terms of Clifford products in order to then find whether they have inverses.

Of great interest are the inhomogeneous multivectors that are equal to their squares. Called idempotents, they generate ideals. Members of these ideals are spinors.

In addition to the mathematical pre-eminence of Clifford algebra over exterior algebra (and similarly for corresponding calculi), there are physical advantages. Whereas exterior algebra is almost sufficient for differential geometry, quantum mechanics requires Clifford algebra.

Chapter 4

EXTERIOR
DIFFERENTIATION

4.1 Introduction

A rigorous treatment of all things pertaining to scalar-valued differential forms
and the exterior calculus is presented in a dedicated long chapter in a future book
whose contents is described in the present chapter 13. For present purposes,
we are simply interested in motivating the exterior derivative and providing
the concepts needed to be functional in the case of scalar-valued differential
forms, without providing proof of everything that a mathematician would find
necessitating proof. We have found pertinent to treat the subject of this chapter
rather expeditiously in order not to blur the focus on vector-valuedness of this
book. Since the next one deals almost exclusively with scalar-valuedness, one
would expect that the order in the publication of these books should be the
opposite of what it is going to be. However, whereas book 2 will deal with
exterior-interior calculus, here we only deal with exterior calculus.

4.2 Disguised exterior derivative

The concept of exterior derivative is central to the Cartan calculus, also known
as exterior calculus. Readers who know the theorems of Gauss and Stokes already have some knowledge of exterior derivatives and of an important theorem
about these derivatives, but without perhaps being aware of it all. We shall
introduce the concept from the perspective of those well-known theorems.

Suppose that we ask ourselves what should go inside the integrals on the
right side of the incomplete equalities

$$\oint\oint v(x,y,z)dy \wedge dz = \iiint ?, \tag{2.1}$$

$$\oint v(x, y, z)dx = \iint ?. \qquad (2.2)$$

The domains of integration on the left are respectively a closed surface and a closed line in simply connected domains. They are the boundaries of the domains of integration of the integrals on the right. We can use the theorems of Gauss and Stokes with $(v_x, v_y, v_z) = [v(x, y, z), 0, 0]$ to answer our question. We thus obtain:

$$\oint\oint v(x, y, z,)dy \wedge dz = \iiint \frac{\partial v(x, y, z)}{\partial x} dx \wedge dy \wedge dz \qquad (2.3)$$

and

$$\oint v(x, y, z)dx = \iint \left[\frac{\partial v}{\partial z}dz \wedge dx - \frac{\partial v}{\partial y}dx \wedge dy \right]. \qquad (2.4)$$

The exterior derivative is precisely the concept that permits us to respond to our question in both cases. It is defined as the operation that maps the integral of an r−form extended to a closed domain into the integral of an $(r + 1)$-form extended to the domain S delimited by the former domain, to be denoted as ∂S.

Let ω be a differential r−form, understood here to be the integrand of an r-integral ($r = 1, 2, 3...$ for a line, surface, volume,...). From a perspective of structural simplicity, we would define its exterior derivative $d\omega$ as the differential $(r + 1)$-form that satisfies

$$\int_{\partial S} \omega = \int_S d\omega, \qquad (2.5)$$

assuming that $d\omega$ exists independently of S. In the literature, (2.5) is referred to as the generalized Stokes theorem or simply Stokes theorem. From the perspective just mentioned, it is not a theorem but an implicit definition.

It is known that $d\omega$ exists, but that would be a tall order to prove. We shall proceed in a different, simpler way, like virtually every author on this subject does. But we wanted to advance here the perspective of the previous paragraph, since it brings to the fore what is going to be achieved. It is worth pointing out that, whereas $d\omega$ is uniquely defined, ω, on the other hand, is not unique. We mean to say that different ω's may have the same $d\omega$, which is reminiscent of the fact that primitives in the calculus of one variable are determined up only to an additive constant.

Notice from this introduction to the concept of exterior *derivative* that more appropriate nomenclature for it would be that of exterior *differential*, something to be remembered until familiarity with the concept permits readers not to be bothered by this ambiguity of the terminology.

Given ω, how does one find $d\omega$? Equations (2.3) and (2.4) tell us that the forms on their right hand sides are, according to the characterization just given, the exterior derivatives of the forms on the left hand sides. More specifically, Stokes theorem, viewed as implicit definition of exterior differentiation, permits us to find that the exterior derivative of $v_i dx^i$ is $(\partial v^i/\partial x^j)dx^j \wedge dx^i$. Similarly, Gauss theorem, again viewed as a definition, permits one to state that the exterior derivative of $v_x dy \wedge dz + v_y dz \wedge dx + v_x dx \wedge dy$ in three dimensions, is $(\partial_x v_x + \partial_y v_y + \partial_z v_z) dx \wedge dy \wedge dz$. Finally, the equation

$$f(B) - f(A) = \int_A^B f_{,i} dx^i \qquad (2.6)$$

can be interpreted to mean that the exterior derivative of a scalar function is just the ordinary differential

$$df = f_{,i} dx^i. \qquad (2.7)$$

(In this case, what is the domain of integration on the left hand side of (2.6)?) Those theorems permitted mathematicians to infer the form of $d\omega$ for arbitrary forms ω, on manifolds of arbitrary dimension. Once the rules for obtaining the exterior derivative were found, the argument was reversed. The rules for obtaining exterior derivatives became their definition and one then proved Eq. (2.5) under the name of Stokes theorem. In Section 3, we introduce the new definition and, in Section 4, we prove the theorem.

4.3   The exterior derivative

Any $r-$form can be represented as a sum of terms of the form $f dx^{i_1} \wedge \ldots \wedge dx^{i_r}$ where f is a zero-form or scalar valued function. The exterior derivative d of a sum is defined as the sum of the exterior derivatives. So, what remains is the defining of the exterior derivative of simple forms of ranks one and greater. Given $\tau = f dx^{i_1} \wedge \ldots \wedge dx^{i_r}$, its exterior derivative is defined as:

$$d(f dx^{i_1} \wedge \ldots \wedge dx^{i_r}) \equiv df \wedge dx^{i_1} \wedge \ldots \wedge dx^{i_r} = f_{,j} dx^j \wedge dx^{i_1} \wedge \ldots \wedge dx^{i_r}. \quad (3.1)$$

One readily shows that $ddf = 0$ since

$$ddf = d(f_{,i} dx^i) = f_{,i,j} dx^j \wedge dx^i = (f_{,i,j} - f_{,j,i})(dx^j \wedge dx^i), \qquad (3.2)$$

which is zero because of the equality of the partial derivatives. In particular, $ddx^i = 0$ (Just for the record: Some mathematicians might disagree with this approach and state that $ddx^i = 0$ must be considered as a postulate; they might be right). Notice that the exterior derivatives of forms of the highest rank, n, are also zero, though for a different reason which should be obvious.

One can show that the definition (3.1) is independent of coordinate system, i.e. one obtains the same result by performing a change of coordinate system and then exterior differentiating as when one exterior differentiates and then changes coordinates.

We finally develop a very useful expression for the exterior product of forms of any rank. Let us start with the exterior product of just two of them. This product can always be expressed as a sum of terms of the form $\rho \wedge \sigma$ where

$\rho = R dx^{i_1} \wedge \ldots \wedge dx^{i_r}$ and $\sigma = S dx^{j_1} \wedge \ldots \wedge dx^{j_s}$. Then

$$
\begin{aligned}
d(\rho \wedge \sigma) &= d(RS) \wedge dx^{i_1} \wedge \ldots \wedge dx^{i_r} \wedge dx^{j_1} \wedge \ldots \wedge dx^{j_s} \\
&= (dR \wedge dx^{i_1} \wedge \ldots \wedge dx^{i_r}) \wedge (S dx^{j_1} \wedge \ldots \wedge dx^{j_s}) \\
&\quad + (-1)^r (R dx^{i_1} \wedge \ldots \wedge dx^{i_r}) \wedge (dS \wedge dx^{j_1} \wedge \ldots \wedge dx^{j_s}) \\
&= d\rho \wedge \sigma + (-1)^r \rho \wedge d\sigma,
\end{aligned}
\tag{3.3}
$$

where the factor $(-1)^r$ is due to the fact that we have moved the dS $1-$form from the front to the right of $dx^{i_1} \ldots dx^{i_r}$. By using the associative property of the exterior product, one gets

$$
d(\rho \wedge \sigma \wedge \tau) = d\rho \wedge \sigma \wedge \tau + (-1)^r \rho \wedge d\sigma \wedge \tau + (-1)^{r+s} \rho \wedge \sigma \wedge d\tau
\tag{3.4}
$$

and so on. An immediate consequence of $ddf = 0$ is that the second derivative $dd\rho$ of any $r-$form is zero since $ddx^m = 0$, and

$$
dd(f dx^i \wedge \ldots \wedge dx^j) = d(df \wedge dx^i \wedge \ldots \wedge dx^j) = (ddf) \wedge dx^i \wedge \ldots \wedge dx^i = 0.
\tag{3.5}
$$

The Gauss and (traditional) Stokes theorems are but particular cases of Eq. (2.5), verification of which we leave to readers. Again, $ddf = 0$ may also be viewed as defining the exterior derivative; $ddx^i = 0$ would then be contained in that definition as particular cases.

4.4 Coordinate independent definition of exterior derivative

The definition of exterior derivative of Section 3.2 is very useful for practical calculations. It has the disadvantage, however, that it invokes a coordinate system or patch. Under a change of coordinates from x^i's to y^j's, the simple $r-$form $f dx^i \wedge \ldots \wedge dx^j$ becomes a sum of $r-$forms $f_{k \ldots m} dy^k \wedge \ldots \wedge dy^m$. We can now apply the rules of Section 3.2 to this linear combination of $r-$forms to obtain its exterior derivative. This result should coincide with the result of transforming to the y coordinate system the exterior derivative obtained in the coordinate system x; a brute force proof of the coincidence of the two results is very cumbersome. This problem becomes a non-issue by introducing a coordinate independent definition of exterior derivative, which we do in the following.

Given an exterior product of forms $\alpha \wedge \beta \wedge \ldots \wedge \gamma$ of respective grades $a, b, \ldots c$, the definition

$$
\begin{aligned}
d(\alpha \wedge \beta \wedge \ldots \wedge \gamma) &\equiv d\alpha \wedge \beta \wedge \ldots \wedge \gamma + (-1)^a \alpha \wedge d\beta \wedge \ldots \wedge \gamma \\
&\quad + \ldots + (-1)^{a+b+\cdots} \alpha \wedge \beta \wedge \ldots \wedge d\gamma
\end{aligned}
\tag{4.1}
$$

applies, in particular, to the exterior product of $1-$forms, which can always be expressed in coordinate independent manner. We thus have:

$$
d(f_0 df_1 \wedge \ldots \wedge df_r) = d[(f_0 df_1) \wedge df_2 \wedge \ldots \wedge df_r] = df_0 \wedge df_1 \wedge \ldots \wedge df_r.
\tag{4.2}
$$

This definition is explicitly independent of coordinate system. It takes the form of Eq. (3.1) when the differential form to be differentiated is first written as $f dx^{i_1} \wedge \ldots \wedge dx^{i_r}$.

4.5 Stokes theorem

In this section we shall prove Stokes theorem, namely Eq. (2.5). Some readers may perhaps wish to just glance at this section and proceed to the next one. We need to first introduce the concept of pull-back of a form, or pull-back for short. The reader has surely come across pull-backs in the calculus with several variables, where pull-backs occur when we perform line and surface integrals.

A (parametrized) surface in R^3 is a map $U \to R^3$ where $U \subset R^2$. The points of U will be represented by the coordinates (u, v) and the points R^3 by the coordinates (x, y, z). As an example, the upper hemisphere H of unit radius is a map $(u, v) \to (x, y, z)$ defined by

$$x = \sin u \cos v$$
$$y = \sin u \sin v$$
$$z = \cos u,$$

where the domain U is $(0 \leq u \leq \pi/2, \ 0 \leq v < 2\pi)$. Suppose now that we want to integrate the form $\omega = z^2 dx \wedge dy$ over the hemisphere H,

$$\int_H \omega, \qquad \omega = z^2 dx \wedge dy$$

The best way to perform the integral is to first obtain dx and dy in terms of du and dv and perform the substitutions in ω. We thus obtain

$$\omega = \cos^2 u \ (\cos u \cos v \ du - \sin u \sin v \ dv) \wedge (\cos u \sin v \ du + \sin u \cos v \ dv)$$
$$= \sin u \cos^3 u \ du \wedge dv \tag{5.2}$$

and finally,

$$\int_H \omega = \int_U \sin u \cos^3 u \ du \wedge dv. \tag{5.3}$$

To make a point, we have been careless with the notation since (5.2) and (5.3) together imply

$$\int_H \omega = \int_U \omega, \tag{5.4}$$

which is not correct. The expression (5.4) is incorrect because ω is not defined on U. The form $z^2 dx \wedge dy$ is defined in R^3, and we denote it as ω. The form $\sin u \cos^3 u \ du \wedge dv$ is defined in $U \subset R^2$, and we shall denote it as $S^*\omega$. This is a so called pull-back of the form ω. The concept of pull-back can be made mathematically rigorous. The important thing to realize is that here it simply means that we move the differential form from one manifold to another.

Instead of (5.4), we should be stating that we are saying that

$$\int_H \omega = \int_U S^* \omega. \tag{5.5}$$

It happens that, to start with, we should write

$$\begin{aligned}
S^* \omega &= (S^* z^2)(S^* dx) \wedge (S^* dy) \\
S^* z &= \cos^2 u \\
S^* dx &= \cos u \cos v \; du - \sin u \sin v \; dv \\
S^* dy &= \cos u \sin v \; du + \sin u \cos v \; dv.
\end{aligned}$$

In physics, however, we want to get to results as fast as possible disregarding sometimes the use of appropriate notation, which distracts us. For the derivation of Stokes theorem, we need more rigor that we are used to. Let us formulate the above considerations in mathematical language.

Let A and M be two differentiable manifolds. Let U be a subset of A and consider a map $S : U \to M$. Given a function f on M, we define function g on U by means of $g(Q) = f(P)$ where $Q \in U$ and where $P = S(Q)$. The "pull-back" $S^* \omega$ of the form $\omega = f_0 df_1 \wedge \ldots \wedge df_r$ on M is defined as the form $g_0 dg_1 \wedge \ldots \wedge dg_r$. Thus the function S^* transforms $r-$forms on M into $r-$forms on A. One can show that

$$S^*(\omega_1 \wedge \omega_2) = (S^* \omega_1) \wedge (S^* \omega_2) \tag{5.6}$$

and that

$$S^*(d\omega) = d(S^* \omega). \tag{5.7}$$

We proceed to prove Stokes theorem. We write ω as $f_0 df_1 \wedge \ldots \wedge df$. Thus $d\omega$ is $df_0 \wedge df_1 \wedge \ldots \wedge df_r$. Let S be a parameterized $(r+1)$ -surface, $S : U \to M$, i.e. a smooth function from a closed, bounded region $U \subset R^{r+1}$. Assume also that U is convex, i.e. contains all line segments between each pair of its points. Let $\alpha = S^* \omega$. Then

$$\int_S d\omega = \int_U d\alpha \tag{5.8}$$

and

$$\int_{\partial S} \omega = \int_{\partial U} \alpha \tag{5.9}$$

where ∂U is the boundary of U. If we prove that

$$\int_U d\alpha = \int_{\partial U} \alpha, \tag{5.10}$$

we shall have proved that

$$\int_S d\omega = \int_{\partial S} \omega.$$

If $(u_0, u_1, \ldots, u_r)$ are the parameters of S, we can write α as

$$\alpha = a_0 du^1 \wedge \ldots \wedge du^r + a_1 du^0 \wedge du^2 \wedge \ldots \wedge du^r + \ldots + a_r du^0 \wedge du^1 \wedge \ldots \wedge du^{r-1}. \tag{5.11}$$

We shall prove that

$$\int_U d(a_0 du^1 \wedge \ldots \wedge du^r) = \int_{\partial U} a_0 du^1 \wedge \ldots \wedge du^r. \tag{5.12}$$

(The other terms are done analogously). If the boundary is convex, it has an upper and a lower part, respectively $u^0 = f_2(u^1, \ldots, u^r)$ and $u^0 = f_1(u^1, \ldots, u^r)$ with $f_1 \leq f_2$. We thus have, with $du^{1 \cdots r} \equiv du^1 \wedge \ldots \wedge du^r$,

$$\int_U d(a_0 du^{1 \cdots r}) = \int_{f_1}^{f_2} \frac{\partial a_0}{\partial u^0} du^0 \wedge du^{1 \cdots r} = \int [a_0]_{f_1}^{f_2} du^{1 \cdots r}$$

$$= \int [a_0(f_2, u^1, \ldots, u^r) - a_0(f_1, u^{1 \cdots r})] du^{1 \cdots r} \tag{5.13}$$

$$= \int_{\partial U} a_0 du^{1 \cdots r}.$$

Q.E.D.

Readers will have noticed that the proof is essentially the same as for the usual Stokes theorem involving two-dimensional surfaces in 3-D Euclidean space. We have left out some matters related to orientation of the forms, which are not unlike the problem of orientation of the bounding curve in the usual Stokes theorem.

Suppose finally that there were some other differential form μ, i.e. other than $d\omega$, that satisfied

$$\int_{\partial S} \omega = \int_S \mu. \tag{5.14}$$

We would then have

$$\int_S d\omega = \int_S \mu \tag{5.15}$$

for any applicable integration domain. Since these can be as small as possible, the two differential forms $d\omega$ and α would have the same coefficients at any point. They would thus be the same differential form. In other words, the exterior derivative is uniquely defined.

4.6 Differential operators in language of forms

Our three-space of everyday life (three-dimensional Euclidean space) is a very special case of differentiable manifold; it has much additional structure (Chapters 6). If this space were not endowed with a dot product, the concepts of gradient, curl, divergence and Laplacian (divergence of gradient) could not be defined within the vector calculus, which is one of its deficiencies. Gradient and curl should not require such a product, thus a so called metric structure. Our computations will make that clear. The divergence does but, fortunately, one could easily pretend that it does not in Euclidean spaces. We shall define those operations with the roles of vector fields replaced by differential forms.

Let M be an $n-$dimensional differentiable manifold. The exterior derivative of a scalar function (zero form) will be called its gradient differential form

$$df = f_{,i}dx^i. \tag{6.1}$$

Let now α be a $1-$form, $\alpha \equiv a_i dx^i$. Its exterior derivative

$$d\alpha = d(a_i dx^i) = a_{i,j}dx^j \wedge dx^i = (a_{i,j} - a_{j,i})(dx^j \wedge dx^i) \tag{6.2}$$

will be called its curl. These definitions are valid on any differentiable manifold and, since they are invariants, in any coordinate system. It is now a matter of specializing to whatever basis we wish to use.

Divergences and Laplacians are a different story, as the metric structure is then essential. Their detail treatment will take place when we shall introduce that structure in a later chapter. Fortunately, we do not need to wait until then to present in Euclidean 3-D a shortcut to the concept of divergence. Given a vector field

$$\mathbf{b} \equiv b^1\mathbf{i} + b^2\mathbf{j} + b^3\mathbf{k} \tag{6.3}$$

in 3-D Euclidean space, we associate with it in Cartesian coordinates the differential 2$-$form

$$\beta = b^1 dy \wedge dz + b^2 dy \wedge dz + b^3 dx \wedge dy \tag{6.4}$$

but only in Cartesian coordinates (it takes other forms in other systems of coordinates, as we shall later see). Roughly, the divergence of the vector field $\mathbf{b}$ can be replaced with the exterior derivative of β,

$$d\beta = \left[\frac{\partial b^1}{\partial x} + \frac{\partial b^2}{\partial y} + \frac{\partial b^3}{\partial z}\right] dx \wedge dy \wedge dz = b^i_{,i}dx \wedge dy \wedge dz, \tag{6.5}$$

to great advantage (Indeed, readers will identify the contents of the square bracket as the divergence of $b^1\mathbf{i} + b^2\mathbf{j} + b^3\mathbf{k}$ in Cartesian coordinates. Divergences are identified with densities, which are to be integrated. The right hand side of (6.5) is a ready to use integrand.

We might wish to relate the gradient and curl differential forms to the gradient and curl vector fields. In the vector calculus, the gradient is

$$grad \; f = f_{,i}\mathbf{e}^i. \tag{6.6}$$

Notice that the basis has to go with superscripts to indicate its appropriate transformation properties so that the gradient will be an invariant. The problem is: what is $\mathbf{e}^i$? Given a coordinate system in a differentiable manifold, a corresponding basis vector field $\mathbf{e}_i$ is defined even if the tangent vector spaces to the manifold are not Euclidean (absence of metric structure). A dot product, equivalently metric structure, is needed in a differentiable manifold in order to define $\mathbf{e}^i$, as we saw in the previous chapter. But why should we invoke structure that is not needed in order to achieve the same objectives?

For the moment let us recall what every reader knows about metrics in Euclidean spaces. They are the expressions

$$ds^2 \equiv g_{ij}dx^i dx^j, \tag{6.7}$$

examples being in 3-D Euclidean space

$$ds^2 = dr^2 + r^2 d\theta^2 + r^2 \sin^2 \theta d\phi^2 \tag{6.8}$$

in spherical coordinates and

$$ds^2 = d\rho^2 + \rho^2 \sin^2 \theta \, d\phi^2 + dz^2 \tag{6.9}$$

in cylindrical coordinates. Orthogonal coordinate systems are those where the metric (to be defined in chapter 6 for Euclidean space and in chapters 9 for generalizations of Euclidean spaces) takes the form

$$ds^2 \equiv \sum_{i=1}^{i=n} (\omega^i)^2, \tag{6.10}$$

with

$$\omega^i \equiv h^i(x)dx^i \qquad \text{(no sum)}. \tag{6.11}$$

Every metric can be written as (6.10), in an infinite number of ways, since once we have found one solution, we can find other solutions by applying, in Euclidean 3-D, arbitrary rotations. But not every $g_{ij}dx^i dx^j$ admits ω^i's of the type (6.11). Orthogonal coordinate systems do, and are easily recognizable since ds^2 then take the form

$$ds^2 = (h^1 dx^1)^2 + (h^2 dx^2)^2 + (h^3 dx^3)^2 + ..., \tag{6.12}$$

which embodies (6.10) and (6.11).

Electrical engineers and certain types of physicists will need to have the differential operators in terms of orthonormal vector basis fields. The gradient vector field does not take the form (6.6) in general orthonormal frame fields.

Computing in terms of orthonormal basis fields is equivalent to computing in terms of bases ω^i's of differential forms that satisfy (6.10). But (6.11) will not be satisfied in general. we naturally have

$$df = f_{,i}dx^i \equiv f_{/i}\omega^i. \tag{6.13}$$

For our present purposes, one may think of the $f_{/i}$ as if they were the components of a vector field in terms of an orthonormal basis field.

We now specialize the coordinate and basis independent equations (6.1)-(6.2) and the Cartesian specific equation (6.5) to orthonormal vector basis fields associated with the spherical and cylindrical coordinates. It is obvious from (6.11) and (6.13) that we can rewrite (6.1) as

$$df \equiv [(h^i)^{-1} f_{,i}](h^i dx^i) = [(h^i)^{-1} f_{,i}]\omega^i. \tag{6.14}$$

The components of df relative to ω^i will be familiar to engineers and physicists using the vector calculus with orthogonal coordinates systems, like the cylindrical and spherical ones.

Some readers may wonder whether we have lost the information provided by the gradient vector field, namely the direction of maximum change of a scalar quantity and the magnitude of such a change per unit length. One may speak of change per unit length only if length is defined in the first place, which a metric does. Hence the gradient differential form does not provide that information if the differentiable manifold is not endowed with metric structure; but the gradient vector field is not even defined. If a metric is given, we may define a vector field with components given by the square brackets in (6.14).

Consider now the curl, i.e. (6.2). The F_j's defined by

$$\alpha \equiv a_j dx^j \equiv F_j \omega^j = F_j h^j dx^j \tag{6.15}$$

will play the role of the components of a vector field in terms of a reciprocal basis. Hence, from (6.15), we get

$$a_j = F_j h^j \qquad \text{(no sum)}, \tag{6.16}$$

which we take to (6.2) to obtain

$$\begin{aligned} d\alpha &= [(F_i h^i)_{,j} - (F_j h^j)_{,i}](dx^j \wedge dx^i) \\ &= (1/h_i h_j)[(F_i h^i)_{,j} - (F_j h^j)_{,i}](\omega^j \wedge \omega^i). \end{aligned} \tag{6.17}$$

End of story. The components of this 2−form are the same ones as those that go by the name curl vector field in the literature

$$\begin{vmatrix} \boldsymbol{i}/h^2 h^3 & \boldsymbol{j}/h^1 h^3 & \boldsymbol{k}/h^1 h^2 \\ \partial/\partial x^1 & \partial/\partial x^2 & \partial/\partial x^3 \\ F_1 h^1 & F_2 h^2 & F_3 h^3 \end{vmatrix}, \tag{6.18}$$

in terms of orthonormal basis fields associated with orthogonal coordinate systems (We have used $F^i = F_i$). The matching of components a_{ij} of a differential 2−form and components of a vector field is a peculiarity of three dimensions. The metric is used (we do not say how at this point) to associate the differential 2−form with the vector field, not to define a concept of curl differential form.

The case of the divergence is different. Here the metric is essential even in dealing with differential forms. It has to do with something which, in Kähler's generalization of the exterior calculus of differential forms, goes by the name of interior derivative. There are structures, like Euclidean spaces, where the divergence of an object of grade r can be replaced with the exterior derivative of an associated differential form of grade $n - r$. Recall that we went from the vector field (6.3), which is of grade one, to the differential form (6.4), which is of grade 3-1.

Corresponding to orthonormal bases, we define new components of β by

$$\beta = F^1 \omega^2 \wedge \omega^3 + F^2 \omega^3 \wedge \omega^1 + F^3 \omega^1 \wedge \omega^2. \tag{6.19}$$

Clearly $b^i = F^i h^j h^k$, with the three indices making a cyclic permutation of (1,2,3). In (6.5), we replace these expression for b^i and use (6.11) to get

$$
\begin{aligned}
d\beta &= \left[\frac{\partial(F^1 h^2 h^3)}{\partial x^1} + \frac{\partial(F^2 h^1 h^2)}{\partial x^2} + \frac{\partial(F^3 h^1 H^2)}{\partial x^3} \right] dx^1 \wedge dx^2 \wedge dx^3 \\
&= \frac{1}{h^1 h^2 h^3} \left[\frac{\partial(F^1 h^2 h^3)}{\partial x^1} + \frac{\partial(F^2 h^1 h^3)}{\partial x^2} + \frac{\partial(F^3 h^1 h^2)}{\partial x^3} \right] \omega^1 \wedge \omega^2 \wedge \omega^3.
\end{aligned}
\tag{6.20}
$$

Readers will recognize the coefficient of this 3−form as the divergence of a vector field of components F^i in terms of an normalized basis field dual to an orthogonal system of coordinates.

4.7 The conservation law for scalar-valuedness

A direct consequence of Stokes theorem is the fact that, if the exterior differential $d\alpha_r$ of an exterior r−form α_r is zero on a simply connected region of a manifold, we have

$$
\int_{\partial R} \alpha_r = 0 \tag{7.1}
$$

on any closed r-surface ∂R in that region. Equation (7.1) is the statement of the conservation law of differential forms. We shall see below that this law comprises (familiar forms of) more specialized conservation laws. The specialization takes place sometimes by splitting ∂R into pieces, usually two (like the two semispheres of a sphere) or three (like the two bases and the lateral face of a cylinder in spacetime).

The conservation law in the case of scalar-valued differential forms is the generalization to those objects of the statement that, if df is null, f is a constant. Equation (7.1) does not explicitly say that, if $d\alpha_r$ is zero, α_r is a constant differential form. We have not even defined what "constant differential form" could possibly mean. The constancy of f if $df = 0$ is a peculiarity of $d\alpha$ being a 1−form, i.e. a function of curves. α is then a function of points. More precisely, the boundary of an open curve is a pair of points. Evaluating a constant as if it were a zero-dimensional integrand is then interpreted to mean

$$
f(B) - f(A) = 0. \tag{7.2}
$$

This applies to "any curve within our curve" (i.e. to any pair of its points) and to any curve between any two points on simply connected regions of manifold. The constancy of f follows. There is the issue of why the minus sign in Eq. (7.2). This has to do with orientation, a constant theme in the examples that follow.

As we indicated, we shall split ∂R into pieces. How we break it depends on the rank of the differential form, especially in relation to the dimensionality of the manifold (thus, for example, on whether we consider the electric field as a piece of a differential 2−form in spacetime or as a differential 1−form in

3-space). The signature also matters. For instance, if we consider 3-D cylinders in spacetime, we take them with axis along the time dimension.

Consider the differential $1-$form $E_i dx^i$. If $d(E_i dx^i)$ is zero, Stokes theorem implies that $\int_{\partial R} E_i dx^i$ is zero on closed curves. Let A and B be two points on one such curve. Notice how a minus sign appears:

$$0 = \int_{\partial R} E_i dx^i = \int_A^B E_i dx^i + \int_B^A E_i dx^i = \int_{A;\,(1)}^B E_i dx^i - \int_{A;\,(2)}^B E_i dx^i, \quad (7.3)$$

where we have used (1) and (2) to denote two different paths between A and B.

Consider next the $2-$form $B_i dx^j \wedge dx^k$, where we sum over cyclic permutations of $(1, 2, 3)$. Assume that $d(B_i dx^j \wedge dx^k)$ is zero. Then

$$\int_{\partial R} B_i dx^j \wedge dx^k = 0. \qquad (7.4)$$

We divide ∂R into two open surfaces: $\partial R = ① \oplus ②$. Then

$$0 = \int_{①;\,out} B_i dx^j \wedge dx^k + \int_{②;\,out} B_i dx^j \wedge dx^k. \qquad (7.5)$$

Suppose that the two surfaces ① and ②, which are bounded by the same curve, were very close to each other. It is clear that "out in one case" means the opposite of "out in the other case." Hence, for common orientation, there must be a negative sign between the two integrands on the right, and their equality follows. The integration of the B differential form is the same for all open surfaces with a common boundary provided that they are considered all at the same instant of time and equally oriented. This caveat means that the B field is to be considered as a differential $2-$form.

E and B together constitute the differential form

$$F = E_i dt \wedge dx^i - B_i dx^j \wedge dx^k, \qquad (7.6)$$

which has the conservation law

$$0 = \int_{\partial R} E_i dt \wedge dx^i - B_i dx^i \wedge dx^k \qquad (7.7)$$

associated with it, where ∂R is the boundary of a 3-surface, and is, therefore, a closed spacetime surface.

In spacetime, consider a ∂R at constant time, i.e. a purely spatial closed surface. The flux of B over ∂R is zero. In order to study the evolution with time of integrals at constant time, we integrate on a rectilinear cylinder with equal spatial bases at times t_1 and t_2. We have six terms to take care of: the integration of both $E_i dx^0 \wedge dx^i$ and of $-B_i dx^j \wedge dx^k$ over the two bases and over the lateral surface. The E-integrals over the bases and the B-integral over the lateral surface are zero. One thus obtains

$$0 = \int_{t_1}^{t_2} dt \oint E_i dx^i + \int_{out\ at\ t_1} B_i dx^j dx^k + \int_{out\ at\ t_2} B_i dx^j dx^k. \qquad (7.8)$$

One differentiates with respect to t_2 at constant t_1, and obtains Faraday's law. That is what having a conserved 2−form in spacetime means in terms of quantities at constant time.

The other conservation law of electrodynamics involves the current 3−form:

$$j = j_{123}dx^1 \wedge dx^2 \wedge dx^3 + j_{023}dt \wedge dx^2 \wedge dx^3 + j_{031}dt \wedge dx^3 \wedge dx^1$$
$$+ j_{012}dt \wedge dx^1 \wedge dx^2. \tag{7.9}$$

The conservation law that is a consequence of $dj = 0$ reads

$$\int_{\partial R} j = 0. \tag{7.10}$$

The j_{123} component is the charge density, ρ. We take as ∂R the boundary of a 4-dimensional cylinder whose lateral "surface" (three-dimensional!) is pushed to infinity, with bases at constant time. The result is the conservation of charge. It is seldom mentioned that one does not need to integrate at constant time to obtain the amount of charge. One also obtains it if one integrates j over any section of the 4-dimensional tube of world lines of all charges in the volume. Hence the components j_{0lm} (equivalently, the components of what is considered a vector field j in the literature) are instrumental not only in obtaining the flux of charge through a surface but also in computing the total change using a section not at constant time (see "Leçons sur les invariants integraux" by É. Cartan [9]).

After the experience gained with orientation, we return to when α is a 0−form. The oriented boundary ∂R now is the pair of points (A, B) together with "direction away from the curve." Direction away from the curve assigns opposite signs to the evaluation of f at the two ends, as becomes obvious by making the curve smaller and smaller until it reduces to a point. Once again if R is a curve,

$$0 = \int_R df = \int_{\partial R} f = f(\partial R) = f(B) - f(A), \tag{7.11}$$

which thus is seen as one more example of application of (7.1).

4.8 Lie Groups and their Lie algebras

The title of this section is meant to de-emphasize the rather abstract concept of Lie algebra and focus instead on the easier to understand concept of Lie algebra of a Lie group, which is of interest in this book.

Most physicists will avoid subtleties when referring to a Lie group and define it as just a continuous group. A mathematician will rightly say that this is not good enough, for two reasons. First, we want to have more than continuity, namely differentiability, and that is what differentiable manifolds are for. Thus a Lie group, G, is a group that is also a differentiable manifold. A second requirement is a compatibility condition between the two structures of group and differentiable manifold. But we shall not enter into that since the concept is satisfied in the theory that we shall be developing. It takes a skilled mathematician to think of examples where the condition is not satisfied.

Although, as we said, we are not interested in Lie algebras in general, let us state that they are modules in which we have or can introduce a multiplication of a type called Lie product. If we kept on, we would be speaking of Lie brackets, but this would take us into unnecessary and distracting considerations on vector fields that we wish to avoid.

(Readers who might have difficulty in the following with our use of terms like planes in the present context should jump now to the opening lines of section 7.2, and come back.)

Let g denote the general element of a group of transformations. Let dg be its exterior derivative. Let $u^i, (i = 1, ..., m)$, be a complete set of parameters in the group, equivalently a coordinate system in the manifold. dg is a linear combination

$$dg = g_{,i}\, du^i \equiv f_i du^i. \tag{8.1}$$

If G were a group of matrices, the $f_i du^i$ would be matrices with entries that are differential 1-forms. Let (Δu^i) denote $m-$tuples of real numbers. Usually G will be a proper subgroup of the group of regular matrices $n \times n$. The set of matrices $f_i \Delta u^i$ then is an m-dimensional hyperplane in the manifold of such $n \times n$ matrices. One might think that $f_i \Delta u^i$ is the tangent plane at g. It is not quite that; it simply is parallel to that tangent plane. This will become easier to understand in the following.

Consider the differential form

$$\omega \equiv (dg)g^{-1} = (f_i du^i)g^{-1} = (f_i g^{-1})du^i, \tag{8.2}$$

called Maurer-Cartan form of the group. Correspondingly, we shall have a plane

$$(f_i g^{-1})\Delta u^i. \tag{8.3}$$

The unit element e of the group is a point of the manifold. Similarly, any other element of the group is a point of the manifold. One can make the following statement: g in the group takes e in the manifold to g in the manifold. So, g^{-1} takes g, its neighborhood and its tangent plane respectively to e, its neighborhood and its tangent plane. Hence $(f_i g^{-1})\Delta u^i$ is independent of g. What is it?

$(f_i g^{-1})\Delta u^i$ would be the tangent plane at e if $f_i \Delta u^i$ were the tangent plane at g. Though $(f_i g^{-1})\Delta u^i$ and $f_i \Delta u^i$ look like such tangent planes, they are not quite so. Here is why. $(f_i g^{-1})\Delta u^i$ and $e + (f_i g^{-1})\Delta u^i$ are parallel planes. So, they do not have any point in common. Since e is in $e + (f_i g^{-1})\Delta u^i$, it cannot be in $(f_i g^{-1})\Delta u^i$. Clearly, $(f_i g^{-1})\Delta u^i$ is the plane at the null matrix that is parallel to the tangent plane at the unit matrix, i.e. at the unit element.

The Maurer-Cartan form of the group satisfies the so called Maurer-Cartan equation of structure:

$$d\omega = \omega \wedge \omega. \tag{8.4}$$

We shall have the opportunity of seeing this equation emerge naturally in several cases of interest for us. Hence, we do not insist on this at this point.

In order to have an algebra, we need a sum and a product. The sum is the sum of matrices. But the product is not the product of matrices, but their anti-symmetrized product. This again, we shall see later in detail.

Let us now practice a little bit with some very simple ω's and corresponding Lie algebras. Consider the matrices for the one dimensional group of rotations. We have

$$\omega \equiv (dg)g^{-1} = \begin{bmatrix} -\sin\phi & -\cos\phi \\ \cos\phi & -\sin\phi \end{bmatrix} \begin{bmatrix} \cos\phi & \sin\phi \\ -\sin\phi & \cos\phi \end{bmatrix} d\phi = \begin{bmatrix} 0 & -1 \\ 1 & 0 \end{bmatrix} d\phi, \tag{8.5}$$

briefly written as $\omega = \mathbf{a}d\phi$, with

$$\mathbf{a} \equiv \begin{bmatrix} 0 & -1 \\ 1 & 0 \end{bmatrix}. \tag{8.6}$$

Consider now the group of translations

$$x' = x + u \tag{8.7}$$

in one dimension. This is not a linear transformation, since translations in dimension n are not given by matrices in the same number of dimensions. n-translations can, however, be represented by matrices in $n+1$ dimensions. Thus (8.7) can be represented as

$$\begin{bmatrix} x' \\ 1 \end{bmatrix} = \begin{bmatrix} 1 & u \\ 0 & 1 \end{bmatrix} \begin{bmatrix} x \\ 1 \end{bmatrix} = \begin{bmatrix} x+u \\ 1 \end{bmatrix}. \tag{8.8}$$

ω is then given by

$$\omega = \begin{bmatrix} 0 & du \\ 0 & 0 \end{bmatrix} \begin{bmatrix} 1 & -u \\ 0 & 1 \end{bmatrix} = \begin{bmatrix} 0 & 1 \\ 0 & 0 \end{bmatrix} du = \mathbf{b}du, \tag{8.9}$$

with

$$\mathbf{b} \equiv \begin{bmatrix} 0 & 1 \\ 0 & 0 \end{bmatrix}. \tag{8.10}$$

The group of rotations is compact and the group of translations is non compact but, in one dimension, they are locally isomorphic, which is to say that their Lie algebras are isomorphic. The composition law in the algebra is (abstraction made of $\mathbf{a}$ and $\mathbf{b}$), $\Delta\phi^1 + \Delta\phi^2$ and $\Delta u^1 + \Delta u^2$ respectively.

When one knows that there is a better option, one realizes by comparison that matrices are cumbersome to work with. In the case of rotations and of Lorentz transformations, there is a much better mathematical tool than matrix algebra. It is Clifford algebra.

Part III

TWO KLEIN
GEOMETRIES

Chapter 5

AFFINE KLEIN GEOMETRY

We leave for chapter 10 the explanation of why we use the term Klein geometry in the title of this part of the book. We shall then have the perspective of having dealt with generalizations of Klein geometries that remain "Klein" in some sense, and shall obtain the perspective of generalizations that cease to be Klein even in that limited sense. For the moment and to keep going, Klein means flat. Better yet, affine Klein geometry and Euclidean Klein geometry simply mean affine and Euclidean geometry.

5.1 Affine Space

The concept of affine connection is of utmost importance in differential geometry. It derives from the concept of affine space Af^n. Geometry in affine space is called affine geometry. In its Kleinean stage, affine geometry is technically defined as the pair (G, G_0) constituted by the affine group, G, and its linear subgroup, G_0. In this presentation, these groups will emerge from the concept of affine space.

A rather intuitive definition of Af^n is the following: a set of points $(A, B, \ldots)$ such that to each ordered pair of points (A, B) we can associate a vector $\boldsymbol{AB} \in V^n$ with the following properties:

(a) $\boldsymbol{AB} = -\boldsymbol{BA}, \qquad \forall A, B \in Af^n$;

(b) $\boldsymbol{AB} = \boldsymbol{AC} + \boldsymbol{CB}, \qquad \forall A, B, C \in Af^n$;

(c) Given an arbitrary point A and vector $\mathbf{v}$, a point B exists such that $\boldsymbol{AB} = \mathbf{v}$.

The vector space is called the associated vector space. In this book, the associated vector spaces will be vector spaces over the reals. To put it slightly differently, affine space is a pair of a vector space and an associated set of points related to it in the aforementioned way; a vector $\mathbf{v}$ acting on a point $\mathbf{A}$ yields a point $\mathbf{B}$.

Let us make $A = B = C$ in the defining property (b). It implies that $\mathbf{AA}$ is the zero vector. But this also applies to any other point. Hence, in principle, any point can be associated with the zero vector. There is a one to one correspondence between the points of affine space and the vectors of a vector space only after one particular point has been arbitrarily associated with the zero vector. Let us call that point O. A point Q can actually be given as $\mathbf{Q}$ ($=\mathbf{OQ}$). From now on, we simply eliminate the point $\mathbf{O}$ from the notation.

A reference system in affine space is constituted by a point $\mathbf{Q}$ and a vector basis at $\mathbf{Q}$ of the associated vector space, which is equivalent to saying that we have a basis $(\mathbf{a}_i)$ at $\mathbf{Q}$. With these conventions, we can write for arbitrary point $\mathbf{P}$ and specifically chosen point $\mathbf{Q}$

$$\mathbf{P} = \mathbf{Q} + A^i \mathbf{a}_i. \tag{1.1}$$

In general, given a coordinate system (x) (also called a chart), the coordinate line x^i is defined as the locus of all points such that x^j is constant for all values of j except i ($j = 1 \ldots i-1, i+1 \ldots$). In particular, the coordinate line A^i has the direction of the basis vector $\mathbf{a}_i$, i.e., the vector associated with any pair of points in this line is proportional to the basis vector $\mathbf{a}_i$. The A^i's are referred to as rectilinear coordinates since they correspond to straight lines.

In Figure 1, the points P and S have the same coordinate A^2; the points P and R have the same coordinate A^1. The vectors $\mathbf{R}$ and $\mathbf{S}$ are respectively equal to $A^1 \mathbf{a}_1$ and $A^2 \mathbf{a}_2$. The rectilinear coordinates are therefore associated with the parallel projections of the vector $\mathbf{P}$ upon the axes, i.e. upon the coordinate lines going through the origin Q.

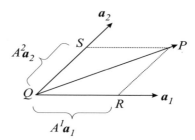

Figure 2: Emergence of rectilinear coordinates

Consider a new frame $(\mathbf{Q}, \mathbf{a}'_1 \ldots, \mathbf{a}'_n)$ at the same point Q of Af^n. In terms of the new basis $(\mathbf{a}'_i)$, we can expand $\mathbf{P}$ as

$$\mathbf{P} = \mathbf{Q} + A'^i \mathbf{a}'_i. \tag{1.2}$$

The relation between the A and A' coordinates are obtained from appropriate substitution in (1.1) or (1.2) of the relation between the bases.

5.2   The frame bundle of affine space

The frame bundle F^n of Af^n is the set of all its frames. Its name reflects its being endowed with the sophisticated structure that will be the subject of this section. Given the particular frame $(\mathbf{Q}, \mathbf{a}_1 ... \mathbf{a}_n)$, any other frame $(\mathbf{P}, \mathbf{a}'_{1...}, \mathbf{a}'_n)$ can be referred to this one through the equations

$$\mathbf{P} = \mathbf{Q} + A^i \mathbf{a}_i, \qquad \mathbf{a}'_i = A^j_{i'} \mathbf{a}_j. \tag{2.1}$$

There is a one-to-one correspondence between the affine frames and the set of $n + n^2$ quantities $(A_{i'}, A^j_{i'})$ with $\det [A^j_{i'}] \neq 0$. Readers familiar by now with the concept of differentiable manifold will easily verify that the frame bundle constitutes a differentiable manifold of dimension $n + n^2$. Other readers need only believe that this a manifold where we can differentiate without worry. The points of the frame bundle manifold are thus the affine frames. The quantities $(A_i, A^j_{i'})$ with $\det [A^j_{i'}] \neq 0$ constitute a system of coordinates on the frame bundle manifold. In other words, they assign labels to frames.

A third frame $(\mathbf{R}, \mathbf{a}''_1, \ldots, \mathbf{a}''_n)$ is given in terms of $(\mathbf{P}, \mathbf{a}'_1, \ldots, \mathbf{a}'_n)$ by

$$\mathbf{R} = \mathbf{P} + A'^i \mathbf{a}'_i, \qquad \mathbf{a}''_i = A^{j'}_{i''} \mathbf{a}'_j. \tag{2.2}$$

Substitution of (2.1) in (2.2) yields $(\mathbf{R}, \mathbf{a}''_1, \ldots, \mathbf{a}''_n)$ in terms of $(\mathbf{Q}, \mathbf{a}_1, \ldots, \mathbf{a}_n)$. The resulting expression remains of the same form. We may view these pairs of equations as transformations. They constitute a group. (Readers, please check that the conditions defining the concept of group are satisfied.) It is called the affine group $Af(n)$. By choosing a frame in F^n as preferred, we have been able to put the members of F^n in a one-to-one correspondence with the members of $Af(n)$. The parameters (A^i, A^j_i) in the group are a system of coordinates in the frame bundle.

In the following, we shall continue to denote a fixed frame as $(\mathbf{Q}, \mathbf{a}_1, \ldots, \mathbf{a}_n)$, and arbitrary frames (said better, frame-valued functions) as $(\mathbf{P}, \mathbf{e}_1, \ldots \mathbf{e}_n)$. Thus, we shall write:

$$\mathbf{P} = \mathbf{Q} + A^i \mathbf{a}_i, \qquad \mathbf{e}_i = A^j_i \mathbf{a}_j, \tag{2.3}$$

for the action of the affine group. The second of these equations represents the action of its maximal linear subgroup. The elements of the matrix inverse to $[A^j_i]$ will be denoted as B^j_i. A formalization of these concepts now follows.

The natural map $\mathbf{P} : F^n \to Af^n$ (i.e. $(\mathbf{P}, \mathbf{e} \ldots \mathbf{e}_n)$ goes to $\mathbf{P}$) is a continuous and onto projection. We define similar projection maps $\mathbf{e}_i$. Af^n is called the base of the frame bundle F^n. The set of all points of the frame bundle that are mapped to $\mathbf{P}$ is called the fiber above $\mathbf{P}$. The fibers of F^n are n^2-dimensional manifolds whose points are in a one-to-one correspondence with the elements of the general linear group $GL(n)$ (i.e. of all linear transformations in n dimensions), once a particular basis at each point $\mathbf{P}$ has been chosen as "preferred". The coordinates in the fibers are the $A^j_{i'}$'s.

We can visualize frame bundles as bundles of fibers. Refer to Figure 2. A name like fiber bundle is more pictorial but less precise than frame bundle. A frame bundle is a fiber bundle where the fibers are made of vector bases; the point they are at makes the bases become frames. But there are also fibers made of vectors, spinors, etc. They are then called vector bundles, spinor bundles, etc. Very precise definitions of fiber bundles, principal fiber bundles, frame bundles, tangent bundles, etc. exist in the literature on differential geometry. We can do without their formal introduction.

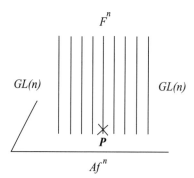

Figure 3: Fibration of F^n over Af^n

Owing to the fact that the frame $(\mathbf{Q}, \mathbf{a}_1, \ldots, \mathbf{a}_n)$ is fixed, we can readily differentiate the vector valued functions (2.3) as follows:

$$d\mathbf{P} = dA^i \mathbf{a}_i, \quad d\mathbf{e}_i = dA_i^j \mathbf{a}_j. \tag{2.4}$$

(some may prefer to think of $-\mathbf{P}$ as the vector field —defined by a fixed point Q— that assigns to each point P the vector from P to Q).

The dA^i's and dA_i^j's span $n-$dimensional and n^2-dimensional modules of $1-$forms. We can think of $d\mathbf{P}$ as $dA^1 \otimes \mathbf{a}_1 + dA^2 \otimes \mathbf{a}_2 + \ldots$ and of $d\mathbf{e}_i$ as $dA_i^1 \otimes \mathbf{a}_1 + dA_i^2 \otimes \mathbf{a}_2 + \ldots$ One thus refers to $d\mathbf{P}$ and $d\mathbf{e}_i$ as vector-valued differential $1-$forms. One should keep in mind, however, that $d\mathbf{e}_i$ is not vector-valued in the same sense that $d\mathbf{P}$ is. The bold faced quantities have subscripts in one case, but not in the other. That is a very important difference whose consequences will become increasingly apparent.

The presence of the constant basis $(\mathbf{a}_i)$ in Eqs. (2.4) is unnecessary and hides the quality of invariance of the concepts with which we are dealing. We proceed to remove that basis. Solving for $\mathbf{a}_i$ in the second equation (2.4) we have

$$\mathbf{a}_i = B_i^j \mathbf{e}_j. \tag{2.5}$$

Substituting (2.5) in (2.4), we obtain

$$d\mathbf{P} = dA^j B_j^i \mathbf{e}_i, \quad d\mathbf{e}_i = dA_i^k B_k^j \mathbf{e}_j, \tag{2.6}$$

which states that $d\mathbf{P}$ and $d\mathbf{e}_i$ are linear combinations of the $\mathbf{e}_i$'s with coefficients $dA^j B_j^i$ and $dA_i^k B_k^j$ that together constitute $n + n^2$ differential $1-$forms in the bundle. We denote these coefficients as ω^i and ω_i^j. In terms of the aforementioned coordinate system, they are given by

$$\omega^i = dA^j B_j^i, \qquad \omega_i^j = dA_i^k B_k^j \qquad (2.7)$$

and are referred to as the left invariant forms of the frame bundle of Af^n. In general, they will look different in terms of other coordinate systems (We speak of left and/or right invariants depending on whether we make groups act on the left or on the right; in this book, it will always be left invariants).

Assume we are given a set of $n + n^2$ independent differential forms that cannot be reduced to the form (2.7) by a change of coordinates. In that case they represent a structure other than an affine space. If they still satisfy certain conditions to be later considered, they may be used as the ω^i and ω_i^j in the system

$$d\mathbf{P} = \omega^i \mathbf{e}_i, \qquad d\mathbf{e}_i = \omega_i^j \mathbf{e}_j, \qquad (2.8)$$

to which Cartan referred to as the connection equations. These forms are then called the connection differential forms.

In the next section, we shall consider conditions to be satisfied by a set (ω^i, ω_i^j) to be reducible to the form (dA^j, dA_i^k) by a set of coordinate transformations. If it is reducible, the frame bundle defined by the (ω^i, ω_i^j) will be the one for an affine space, and the coordinate transformation will take us from curvilinear to rectilinear coordinates. The identification of the actual transformation is an additional problem. Suffice to say that, if the conditions are satisfied, we may integrate the (pull-back of the) differential forms (ω^i, ω_i^j) along any curve between two points and relate any two points of the frame bundle in a path-independent way.

When, going the other way, we change from the rectilinear coordinates A^i to curvilinear coordinates x^i, the ω_k^j will also depend on the x^i. If, in addition, we change the tangent basis field in terms of which we construct the bundle to a non-constant basis field, as we shall do later, the ω_i^j's will also depend on the differentials of those curvilinear coordinates. For the concept of constant basis field, recall for the moment what was said in section 6 of chapter 3. We return to this in section 4.

The connection differential forms are said to be invariant because they are independent of coordinate system, even if one uses coordinates to represent them. This and many other statements made so far will be clarified with the introduction of other coordinate systems in section 5, and with the simpler but structurally similar examples of chapter 6, and also with the simple examples of Cartanian generalizations of chapter 7.

5.3   The structure of affine space

The system of equations (2.4) is obviously integrable to yield (2.3). The system (2.6) also is integrable, since it is a different version of (2.4), and the result of the integration again is (2.3). Consider next the system of differential equations

(2.8) where ω^i and ω_i^j do not take the form (2.7). They may —or may not— be the same differential forms (2.7), though expressed in terms of some other coordinate system. If they are (are not), we say that the system 2.8 is (respectively is not) integrable.

The issue of integrability is resolved by application of a theorem called Frobenius theorem. Its application to the system being considered here results in the following: a necessary and sufficient condition for the integrability of the system (2.8) (with rather arbitrary differential forms ω^i and ω_i^j; more on all this to be found in a later chapter) is that $d(\omega^i \mathbf{e}_i) = 0$, $d(\omega_i^j \mathbf{e}_j) = 0$. The necessity condition is obvious. Suffice to differentiate the right hand side of (2.4):

$$d d\mathbf{P} = d(dA^i \mathbf{a}_i) = ddA^i \mathbf{a}_i + dA^i d\mathbf{a}_i = 0 + 0 = 0, \tag{3.1}$$

$$d d\mathbf{e}_i = d(dA_i^j \mathbf{a}_j) = ddA_i^j \mathbf{a}_j + dA_i^j d\mathbf{a}_j = 0 + 0 = 0. \tag{3.2}$$

We shall not prove the sufficiency condition.

Let us write $dd\mathbf{P} = 0$ and $dd\mathbf{e}_i = 0$ in terms of the invariant forms of the bundle. Differentiating (2.8), we get:

$$0 = d\omega^i \mathbf{e}_i - \omega^j \wedge d\mathbf{e}_j = (d\omega^i - \omega^j \wedge \omega_j^i)\mathbf{e}_i, \tag{3.3}$$

$$0 = d\omega_i^j \mathbf{e}_j - \omega_i^k \wedge d\mathbf{e}_k = (d\omega_i^j - \omega_i^k \wedge \omega_k^j)\mathbf{e}_j. \tag{3.4}$$

Hence, since the $(\mathbf{e}_i)$ is a vector basis, we have:

$$d\omega^i - \omega^j \wedge \omega_j^i = 0, \qquad d\omega_i^j - \omega_i^k \wedge \omega_k^j = 0. \tag{3.5}$$

These are called the equations of structure of affine space.

The relevance of the equations of structure thus lies in the fact that they are the integrability conditions of the connection equations, (2.8). They are the basis for the generalization of affine geometry to differentiable manifolds with an affine connection, which are the Cartanian generalization of affine space. In order to get perspective, we shall now make a small incursion into that generalization, which is the subject of chapter 8.

Differentiable manifolds endowed with affine connection also have associated frame bundles. Assume we were given the ω^i and ω_k^j in the bundle of one such generalization. If we were to perform the computations of the left hand sides of (3.5), we would not get zeros in general. We would get linear combinations of the $\omega^l \wedge \omega^p$, but not of the $\omega^k \wedge \omega_s^r$ and $\omega_m^l \wedge \omega_q^p$ terms, or else we would be dealing with some structure still more general. The generalization of (3.5) to our affine connections will then read:

$$d\omega^i - \omega^j \wedge \omega_j^i = R_{kl}^i \omega^k \wedge \omega^l, \qquad d\omega_i^j - \omega_i^k \wedge \omega_k^j = R_{jlm}^i \omega^l \wedge \omega^m. \tag{3.6}$$

One uses the terms torsion and affine curvature to refer to the right hand sides of equations (3.6) respectively, and represents them with the symbols Ω^i and Ω_j^i :

$$\Omega^i \equiv R_{kl}^i \omega^k \wedge \omega^l, \qquad \Omega_j^i \equiv R_{jlm}^i \omega^l \wedge \omega^m, \tag{3.7}$$

This is typical when one knows the R^i_{kl}'s and R^i_{jlm}'s but not the ω^k and ω^r_s. Sometimes, one refers to the $d\omega^i - \omega^j \wedge \omega^i_j$ and the $d\omega^j_i - \omega^k_i \wedge \omega^j_k$ themselves as the torsion and the curvature. Since computations with the explicit form that ω^j and ω^i_j take in the bundle would be laborious in general, one actually computes in sections of the bundle, a concept which we are about to introduce. The Ω^i and Ω^i_j may be viewed as differential form components of a $(1,0)$−tensor and $(1,1)$−tensor respectively. But this is only just one way of looking at curvature and torsion. Impatient readers can refer to sections 5.9 and 8.13 for their Lie algebra valuedness.

Let us return to affine space. The equations of structure of affine space are then written as

$$\Omega^i = 0, \qquad \Omega^i_j = 0. \qquad (3.8)$$

One then says that affine spaces are particular cases of differentiable manifolds endowed with affine connections whose torsion and curvature are zero.

An interesting exercise at this point with the equations of affine space is the following. Substitute Eqs. (2.7) in (3.5). Use $A^j_i B^k_j = \delta^k_i$ and reach two trivially looking identities, $0 = 0$. In generalized spaces, they will take interesting forms. We shall then use this trivial identity to simplify calculations.

5.4 Curvilinear coordinates: holonomic bases

A section of the affine frame bundle is a continuous map $S : Af^n \to F^n$ such that $\mathbf{P} \circ S$ is the identity. In simpler terms, it is a map that picks one and only one point from each fiber. Since those points are frames, the section is a frame field, which we shall assume to be not only continuous but also differentiable to whichever order may be needed. The concept is equally applicable to the bundles of generalizations of affine and Euclidean spaces.

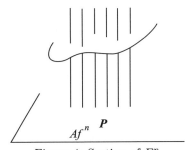

Figure 4: Section of F^n

As we shall see in the next paragraph, sections emerge naturally when one introduces in the base manifold (i.e. affine space in this chapter) curvilinear coordinates through an invertible transformation: $x^i = x^i(A^1 \ldots A^n)$, $i = 1 \ldots n$. Needless to say that we may define new curvilinear coordinates in terms of other curvilinear coordinates rather than directly from the A^i. The x^i are no longer rectilinear if the equations of the transformation are not linear. Inexperienced readers should be aware of the inconsistency in terminology from one area of mathematics to another. One should use the term affine equations to refer to what are usually called linear equations. Linear transformation yields the special linear equations whose independent terms are all zero.

Curvilinear coordinate systems define point-dependent bases of vectors $\mathbf{e}_i$ by means of

$$\mathbf{e}_i \equiv \frac{\partial \mathbf{P}}{\partial x^i}. \tag{4.1}$$

We use the same term $\mathbf{e}_i$ as before since these frame fields are sections of the bundle of frames. One says that the $\mathbf{e}_i$ of (4.1) are pull-backs (to sections) of the $\mathbf{e}_i$ of (2.3). Curvilinear coordinates are not components of $\mathbf{P}$ or some other vector upon some basis $\{\mathbf{b}_i\}$ and, in fact, one first defines those coordinates and then a basis field associated with them.

The basis vectors $\mathbf{e}_i$ and $\mathbf{a}_j$ are related by:

$$\mathbf{e}_i = \frac{\partial \mathbf{P}}{\partial A^j} \frac{\partial A^j}{\partial x^i} = \frac{\partial A^j}{\partial x^i} \mathbf{a}_j. \tag{4.2}$$

The partial derivatives are not constants but functions of the coordinates, unless we were dealing with a transformation between two rectilinear coordinate systems. To be precise, we are dealing here with composite functions, since the A^i are themselves functions that assign numbers to the points of the manifold (Refer to point (c) of section 6 of chapter 1). Because of its importance, we repeat that, when dealing with coordinates, we use the same symbol for coordinate functions and for the value that those functions take. Whether we are considering the ones or the others will be obvious from the context. The nature of the coordinates as functions on the manifold started with the definition of $\mathbf{P}$ as a projection map on Af^n. It follows that rectilinear coordinates are also functions on Af^n. Curvilinear coordinates being functions of the rectilinear coordinates, they also are (composite) functions on Af^n.

The $(\mathbf{e}_i)$ vary from one point to another. Each coordinate system determines a vector basis at each point of the region of affine space that it covers (No basis is determined at the origin of the system of polar coordinates, since this system does not assign a unique pair of numbers to it; the definition (4.2) fails in any case). We are thus dealing here with fields of bases, not just one basis as was the case with $(\mathbf{a}_i)$. But $(\mathbf{a}_i)$ can now be viewed as a constant basis field. This becomes obvious when we obtain a system of rectilinear coordinates from another one in the same way as for curvilinear coordinates.

The vector basis fields generated in the preceding way by the system of curvilinear coordinates is referred to as holonomic or coordinate basis field. In that change from rectilinear to rectilinear, the new basis field $(\mathbf{b}_i)$ appears as a particular case of curvilinear basis field that happens to be the same everywhere in terms of the $(\mathbf{a}_i)$, and thus equal to itself from one point to another. The argument can be reversed, meaning that we could have started the whole argument with the basis $(\mathbf{b}_i)$ and have obtained the basis $(\mathbf{a}_i)$ as a constant frame field also denoted $(\mathbf{a}_i)$.

Non-holonomic basis fields will be introduced in the next section. We leave for then the push-forward (a concept that amounts here to a recovery) of the invariant forms in the bundle from the dx^k and ω_i^j just considered.

The use of a particular coordinate system does not force us to use corresponding holonomic basis fields (unlike in the most common version of the tensor calculus, version where bases are not explicitly used).

From the definition (4.1) of $\mathbf{e}_i$, we have:

$$d\mathbf{P} = dx^i \mathbf{e}_i. \tag{4.3}$$

Along each coordinate line, all the dx^i's but one are zero. Equation (4.3) then shows that the vector is tangent to the coordinate line x^i. In a similar way to the obtaining of Eq. (4.2), we would get

$$\mathbf{a}_i = \frac{\partial \mathbf{P}}{\partial x^j} \frac{\partial x^j}{\partial A^i} = \frac{\partial x^j}{\partial A^i} \mathbf{e}_j. \tag{4.4}$$

This shows that the vectors $(\mathbf{e}_j)_P$ constitute a basis since any vector can be expressed as a linear combination of the $\mathbf{e}_i$'s through the $\mathbf{a}_i$'s (and the number of vectors $\mathbf{e}_j$ is the same as the dimension of the vector space).

Let us practice some of these ideas using polar coordinates in the plane

$$x = \rho \, \cos\phi, \qquad y = \rho \, \sin\phi. \tag{4.5}$$

We use (4.2) with (x, y) and (ρ, ϕ) replacing the A^i's and x^i's respectively. We thus have

$$\mathbf{e}_\rho = \cos\phi \mathbf{a}_1 + \sin\phi \mathbf{a}_2 \qquad \mathbf{e}_\phi = -\rho\sin\phi \mathbf{a}_1 + \rho\cos\phi \mathbf{a}_2 \tag{4.6}$$

where we have used $\mathbf{a}_1$ and $\mathbf{a}_2$ for $\mathbf{i}$ and $\mathbf{j}$. Notice that, with this definition of $\mathbf{e}_j$, $d\mathbf{P}$ is not given by the familiar $d\mathbf{P} = d\rho\mathbf{e}_\rho + \rho d\phi\mathbf{e}_\phi$, which corresponds to orthonormal bases, but rather by

$$d\mathbf{P} = d\rho\mathbf{e}_\rho + d\phi\mathbf{e}_\phi, \tag{4.7}$$

in accordance with Eq. (4.3). The concept of orthonormal bases does not even enter here since a dot product or a metric has not been defined. Thus, although ρ and ϕ have metric connotation, we ignore it here and consider those functions of (x, y) as any other functions without such connotation.

From (4.6), we read $A_1^1 = \cos\phi$, $A_1^2 = \sin\phi$, $A_2^1 = -\rho\sin\phi$, $A_2^2 = \rho\cos\phi$ for the vector basis (on the section) that is above the point (ρ, ϕ) of Af^2.

Exercise. Obtain $\mathbf{e}_i$ in terms of $\mathbf{a}_j$, and vice versa, for the spherical coordinates:

$$r = \sqrt{x^2 + y^2 + z^2}, \quad \phi = \arctan\frac{y}{x}, \quad \theta = \arctan\frac{\sqrt{x^2 + y^2}}{z}. \tag{4.8}$$

It is not convenient to invert the change of basis by inverting the matrix that gives this change if the inverse coordinate transformation is remembered. Thus, in this case, we would proceed in the same way but starting from the transformations

$$x = r\sin\theta\cos\phi, \quad y = r\sin\theta\sin\phi, \quad z = r\cos d\theta. \tag{4.9}$$

Notice that we do not need to draw figures to relate different vector bases, which may not be easy to do for unusual coordinate systems.

We next summarize much of what has been done in this section. We arbitrarily chose a point of affine space and introduced an arbitrary basis $(\mathbf{a}_i)$. Rectilinear coordinates (A^i) were obtained as components of $\mathbf{P}$ relative to that basis. This was followed by the introduction of curvilinear coordinates as functions of the rectilinear ones. We then generated basis fields associated with the coordinate system. We completed the process by finally viewing the rectilinear coordinates as the particular cases of curvilinear coordinates that generate constant vector basis fields $\{\mathbf{a}_i\}$.

The use of sections —thus of a specific vector basis at each point of affine space— has the implication that the ω_i^j's are now linear combinations

$$\omega_i^j = \Gamma_{ik}^j dx^k \tag{4.10}$$

of the differentials of the coordinates in affine space. We thus have

$$d\mathbf{P} = dx^i \mathbf{e}_i \tag{4.11a}$$

$$d\mathbf{e}_i = \omega_i^j \mathbf{e}_j = \Gamma_{ik}^j \mathbf{e}_j dx^k. \tag{4.11b}$$

Consistently with concepts already advanced, these equations pertain to quantities (on the left) that are said to be pull-backs to sections of quantities in the bundle.

One can show that the pull-backs of exterior derivatives and exterior products are the exterior derivatives and exterior products of the pull-backs of the quantities being exterior differentiated and exterior multiplied. We leave for section 6 the pull-back of the equations of structure to sections of the bundle, whether holonomic sections like those we have just seen, or anholonomic ones, about to be considered.

Finally, the basis fields of two different coordinate patches can be related in a way similar to (4.4). Let (x^i) and (x'^i) be two sets of coordinates. We shall denote by $(\mathbf{e}'^i)$ the basis field for the coordinate patch (x'^i). A change of coordinate vector basis field is then given by

$$\mathbf{e}_i = \frac{\partial\mathbf{P}}{\partial x'^j}\frac{\partial x'^j}{\partial x^i} = \frac{\partial x'^j}{\partial x^i}\mathbf{e}'_j \tag{4.12a}$$

$$\mathbf{e}'_i = \frac{\partial\mathbf{P}}{\partial x^j}\frac{\partial x^j}{\partial x'^i} = \frac{\partial x^j}{\partial x'^i}\mathbf{e}_j. \tag{4.12b}$$

5.5 General vector basis fields

A constant basis field $(\mathbf{a}_i)$ together with a matrix-valued function $A^i_{j'}$ of the coordinates A^k (with det $A^i_{j'} \neq 0$) defines a field of bases by

$$\mathbf{e}_j = A^i_{j'}(A^k)\mathbf{a}_i. \tag{5.1}$$

We define corresponding differential forms ω^j by using (5.1) in

$$d\mathbf{P} = \omega^j \mathbf{e}_j = dA^i \mathbf{a}_i \tag{5.2}$$

and comparing coefficients. Inverting these relations yields the ω^j's in terms of the dA^i's. Let us denote these coefficients as $A^{j'}_i$:

$$\omega^j = A^{j'}_i dA^i. \tag{5.3}$$

This reduces to the case in the previous section if ω^j is dx^j. But this is possible if and only if the differential system

$$dx^j = A^{j'}_i dA^i. \tag{5.4}$$

is integrable. The integrability conditions are:

$$\frac{\partial A^{j'}_i}{\partial A^k} = \frac{\partial A^{j'}_k}{\partial A^i} \tag{5.5}$$

for all values of the indices. If this condition is not satisfied, the basis field $(\mathbf{e}'_i)$ is called anholonomic or non-holonomic.

The preceding presentation of anholonomic basis fields could equally have taken place by defining the ω^j in terms of the differentials of curvilinear coordinates (x^j), rather than rectilinear ones. Instead of (5.1), we would have

$$\mathbf{e}'_j = A^i_{j'}(x^k)\mathbf{e}_i \tag{5.6}$$

Again for the purpose of contrast of affine space with its Cartanian generalizations, let us remark the following. In the generalizations, no basis field can be given in the form (4.1) because there is no $\mathbf{P}$ to be differentiated. If and when we use $d\mathbf{P}$ in that more general context, it will be just notation for some vector-valued differential form, not the result of differentiating some vector field.

As an example of introduction of a non-holonomic frame field we have the orthonormal frame field with bases $(\hat{\mathbf{e}}_\rho, \hat{\mathbf{e}}_\phi)$ associated with the polar coordinates. In this case, Eqs. (5.1) now read

$$\hat{\mathbf{e}}_\rho = \cos\phi\,\mathbf{a}_1 + \sin\phi\,\mathbf{a}_2 \qquad \hat{\mathbf{e}}_\phi = -\sin\phi\,\mathbf{a}_1 + \cos\phi\,\mathbf{a}_2. \tag{5.7}$$

We then write $d\mathbf{P}$ as

$$d\mathbf{P} = d\rho\,\hat{\mathbf{e}}_\rho + \rho d\phi\,\hat{\mathbf{e}}_\phi. \tag{5.8}$$

The vector basis field corresponding to $\omega^1 = d\rho$, $\omega^2 = \rho d\phi$ is anholonomic. The coordinates $A^j_{i'}$ of intersection of this section with the fiber at (ρ, ϕ) become the following functions $A^1_1 = \cos\phi$, $A^1_2 = -\sin\phi$, $A^2_1 = \sin\phi$, $A^2_2 = \cos\phi$. Needless to say that we could have chosen a system of coordinates in the fibers such that the coordinates of $(\hat{\mathbf{e}}_\rho, \hat{\mathbf{e}}_\phi)$ were $A^1_1 = 1$, $A^1_2 = 0$, $A^2_1 = 0$, $A^2_2 = 1$.

Once again, the coefficients of the $d\mathbf{e}_i$'s in terms of the basis field itself will be denoted as ω^j_i:

$$d\mathbf{e}_i = \omega^j_i \mathbf{e}_j, \tag{5.9}$$

i.e. as for holonomic basis fields. Correspondingly, we could express ω^j_i as a linear combination

$$\omega^j_i = \Gamma^j_{ik}\omega^k. \tag{5.10a}$$

However, for the purpose of exterior differentiating ω^j_i it is better to have it expanded as

$$\omega^j_i = \Gamma'^j_{ik}dx^k, \tag{5.10b}$$

even if the ω^j_i pertain to a non-holonomic section of the bundle.

Both the Γ^j_{ik} and the Γ'^j_{ik} are functions of only the coordinates of the base space, i.e of affine space, since we are dealing with a section of the bundle rather than on the fibers or on the whole bundle.

The Γ'^j_{ik} will be called mixed components of the connection. The reason does not have to do with having superscripts and subscripts, but rather that, for the $\mathbf{e}_j$ that goes together with ω^j_i in (5.9), we have $d\mathbf{P} = \omega^j\mathbf{e}_j$, but not $dx^j\mathbf{e}_j$. We state this fact by saying that dx^j and $\mathbf{e}_j$ are not dual of each other, or that they do not correspond to each other.

Be aware of the fact that in this paragraph the use of primed quantities will not be the same as in the previous one, in particular in the case of Γ'^j_{ik}. The relation between the ω^j_i in two different sections is obtained by differentiating (5.6). We get

$$d\mathbf{e}'_i = A^r_{i'}d\mathbf{e}_r + dA^l_{i'}\mathbf{e}_l = A^r_{i'}\omega^l_r\mathbf{e}_l + dA^l_{i'}\mathbf{e}_l. \tag{5.11}$$

After writing $d\mathbf{e}'_i$ as $\omega'^j_i\mathbf{e}'_j$ and $\mathbf{e}_l$ as $A^{j'}_l\mathbf{e}'_j$, we proceed to equate coefficients to obtain

$$\omega'^j_i = A^r_{i'}A^{j'}_l\omega^l_r + dA^l_{i'}A^{j'}_l. \tag{5.12}$$

Because of the last term, ω^i_j does not "transform" linearly. The same comment applies to the Γ^j_{ik}, since (5.12) in turn yields the following transformation of connection coefficients

$$\Gamma'^j_{ik} = A^r_{i'}A^{j'}_l A^m_{k'}\Gamma^l_{rm} + A^l_{i'/k}A^{j'}_l, \tag{5.13}$$

where, recalling our previous use "/" of a subscripts we have defined $A^l_{i'/k}$ through $dA^l_{i'} = A^l_{i'/k}\omega'^k$, and where we have further used that $\omega^l_r = \Gamma^l_{rm}\omega^m = \Gamma^l_{rm}A^m_{k'}\omega'^k$ and that $\omega'^j_i = \Gamma'^j_{ik}\omega'^k$.

Exercise. Verify that, if the two bases are holonomic, (5.13) is equivalent to

$$\Gamma^j_{ik} = \frac{\partial x'^l}{\partial x^i} \frac{\partial x^j}{\partial x'^m} \frac{\partial x'^q}{\partial x^k} \Gamma'^m_{lq} + \frac{\partial^2 x'^l}{\partial x^i \partial x^k} \frac{\partial x^j}{\partial x'^l}. \tag{5.14}$$

Given a basis field $\{\mathbf{e}\}$, the affine frame bundle is represented by $(\mathbf{P}, A\{\mathbf{e}\}_P)$ for all $\mathbf{P}$ and all the non-singular $n \times n$ matrices A. Curly brackets are used for column matrices. We could have chosen another field and have built the bundle in the same way: $(\mathbf{P}, A'\{\mathbf{e}'\}_P)$. Although the two representations use fields and coordinates, they constitute the same (total) set of frames. For any one of them, its coordinates A^j_i depend on the section to which we applied the linear group. The obtaining of the push-forward of the ω^i from a section to the frame bundle should be obvious by now: directly apply the matrices of the full linear group to the ω^i written as a column.

5.6 Structure of affine space on SECTIONS

In section 5.3, $(\mathbf{P}, \mathbf{e}_i)$ represents n+1 projection maps, $\mathbf{P} : F^n \to Af^n \to V^n$ and $\mathbf{e}_i : F^n \to V^n$. These maps are functions on F^n. On the other hand, the vector fields $\mathbf{e}_i$ of section 4.4 are functions on Af^n. They can, therefore, be denoted as $\mathbf{e}_i(\mathbf{P})$. One often uses the not quite correct notation $\mathbf{e}_i(x^j)$, or simply $\mathbf{e}_i(x)$.

Since the pull-back of the derivative is the derivative of the pull-back, we have

$$d\mathbf{P}(x) = \omega^i(x)\mathbf{e}_i(x), \tag{6.1}$$

$$d\mathbf{e}_i(x) = \omega^j_i(x)\mathbf{e}_j(x). \tag{6.2}$$

And, since the pull-back of the exterior product is the exterior product of the pull-backs of the factors, we can write the pull-back of the equations of structure in the bundle, (3.5), as

$$d\omega^i(x) = \omega^j(x) \wedge \omega^i_j(x), \tag{6.3}$$

$$d\omega^j_i(x) = \omega^k_i(x) \wedge \omega^j_k(x). \tag{6.4}$$

These equations thus give us the structure of affine space from a section's perspective. They will be written without explicit indication of the domains of the functions, i.e. as in (3.5). Again, we shall know from the context whether we are working in the frame bundle or on a particular section.

When a differentiable manifold endowed with a connection also has a metric defined on it, there are some very useful non-holonomic basis fields. Whether one uses the ones or the others depends on what one is going to do. Even if a section is non-holonomic, it is preferable, as we already said, to use $\Gamma'^j_{ik}dx^k$ rather than $\Gamma'^j_{ik}\omega^k$ if one is going to exterior differentiate ω^j_i, since $ddx^k = 0$. But one should do so only temporarily and then return to the non-holonomic bases. We proceed to explain.

An important disadvantage of using holonomic sections is that $d\omega^i = ddx^i = 0$ and Eq. (6.3) then becomes

$$dx^j \wedge \Gamma^i_{jk} dx^k = (\Gamma^i_{jk} - \Gamma^i_{kj})(dx^j \wedge dx^k) = 0, . \qquad j < k, \qquad (6.5)$$

which in turn implies

$$\Gamma^j_{ik} = \Gamma^j_{ki}. \qquad (6.6)$$

This type of simplicity masks the fact that we are dealing with the first equation of structure, rather than just a property of the connection forms.

The advantage has to do with the use one may make of them when exterior differentiating, even if one is dealing with non-holonomic tangent basis fields. Let ω^j_i belong to a non-holonomic section. Nothing impedes us from expressing ω^j_i as $\Gamma'^j_{ik} dx^k$ in the second equation of structure. We then have:

$$d\omega^j_i = d(\Gamma'^j_{ik} dx^k) = (\Gamma'^j_{ik,l} - \Gamma'^j_{il,k})(dx^l \wedge dx^k). \qquad (6.7a)$$

$$-\omega^m_i \wedge \omega^j_m = (\Gamma'^m_{ik}\Gamma'^j_{ml} - \Gamma'^m_{il}\Gamma'^j_{mk})(dx^l \wedge dx^k) \qquad (6.7b)$$

$$0 = \Omega^j_i = (\Gamma'^j_{ik,l} - \Gamma'^j_{il,k} + \Gamma'^m_{ik}\Gamma'^j_{ml} - \Gamma'^m_{il}\Gamma'^j_{mk})(dx^l \wedge dx^k), \qquad (6.7c)$$

where we have used (3.8). We thus have

$$R'^j_{ilk} = \frac{1}{2}(\Gamma'^j_{ik,l} - \Gamma'^j_{il,k} + \Gamma'^m_{ik}\Gamma'^j_{ml} - \Gamma'^m_{il}\Gamma'^j_{mk}) = 0, \qquad (6.8)$$

with R'^j_{ilk} defined by

$$\Omega^i_j \equiv R'^i_{jlm}(dx^l \wedge dx^m). \qquad (6.9)$$

The components R'^i_{jlm} are of the type we have called mixed. They may easily prompt one to make errors prompt if one is not using explicitly bases for the tensor valuedness.

We may now return to when all four indices pertain to non-holonomic bases by substituting the dx^k in terms of the ω^s in the equation

$$R'^i_{jlm} dx^l \wedge dx^m = R^i_{jqr}\omega^q \wedge \omega^r. \qquad (6.10)$$

We find not only the R^i_{jqr} but also a linear transformation of components of the curvature that affects only two indices, since we do not change the basis pertaining to the other two indices.

Of course, we also have

$$\Gamma^j_{ik,l} - \Gamma^j_{il,k} + \Gamma^m_{ik}\Gamma^j_{ml} - \Gamma^m_{il}\Gamma^j_{mk} = 0 \qquad (6.11)$$

in affine space. This obviously holds in any tangent basis field and concomitant basis of differential forms.

5.7 Differential geometry as calculus of vector-valued differential forms

Tensor calculus and modern versions of differential geometry fail to show that classical differential geometry is little else than calculus of vector-valued differential forms.

In order to eliminate some clutter, the term vector-valued differential form means here vector-field-valued differential form, since they are defined over curves, surfaces, etc., not at points. Let a vector field be represented as

$$\mathbf{v} = V^i \mathbf{a}_i = v^l \mathbf{e}_l, \tag{7.1}$$

where $(\mathbf{a}_i)$ but not $(\mathbf{e}_l)$ is a constant basis field. Differentiation of $V^i \mathbf{a}_i$ is trivial.
Exercise. Justify the Leibniz rule to differentiate $v^l \mathbf{e}_l$, using $\omega_l^i \mathbf{e}_i$ for $d\mathbf{e}_l$. Hint: in $v^l \mathbf{e}_l$ replace $\mathbf{e}_l$ in terms of $\mathbf{a}_i$, differentiate, express the result in terms of $\mathbf{e}_l$ and compare with the result of applying the Leibniz rule.

Based on the Leibniz rule, we have

$$d\mathbf{v} = dv^i \mathbf{e}_i + v^k d\mathbf{e}_k = (dv^i + v^k \omega_k^i)\mathbf{e}_i. \tag{7.2}$$

Since the expression (7.2) is valid for any basis field and always equal to $d(V^i \mathbf{a}_i)$, we have

$$d\mathbf{v} = (dv^i + v^k \omega_k^i)\mathbf{e}_i = (dv'^i + v'^k \omega_k'^i)\mathbf{e}_i' = (dv''^i + v''^k \omega_k''^i)\mathbf{e}_i'' = \ldots \tag{7.3}$$

This shows that the quantities $(dv^i + v^k \omega_k^i)$ transform like the components of a vector field. The so called covariant derivatives $v_{;k}^j$,

$$v_{;k}^j \equiv v_{/k}^j + v^l \Gamma_{lk}^j, \tag{7.4}$$

are the components of the vector-valued differential $1-$form $d\mathbf{v}$,

$$d\mathbf{v} = v_{;k}^j \omega^k \mathbf{e}_j. \tag{7.5}$$

By virtue of (7.3), $(dv^i + v^k \omega_k^i)\mathbf{e}_i$ is an invariant, but $dv^i \mathbf{e}_i$ and $v^k \omega_k^i \mathbf{e}_i$ are not. This is obvious because $v^k \omega_k^i \mathbf{e}_i$ is zero in constant basis fields but not in other ones, since ω_k^i does not transform linearly, which impedes that ω_k^i also be zero in other frame fields.

Differentiating again $d\mathbf{v}$, we obtain a vector-valued differential $2-$form, $dd\mathbf{v}$. Its components are not what the tensor calculus calls covariant derivatives. For that, we would have to differentiate $v_{;k}^j \phi^k \mathbf{e}_j$. Resorting to (7.2), we obtain five terms, two from the differentiation of $dv^i \mathbf{e}_i$, and three more from the differentiation of $v^k \omega_k^i \mathbf{e}_i$. The first term, ddv^i is obviously zero. The second and third terms cancel each other out. We thus have

$$dd\mathbf{v} = v^k d(\omega_k^i \mathbf{e}_i) = v^k (d\omega_k^i - \omega_k^j \wedge \omega_j^i)\mathbf{e}_i. \tag{7.6}$$

Taking into account the second of equations (3.6), we get

$$dd\mathbf{v} = v^i R^{j}_{i\,kl}(\omega^k \wedge \omega^l)\mathbf{e}_j. \tag{7.7}$$

The components of the vector-valued differential 2−forms $dd\mathbf{v}$ are $v^i R^{j}_{i\,kl}$. We define an object whose coefficients are the R^{j}_{ikl} as follows:

$$\mho \equiv R^{j}_{i\,kl}(\omega^k \wedge \omega^l)\mathbf{e}_j \phi^i, \tag{7.8}$$

where (ϕ^i) is the field of bases of linear 1−forms dual to the basis field $(\mathbf{e}_j)$. $\mho$ is an invariant object because its action on an arbitrary vector field $\mathbf{v}$ yields the invariant $dd\mathbf{v}$:

$$\mho \llcorner \mathbf{v} = R^{j}_{i\,kl}(\omega^k \wedge \omega^l)\mathbf{e}_j \phi^i \llcorner v^m \mathbf{e}_m = v^i R^{j}_{i\,kl}(\omega^k \wedge \omega^l)\mathbf{e}_j = dd\mathbf{v}, \tag{7.9}$$

where we have used that $\phi^i \llcorner \mathbf{e}_m = \delta^i_m$. Although, following custom, we have used the term affine curvature to refer to Ω^j_i, we should have reserved this name for $\mho$, which, being given by

$$\mho \equiv \Omega^j_i \phi^i \mathbf{e}_j, \tag{7.10}$$

is a (1,1)-tensor-valued differential 2−form. It can also be viewed as a (1,1)-tensor whose components are differential 2−forms. We prefer the first of these two options, i.e. differential form being the substantive and tensor-valued being the adjective. Notice that the covariant valuedness has been induced and accompanied by the contravariant vector-valuedness.

Although this is the first time we have mentioned covariant vector-valuedness, it had already appeared in ω^j_i. But we refrained from mentioning it until now in order to avoid prompting inexperienced readers into thinking that $\omega^j_i \phi^i \mathbf{e}_j$ is an invariant, which is not. Readers can start figuring out why in view of the last part of section 5. We shall consider again the nature of ω^j_i in several times in the future.

One obtains that $\mho$ is zero in affine space by simply differentiating $V^i \mathbf{a}_i$ twice. Once again, we shall later introduce structures where $dd\mathbf{v}$ is no longer zero and where, in general, the rule that dd is zero does not apply to differential forms that are not scalar-valued.

$\mathbf{e}_i$ and $dd\mathbf{e}_i$ are not invariant under a change of section. The reason is obvious: $\mathbf{e}_i$ for given value of i represents different vectors in different sections. $dd\mathbf{e}_i$ will be considered in a later chapter, in the context of non-zero curvature.

Let us compare $d\mathbf{v}$ with another very significant vector-valued differential 1−form, $d\mathbf{P}$. We have

$$d\mathbf{P} = \omega^i \mathbf{e}_i = \delta^i_j \omega^j \mathbf{e}_i. \tag{7.11}$$

Whereas the differentiation of $d\mathbf{P}$ yields the torsion, $\Omega^i \mathbf{e}_i$, the differentiation of $d\mathbf{v}$ leads us to the affine curvature. This might look surprising if one were to view (7.11) as a particular case of (7.5), but whereas the components δ^j_k are connection-independent, the components $v^j_{;k}$ are connection-dependent.

For further practice, let us deal with the differentiation of $v^j_{;k} \phi^k \mathbf{e}_j$. First, and just in case, let us emphasize that $\phi^k \mathbf{e}_j$ is a tensor product, the symbol for such product being ignored as in most of the literature when we multiply copies of different spaces. Observe also that one writes $\omega^i \mathbf{e}_i$ rather then $\omega^i \otimes \mathbf{e}_i$, even though that is what one means.

We apply the Leibniz rule for differentiation of tensor products and obtain, using $\phi^i(\mathbf{a}_j) = \delta^i_j$,

$$d\phi^i = -\omega^i_j \phi^i. \tag{7.12}$$

This meager treatment of $d\phi^i$ is an excuse to bring to your attention the following remarks. Euclidean and pseudo-Euclidean geometry, whether generalized or not, is not only our true interest but also the main subject of geometric interest. It is customarily but not necessarily dealt with in the literature as an outgrowth of affine geometry. This is due to historical reasons. In order to truly address Euclidean geometry with a Euclidean perspective, one should use ab initio algebra specific to Euclidean spaces, which is Clifford algebra. This algebra has not yet occupied the role it should have in mathematical teaching. But we shall at least be able to replace the evaluation "$\phi^i(\_)$" with the product "$\mathbf{e}^i$." when we shall do Euclidean geometry in the next chapter, where the replacement for (7.12) best fits (See section 3 of that chapter, and specifically equation 3.3).

As a concession to tensor calculus because of its dominance of the physics literature, let us differentiate $v^j_{;k} \phi^k \mathbf{e}_j$. We use the Leibniz rule as it applies to tensor products and obtain

$$d(v^j_{;k} \phi^k \mathbf{e}_j) = dv^j_{;k} \phi^k \mathbf{e}_j - v^j_{;l} d\phi^l \mathbf{e}_j + v^l_{;k} \phi^k d\mathbf{e}_l. \tag{7.13}$$

Readers are invited to substitute pertinent expressions for $d\phi^l$ and $d\mathbf{e}_l$ in (7.13) and obtain what, in the tensor literature, goes by the name of second covariant derivatives: the components of the (1,1)-valued differential $1-$form $d(v^j_{;k} \phi^k \mathbf{e}_j)$, not of $dd\mathbf{v}$, which is zero in this case. In the dedicated section 12 of chapter 8, we shall see why covariant derivatives are horrible, except for (first) covariant derivatives of (p,q)-valued $0-$forms, i.e. of tensor fields.

The structures introduced in this chapter allow us already to address more involved computational cases (In order not to overwhelm readers with calculations, in detriment of structural considerations, we shall introduce them little by little in different chapters). After all, equations to be introduced later on manifolds endowed with affine connections also apply here, but setting everywhere to zero both torsion and affine curvature.

5.8 Invariance of connection differential FORMS

The term connection differential forms has restricted and comprehensive meaning's, namely to refer to (ω^j_i) and (ω^k, ω^j_i), respectively. Here we shall refer to the second of these options. Both (ω^j_i) and (ω^k, ω^j_i) are invariants in the bundle. But, what does it mean?

We start by making abstraction of (i.e. ignoring) the rich structure of frame bundles and of the nature of its elements as frames. What results is called the topological frame bundle. If we refibrate the set of bases of our bundle over the topological frame bundle, the fibers are constituted by just one element, as the group in the new fibers only contains the identity transformation. The connection differential forms are then said to be scalars or invariants because they obviously do not change under the transformations in the fiber, which only comprise the identity transformation.

Readers who find this second bundle too trivial to be helpful need consider a third fibration, intermediate between the standard fibration and the one we have just introduced. It is the Finslerian fibration. However, given that linear transformations have less intuitive geometric interpretations than rotations, we are going to provide the Finslerian refibration of bundles of orthonormal frames. Orthonormal frames (i.e. made with orthonormal bases) belong to the next chapter and some readers may consider postponing further reading of this section until later on. For many other readers, knowledge of special relatively (SR) will allow them to deal at this point with a bundle that we have not yet introduced.

Consider the set of all the inertial bases at all points of relativistic flat spacetime. The bases at each point are related among themselves by the action of the (homogeneous!) Lorentz group in four dimensions. These fibers constitute differentiable manifolds of dimension six, all of them identical. The coordinates on the fibers can be taken to be three velocity components v^i and three rotation parameters ϕ^i. These six coordinates together with the four spacetime coordinates label all the frames in our set. The requirement we made that, from a physical perspective, the spatial (sub)bases have to be inertial is not essential; it has been made in order not to get distracted with other issues.

The aforementioned rotations constitute a three dimensional subgroup of the 10-dimensional Poincaré group. The quotient of the Poincaré group by that rotation group is a seven dimensional set labelled by spacetime and velocity coordinates. It can be taken as a base manifold for a "Finslerian" refibration of the 10-dimensional set of frames, with the three-dimensional rotation group acting on the fibers. We now need to consider how the concept of tensoriality applies to sections of this bundle.

We say that a set of quantities transforms vectorially under a group of transformations of the bases of a vector space if, under an element of the group, they transform like the components of vectors, whether contravariant or covariant. We say that some quantity is a scalar if it is an invariant under the transformations of a group. A quantity (respectively, a set of quantities) may be scalar (respectively vectorial) under the transformation of a group, while not being so under the transformations of another group. Thus the sum $\sum_i v^i v^i$ is a scalar under the transformations of the rotation group, but not of the linear group. The invariant sum $\sum_{i,j} g_{ij} v^i v^j$ (see next chapter) is the *affine extension by the linear group* of the expression $\sum_i v^i v^i$, which is left invariant by the (largest) rotation subgroup of the linear group in question.

Before we return to the issue of refibrations, we advance one more concept about the set of frames of SR. Let μ be $(0, i)$ with $i = 1, 2, 3$. Unlike in four-dimensional affine space, there are now only six independent ω^ν_μ's, namely

(ω_0^i, ω_l^m) with $m > l$, which is the same number as the dimensionality of the fibers. The reason is that, as we shall see in the next chapter, $\omega_i^0 = \omega_0^i$ and $\omega_m^l = -\omega_l^m$.

In the Finslerian refibration of the frames of SR, the ω_l^m (as opposed to the ω_λ^μ; $\lambda = 0, l$; $\mu = 0, m$) play the role of the connection, and the ω_0^i live in the new base manifold, i.e. the one to which we have assigned the coordinates (x^μ, v^i). The ω_0^i are like the ω^i in that they do not depend on the differentials $d\phi^i$ of the coordinates on the fiber, and that they transform like the components of a vector under the rotation group in three dimensions. ω^0 is invariant, also called scalar, under the same group.

We return to the statement of invariance of the ω^μ's and ω_μ^ν's when the set of frames is viewed as a bundle over itself. All the differential forms ω^μ and ω_μ^ν are scalars because the group in the fibers only has the unit element. We may legitimately ask: is the refibration of the bundle "over itself" just a trick to create such invariance out of thin air? No, it is not. We are about to see in the next section and in the next chapter that the $(\omega^\mu, \omega_\nu^\lambda)$ are the components of a differential $1-$form taking values in the Lie algebra of the affine or the Euclidean (or Poincaré) group, as the case may be. The Lie algebra is of the same dimension as the group G in the pair (G, G_0) that constitutes the Klein geometries to which those examples refer. We need, therefore, not view the connection as some entity that fails to belong to a tensor structure, but as an entity that belongs to some other structure, a Lie algebra The tensorial approach to geometry —which has very little to do with the groups G and only marginally more with the group G_0 in standard presentations for physicists of the tensor calculus— is a historical accident which should be recognized as such.

Once the set of all frames of a geometry is viewed as the arena where the action takes place in differential geometry, we change coordinates in it, say from y^A to z^A. The index A goes from one to $n + n^2$ in affine space, and from one to $n + [n(n-1)/2]$ in spaces endowed with a dot product. Take, for example, ω_1^2. We have

$$\omega_1^2 = a_1{}^2{}_A dy^A = b_1{}^2{}_A dz^A = \ldots,$$

where y and z are coordinate systems in the bundle and where we sum over all its coordinates. That is the form that the statement of invariance of ω_1^2 takes.

5.9 The Lie algebra of the affine group

The Lie algebra of the affine group is no big deal. It is implicit in the connection equation $\{d\mathbf{e}_i\} = [\omega_i^j]\{\mathbf{e}_j\}$ and what we do with it. The idea is simply to understand what is meant by affine Lie algebra valuedness of the connection differential form ω_i^j. For this author, the usefulness of dealing here with this concept has to do with correcting the wrong impression that whereas Yang-Mills theory has to do with bundles and Lie algebras, traditional differential geometry does not. It does.

An affine group comprises a corresponding subgroup of translations. They are not linear transformations on vectors since

$$T_{\mathbf{w}}(\mathbf{u} + \mathbf{v}) \neq T_{\mathbf{w}}(\mathbf{u}) + T_{\mathbf{w}}(\mathbf{v}), \tag{9.1}$$

where $T_{\mathbf{w}}$ denotes a translation by the vector $\mathbf{w}$. They cannot, therefore, be represented by $n \times n$ matrices. They can, however, be represented by $(n + 1) \times (n + 1)$ matrices, namely

$$\begin{bmatrix} 1 & A^i \\ 0 & I \end{bmatrix}, \tag{9.2}$$

where I is the unit $n \times n$ matrix, and where the A^i are the components of the translation relative to a basis $\mathbf{a}_i$ (Our Latin indices take again the values 1 to n; they momentarily took the values 1 to $n - 1$ in the previous section). More concretely, we have

$$\begin{bmatrix} \mathbf{P} \\ \mathbf{e}_i \end{bmatrix} = \begin{bmatrix} 1 & A^j \\ 0 & I \end{bmatrix} \begin{bmatrix} \mathbf{Q} \\ \mathbf{a}_j \end{bmatrix} \tag{9.3}$$

for a translation and

$$\begin{bmatrix} \mathbf{P} \\ \mathbf{e}_i \end{bmatrix} = g \begin{bmatrix} \mathbf{Q} \\ \mathbf{a}_j \end{bmatrix} = \begin{bmatrix} 1 & A^j \\ 0 & A^j_i \end{bmatrix} \begin{bmatrix} \mathbf{Q} \\ \mathbf{a}_j \end{bmatrix} \tag{9.4}$$

for a general affine transformation g, with $\mathbf{Q}$ and $\mathbf{a}_j$ representing a fixed point and a fixed basis. Equations (2.6) are now written as

$$\begin{bmatrix} d\mathbf{P} \\ d\mathbf{e}_i \end{bmatrix} = dg \cdot g^{-1} \begin{bmatrix} \mathbf{P} \\ \mathbf{e}_j \end{bmatrix} = \begin{bmatrix} 0 & \omega^j \\ 0 & \omega^j_i \end{bmatrix} \begin{bmatrix} \mathbf{P} \\ \mathbf{e}_j \end{bmatrix} \tag{9.5}$$

with ω^j and ω^j_i given by (2.7).

A Lie group is a group that is at the same time a differentiable manifold. The Lie algebra of a Lie group is nothing but $dg \cdot g^{-1}$; if g is the general element of an affine, linear, Euclidean, rotation group for a given dimension, $dg \cdot g^{-1}$ is the respective affine, linear, Euclidean, rotation Lie algebra. They are matrices with entries that are differential 1−forms In fact Lie algebras are sometimes defined as matrix algebras that so and so. One thus says that the connection (ω^i, ω^k_j) is a differential 1−form valued in the Lie algebra of the affine group.

The Lie algebra of the linear group is read from

$$\{d\mathbf{e}_i\} = [\omega^j_i]\{\mathbf{e}_j\}, \tag{9.6}$$

thus valued in the Lie algebra of the linear subgroup of the affine group.

Notice that the square matrix in (9.6) results from removing the first column and the first row of the square matrix in (9.5). Never mind that ω^j_i is a linear combination of the dx^l and the dA^m_l. In one case we are dealing with the manifold (the bundle) where the differential forms are defined and another one is where they take their values (the Lie algebra). If (9.6) is obtained by considering the linear group for dimension n on its own, the ω^j_i depend only on the coordinates A^m_l and their differentials. But if (9.6) is extracted from (9.5) by eliminating the first row and column, the ω_i^j depend also on the differentials dA^i.

As usual, we differentiate (9.5) and obtain

$$
\begin{bmatrix} \Omega^i \mathbf{e}_i \\ \Omega_i^j \mathbf{e}_j \end{bmatrix} = \begin{bmatrix} 0 & d\omega^j \\ 0 & d\omega_i^j \end{bmatrix} \begin{bmatrix} \mathbf{P} \\ \mathbf{e}_j \end{bmatrix} - \left\{ \begin{bmatrix} 0 & \omega^k \\ 0 & \omega_i^k \end{bmatrix} \wedge \begin{bmatrix} 0 & d\omega^j \\ 0 & d\omega_k^j \end{bmatrix} \right\} \begin{bmatrix} \mathbf{P} \\ \mathbf{e}_j \end{bmatrix}, \quad (9.7)
$$

i.e. torsion and curvature. Notice that we do not get zero in the first row of the column matrix on the right. The exterior product of matrices simply means the skew-symmetrization of the matrix product, which permits us to write

$$
\begin{bmatrix} \Omega^i \mathbf{e}_i \\ \Omega_i^j \mathbf{e}_j \end{bmatrix} = \begin{bmatrix} 0 & d\omega^j - \omega^k \wedge \omega_k^j \\ 0 & d\omega_i^j - \omega_i^k \wedge \omega_k^j \end{bmatrix} \begin{bmatrix} \mathbf{P} \\ \mathbf{e}_j \end{bmatrix}, \quad (9.8)
$$

where we again see the first row if we are to represent $\Omega^i \mathbf{e}_i$ and Ω_i^j jointly in matrix form. This helps emphasize that the first row also plays a role and that, therefore, Ω^i takes values in the Lie algebra of the affine group and not just the linear group. This is due in any case to the fact that $\Omega^i \mathbf{e}_i$ involves the first row of the square matrix

We return to the main argument. With summation over repeated indices, the square matrix in (9.5) can be written as

$$
\omega^j \boldsymbol{\alpha}_j + \sum_{i,j} \omega_i^j \boldsymbol{\alpha}_i^j, \qquad i, j = 1, ..., n. \quad (9.9)
$$

where the $\boldsymbol{\alpha}_j$ and $\boldsymbol{\alpha}_i^j$ are easily identifiable from (9.5) and constitute a basis in the algebra of $(n+1) \times (n+1)$ matrices whose first column is made of zeroes. Each $\boldsymbol{\alpha}_j$ has all its elements zero except for the unit in the first row and $j+1$ column ($j > 0$), and each $\boldsymbol{\alpha}_i^j$ has all its elements zero except for the unity in row $i+1$ and column $j+1$ ($i > 0, j > 0$). The space spanned by ($\boldsymbol{\alpha}_j, \boldsymbol{\alpha}_j^i$) is called the Lie algebra of the affine group. The issue arises of the product that makes said space an algebra. The skew-symetrized product of matrices does it, and also makes torsion and affine curvature belong to the Lie algebra.

5.10 The Maurer-Cartan equations

In his 1937 book "The theory of finite continuous groups and differential geometry treated by the method of the moving frames", Cartan shows that the equations on continuous groups known as Maurer equations reveal the structure of the differential forms $dg\,g^{-1}$ to which those equations refer [22]. That is the reason why, he said, "they are known as equations of structure of É. Cartan". Nowadays, they are called Maurer-Cartan equations.

The Maurer-Cartan equations state that

$$
d\omega^A = C_{BC}^A (\omega^B \wedge \omega^C), \quad (10.1)
$$

where the ω^D represent a basis for the Lie algebra $dg\,g^{-1}$. The purpose of the parenthesis in (10.1), which is related to what values of the indices the sum refers to, is the same as in previous occasions. The relevance of these equations, which would otherwise be trivial, is that the C_{BC}^A are constants, specifically called structure constants. The use of capitals for the indices serves to call attention to the fact that the ω^A refer not only to what we have referred as the ω^μ (or ω^i), but also to the ω_μ^ν. Readers can read the C_{BC}^A for the affine group and its linear subgroup from the already given equations of structure.

The Maurer equations are from 1888, but not exactly as we have written them, since the exterior calculus was not born until 1899. One only had differential 1−forms, called Pfaffian, and something called their bilinear covariants, which are related to their exterior derivatives. We mention it so that readers will see the so called properties of the constants of structure from the new perspective that the exterior calculus brings to them.

Since we are going to differentiate (10.1), it is useful to rewrite it as

$$dw^A = \frac{1}{2}C_{BC}'^A \omega^B \wedge \omega^C. \tag{10.2}$$

The C_{BC}^A do not determine the $C_{BC}'^A$, these (the non-null ones to be precise) being double in number if we add the condition

$$C_{BC}'^A = -C_{CB}'^A. \tag{10.3}$$

Half of them are equal to the C_{BC}^A, and the other half are their opposite. We can now drop the primes.

Differentiation of (10.2) yields zero on the left, and also for the term where we differentiate $C_{BC}'^A$. The last two terms are equal. Substitution of (10.2) in those terms yields:

$$0 = C_{BC}^A C_{DE}^B \omega^D \wedge \omega^E \wedge \omega^C, \tag{10.4}$$

and, therefore,

$$0 = C_{BC}^A C_{DE}^B + C_{BD}^A C_{Ec}^B + C_{BE}^A C_{CD}^B - C_{BC}^A C_{ED}^B - C_{BD}^A C_{CE}^B - C_{BE}^A C_{DC}^B. \tag{10.5}$$

The first, second and third terms are respectively equal to the fourth, fifth and sixth terms. Equations (10.5) can thus be written as

$$0 = C_{BC}^A C_{DE}^B + C_{BD}^A C_{EC}^B + C_{BE}^A C_{CD}^B. \tag{10.6}$$

Whereas $C_{BC}^A = -C_{CB}^A$ is the definition of the symbols on one side in terms of the symbols of the other side, (10.6) is an actual property, which, as presented, hides its inner simplicity, namely

$$dd\omega^A = 0 \tag{10.7}$$

for the invariant forms of the group. We leave it for the reader to figure out the trivial geometric identities that (10.7) correspond to.

Let us now see what difference Cartan's generalization of affine space makes. In the bundle, the ω^l are independent of the ω_i^j. Hence the $R_{ikl}^j \omega^k \wedge \omega^l$ are not of the form $C_{ilp}^{jkm} \omega_k^l \wedge \omega_m^p$, and the ω_i^j are not, therefore, the invariant forms of the linear subgroup. This remark has to be seen in the context that, as we shall see in chapter 8, only the pull-backs to fibers of the ω_i^j's of differentiable manifolds endowed with an affine connection are invariant forms of the linear group, not the ω_i^j themselves.

The case of the first equation of structure is a little bit tricky because the first equation of structure is more remotely connected to the translations than the second equation of structure is related to the linear transformations. In the first place, the Maurer-Cartan equation of the translation group cannot contain the ω_i^j (Notice that, in contrast to the first equation of structure, the second equation of structure of affine space contains only the the ω_i^j, not both the ω_i^j and the ω^l). So the issue of generalization of the first equation of structure of affine space is not an issue of generalization of the Maurer-Cartan equation of the translation group.

5.11 HORIZONTAL DIFFERENTIAL FORMS

The sophisticated concept of horizontality concerns the bundle as a whole, not its sections. After pulling differential forms from a bundle to its sections, the concept can no longer be formulated. Said better, the pull-back to a section of a horizontal form of the bundle is a tensorial object. The concept of horizontality is useful for understanding that something that looks like a tensor may not be one.

In the bundle, we can express a differential $1-$form in the following alternative ways:

$$\alpha = P_m(x, A)dx^m + P_i^k(x, A)dA_k^i = Q_r(x, A)\omega^r + Q_i^k(x, A)\omega_k^i. \tag{11.1}$$

The last of these two expansions of α has an intrinsic significance that the first one lacks, significance that is due to the invariant character of (ω^r, ω_k^i) in the frame bundle. That is paradoxically due to the fact that the ω_k^i is a linear combination of the dx^m and dA_k^i, but the ω^r is not a function of the dA_k^i. The term $P_m(x, A)dx^m$ has contributions not only from ω^r but also from ω_k^i, the latter contribution not being compensated by a contribution to $P_i^k(x, A)dA_k^i$ from $Q_r(x, A)\omega^r$.

One refers to $Q_r(x, A)\omega^r$ as the horizontal part of α. Later in this section, we shall discuss covariant derivatives from this perspective. Who would imagine that in the differentials of the components of a vector field hide (pull-backs of) terms of both types, linear in ω^r and in ω_k^i? They do!

The set of frames of the frame bundle can be expressed as the push-forward from any given section $\{\mathbf{e}\}$. Symbolically, we have

$$\{\mathbf{E}\} = g_0\{\mathbf{e}\}, \tag{11.2}$$

where g_0 is the linear group and where $\{e\}$ is some basis field. Of course, we should have replaced g_0 with G_0, but, as when we wrote $dg_0\,g_0^{-1}$, we are following notational custom in dealing with this topic. Differentiating (11.2), we have:

$$d\{\mathbf{E}\} = dg_0\{e\} + g_0 d\{e\} = dg_0\,g_0^{-1}\{\mathbf{E}\} + g_0\omega\{e\}$$
$$= (dg_0\,g_0^{-1} + g_0\omega g_0^{-1})\{\mathbf{E}\}, \tag{11.3}$$

where ω represents the ω_i^j.

The expression $v^m\mathbf{e}_m$ for a vector field, $\mathbf{v}$, pertains to sections $(\mathbf{e}_m)$. But, like the basis field $(\mathbf{e}_m)$, it can be pushed forward to the bundle, meaning that we can write $\mathbf{v}$ it as

$$\mathbf{v} = v^i\mathbf{e}_i = V^j\mathbf{E}_j. \tag{11.4}$$

The V^j will be a function of the coordinates in the fibers by virtue of its relation to v^i. It will also depend on x because so does v^i in general.

We use (11.4) to define differential forms Dv^m and DV^m:

$$d\mathbf{v} = (dv^m + v^i\omega_i^m)\mathbf{e}_m \equiv (Dv^m)\mathbf{e}_m \tag{11.5a}$$

and

$$d\mathbf{v} = (dV^m + V^i\omega_i^m)\mathbf{E}_m \equiv (DV^m)\mathbf{E}_m, \tag{11.5b}$$

Clearly, (Dv^m) is a pull-back of (DV^m). Equivalently, (DV^m) is the push-forward of (Dv^m). Similarly the ω_i^m in (11.5a) is the pull-back to the same section of the ω_i^m in (11.5b), which pertains to the bundle. In parallel to (11.4), we have

$$d\mathbf{v} = (Dv^m)\mathbf{e}_m = (DV^m)\mathbf{E}_m. \tag{11.6}$$

The relation between (Dv^m) and (DV^m) is implied by the relation between $(\mathbf{e}_m)$ and $(\mathbf{E}_m)$, and is the same as the relation between (v^m) and (V^m), as comparison of (11.4) and (11.6) shows. Equations (11.5)-(11.6) thus imply

$$DV^j = A_i^j Dv^i. \tag{11.7}$$

In addition to depending on A_i^j, DV^j depends on x and dx, since dv^i does, but not on ω_i^j. However, dv^i belongs to a section, where the pull-back of ω_i^j enters Dv^i. We must thus deal with this issue in the frame bundle, where ω^j and ω_i^j are independent.

We rewrite (11.6) as

$$dV^m = DV^m + (-V^i)\omega_i^m. \tag{11.8}$$

In this equation each and every quantity belongs to the bundle. It follows that DV^m is the horizontal part of dV^m, because of the definition of horizontality together with the fact that this decomposition is unique in the bundle. Abusing the language, we may say that the covariant derivatives are the coefficients of the horizontal part of dv^m; we should be referring to the horizontal part of dV^m, not dv^m.

Readers who need to go into greater generality because of their parallel interest in the tensor calculus would have to repeat the foregoing process with tensor fields rather than vector fields. In this book we are not interested in the tensor calculus.

Chapter 6

EUCLIDEAN KLEIN GEOMETRY

6.1 Euclidean space and its frame bundle

By the name of Euclidean point space or simply Euclidean space, $\mathcal{E}^n$, we refer to the affine space whose associated vector space is Euclidean. As we become increasingly conscientious of the role of groups in defining a geometry, we shall start to avoid statements like "*a Euclidean connection is an affine connection that ...*". The pair of groups (G, G_0) involved in defining anything affine is different from the pair of groups (G, G_0) *directly involved* in defining anything Euclidean.

The dot product allows us to define the Euclidean bases, i.e. those satisfying the orthonormality condition

$$\mathbf{a}_i \cdot \mathbf{a}_j = \delta_{ij}. \tag{1.1}$$

The Euclidean frame bundle is the bundle of all Euclidean frames, i.e. pairs of a point and a vector basis. The group G for this bundle is the Euclidean group, and the group G_0 in its fibers is the rotation group, both for the same dimension n. The whole bundle can be obtained from any section by the action of the rotation group for the given number of dimensions. Thus, for instance, we have the section

$$\widehat{\mathbf{e}}_{p(p_0)} = \mathbf{i} \cos\phi + \mathbf{j} \sin\phi, \quad \widehat{\mathbf{e}}_\phi = -\mathbf{i} \sin\phi + \mathbf{j} \cos\phi \tag{1.2}$$

in $\mathcal{E}^2$. The circumflex is used to denote orthonormal bases. By the action of the rotation group in two dimensions, we obtain (or recover, if you will) the full bundle from a section. Do not confuse the point dependent rotation of angle ϕ defining the section (See Eq. (1.13)) with the set of all rotations in the plane to obtain the bundle (See Eqs. (1.11) and (1.12)).

Returning to general dimension, n, let α denote the coordinates in the fibers (i.e. the parameters of the rotation group in n dimensions) and let x denote the (in general) curvilinear coordinates in a section. By the construction of the previous paragraph, we shall have

$$\widehat{\mathbf{e}}_i(x, \alpha) \cdot \widehat{\mathbf{e}}_j(x, \alpha) = \delta_{ij}, \tag{1.3}$$

in Euclidean frame bundles and

$$\widehat{\mathbf{e}}_i'(\alpha) \cdot \widehat{\mathbf{e}}_j'(\alpha) = \delta_{ij} \tag{1.4}$$

in their fibers (meaning that x is no longer a function but numbers, absorbed in the notation).

A natural way to create the bundle of frames of a Euclidean space is to first introduce a basis field by, for example, using vectors tangent to the coordinate lines, then orthonormalizing them and finally applying the corresponding group of rotations at each point of the space in question. It must be clear that the bases in each fiber are related by the group G_0, i.e. by rotations. In the sections and in the bundle, on the other hand, the relation between the frames is given by the group G, i.e. by general displacements, which also include translations.

In Euclidean space, the rule to compare vectors at a distance is implicit in the fact that all vector bases can be referred to just one vector space and thus just one vector basis, $(\mathbf{i}, \mathbf{j}, \mathbf{k}, \ldots)$. Then, retrospectively, this is equivalent to having sections of the bundle of frames where the basis at one point is equal to the basis at any other point. Of course, most sections are not constant. The section (1.2) is not constant in Euclidean space; see, however, section 2 of chapter 7 for a differentiable manifold where it is.

Whether we differentiate equation (1.3) or (1.4), we get, using $d\mathbf{e}_i = \omega_i^l \mathbf{e}_l$,

$$\omega_i^l \delta_{lj} + \omega_j^l \delta_{il} = \omega_{ij} + \omega_{ji} = 0, \tag{1.5}$$

with ω_{ij} defined as $\omega_i^k \delta_{kj}$. Clearly,

$$\omega_i^j = -\omega_j^i, \tag{1.6}$$

which implies that there are now only $n(n-1)/2$ independent differential forms ω_i^j. There were n^2 independent such forms in the affine case.

When the position vector $\mathbf{P}$ is given in terms of orthonormal basis,

$$\mathbf{P} = A^i \widehat{\mathbf{a}}_i, \tag{1.7}$$

the coordinates A^i are called Cartesian (not all rectilinear coordinates are Cartesian). Obviously, we have

$$d\mathbf{P} = dA^i \widehat{\mathbf{a}}_i$$

in these constant sections, and

$$d\mathbf{P} = dA^i \widehat{\mathbf{a}}_i = \omega^i(\alpha, dA^i) \widehat{\mathbf{e}}_i \tag{1.8}$$

in arbitrary orthonormal bases or basis fields $(\widehat{\mathbf{e}}_i)$ of the bundle. The α are the parameters of the rotation group for the given dimension n. The ω^i's and the $\widehat{\mathbf{e}}_i$ are said to be dual to each other.

Had we used an arbitrary section (orthonormal!) to span the bundle, $d\mathbf{P}$ would then take the more general form

$$d\mathbf{P} = \omega^i(x, \alpha, dx)\mathbf{e}_i. \tag{1.9}$$

See, for example, Eq. (1.12), where α' is a coordinate in the fibers. On sections, α is a set of functions $\alpha(x)$ and, therefore,

$$d\mathbf{P} = \omega^i(x, \alpha(x), dx)\mathbf{e}_i = \omega'^i(x, dx)\mathbf{e}_i. \tag{1.10}$$

We proceed to illustrate in the bundle of $\mathcal{E}^2$ that the ω^i are invariant forms, a point made in more general terms in the last section of the previous chapter. In terms of the Cartesian basis (dx, dy) of differential forms, the ω^i of the bundle can be given as

$$\left\{ \begin{matrix} \omega^{1}{}' \\ \omega^{2}{}' \end{matrix} \right\} = \begin{pmatrix} \cos\alpha & \sin\alpha \\ -\sin\alpha & \cos\alpha \end{pmatrix} \left\{ \begin{matrix} dx \\ dy \end{matrix} \right\}, \tag{1.11}$$

for all $0 \le \alpha < 2\pi$. In terms of the basis $(d\rho, \rho d\phi)$, we similarly have

$$\left\{ \begin{matrix} \omega^{1}{}' \\ \omega^{2}{}' \end{matrix} \right\} = \begin{pmatrix} \cos\alpha' & \sin\alpha' \\ -\sin\alpha' & \cos\alpha' \end{pmatrix} \left\{ \begin{matrix} d\rho \\ \rho d\phi \end{matrix} \right\} \tag{1.12}$$

for all $0 \le \alpha' < 2\pi$. The forms (1.11) for the set of all α are the same as the set of forms (1.12) for all α'. This is clear since

$$\left\{ \begin{matrix} d\rho \\ \rho d\phi \end{matrix} \right\} = \begin{pmatrix} \cos\phi & \sin\phi \\ -\sin\phi & \cos\phi \end{pmatrix} \left\{ \begin{matrix} dx \\ dy \end{matrix} \right\}, \tag{1.13}$$

which substituted in (1.12) yields (1.11) with $\alpha = \alpha' + \phi$. Notice the appearance of the coordinate ρ in (1.12). No similar appearance of x or y takes place in (1.11). In general, one should not assume that an equation or statement valid in terms of Cartesian coordinates is also valid when other coordinates or basis fields are involved.

It is clear now that

$$(\mathbf{P}_2 - \mathbf{P}_1)^2 = A^i A_i = \Sigma(A^i)^2, \tag{1.14}$$

where the A^i are in $\mathcal{E}^3$ the (Cartesian) coordinates x, y, z. Needless to say, we have $A_i = A^i$, which we have used in (1.14).

Let us define ds^2, called the metric, as

$$ds^2 \equiv d\mathbf{P} \cdot d\mathbf{P} = \omega^i \omega^j \widehat{\mathbf{e}}_i \cdot \widehat{\mathbf{e}}_j = \omega^i \omega^j \delta_{ij} = \sum_i (\omega^i)^2 = \omega^i \omega_i \tag{1.15}$$

with $\omega_i = \omega^i$ These equations belong to both the bundle and its sections. It is well known from elementary calculus how to obtain the distance on a curve given the metric.

It is worth being aware of the fact that the product in $(\omega^i)^2$ and in $\omega^i\omega_i$ is a tensor product. A large part of the content of the concept of tensor product is linearity in each of the factors (There are also other linear products, like exterior and Clifford, which are not tensor products). Let us consider the simple example of dimension two. A basis for the module of tensors of rank 2 in the tensor algebra of differential forms for 2-dimensional space (whether, affine, Euclidean or otherwise) is given by

$$\omega^1\omega^1, \ \omega^1\omega^2, \ \omega^2\omega^1, \ \omega^2\omega^2.$$

In contrast, and as we already know, it would be just $\omega^1\omega^2$ (written $\omega^1 \wedge \omega^2$) for exterior forms of grade two. We shall not delve further into this since there is a much better way of doing metrics in differential geometry than using tensors. Just for reference: in Finsler bundles, the distance becomes a differential $1-$form [73].

A pseudo-Euclidean space is one where the associated vector space is pseudo-Euclidean. That means that $\widehat{\mathbf{e}}_i \cdot \widehat{\mathbf{e}}_i$ (no summation) is -1 for at least one value of the index. The metric then takes the form

$$ds^2 = \sum_i \epsilon_i(\omega^i)^2 = \omega^i\omega_i, \tag{1.16}$$

where $\epsilon_i = \pm 1$. The term signature is used to refer to the number of indices for which $\epsilon_i = +1$ minus the number of indices for which $\epsilon_i = -1$. With display of actual signs of the ϵ_i, signatures of the types $(1, -1, -1, \ldots, -1)$ and $(-1, 1, \ldots, 1)$ are called Lorentzian. For dimension $n = 4$, the space is called the Lorentz-Einstein-Minkowski space. The equations to follow will remain the same, independently of signature, except when we warn to the contrary. When dealing specifically with the Lorentzian signature, we shall use Greek indices, the index "zero" being reserved for the sign in the minority in the signature. For $(1, -1, -1, -1)$, we have

$$\omega_0 = \omega^0, \ \ \omega_i = -\omega^i, \tag{1.17}$$

$$\omega_0{}^i = -\omega_{0i} = \omega_{i0} = \omega_i^0, \tag{1.18}$$

as follows from defining ω_μ as $\epsilon_\mu\omega^\mu$ (no sum). The G and G_0 groups now are the Poincaré and Lorentz groups, respectively.

6.2   Extension of Euclidean bundle to affine bundle

By the action of the linear group on the fibers of a Euclidean frame bundle, we "recover" the affine frame bundle for the same number of dimensions. It is called the affine extension of the Euclidean frame bundle. Arbitrary bases of this extension satisfy a relation of the form

$$\mathbf{e}_i(x, \alpha) \cdot \mathbf{e}_i(x, \alpha) = g_{ij}(x), \tag{2.1}$$

rather than (1.3). On sections,

$$\mathbf{e}_i(x) \cdot \mathbf{e}_j(x) = g_{ij}(x). \tag{2.2}$$

In 1922, Cartan exploited equations (2.1)-(2.2) to create geometry in bundles. The idea is that the dependence on the parameters α of a group on the left hand side of (2.1) is not present on its right hand side. For a given point of coordinates x, a basis $\mathbf{e}_i(x)$ at that point generates a "fiber of bases" yielding the same components $g_{ij}(x)$ of the metric.

The extension of the metric to the affine frame bundle is given by

$$d\mathbf{P} \cdot d\mathbf{P} = g_{ij}(x)dx^i dx^j, \tag{2.3}$$

and alternatively as

$$d\mathbf{P} \cdot d\mathbf{P} = g'_{ij}(x)\omega^i \omega^j, \tag{2.4}$$

where the ω^i are not holonomic. Non-holonomic bases are rarely if ever used except for $g'_{ij}(x) = \delta_{ij}$, since it is possible to always write the metric as (1.16) through diagonalization. In most cases of practical interest, this diagonalization can be achieved by inspection. If the metric is diagonal, i.e. if it is of the form

$$ds^2 = \sum_i g_{ii}dx^i dx^i, \tag{2.5}$$

we obviously have sections where

$$\omega^i = \sqrt{\epsilon_i g_{ii}}dx^i \qquad \text{(no sum).} \tag{2.6}$$

If the metric is not diagonal, it can be diagonalized in more than one way, as we already saw.

Differentiation of (2.2) yields

$$\omega_i^l g_{lj} + \omega_j^l g_{il} = dg_{ij}, \tag{2.7}$$

and, therefore,

$$dg_{ij} - \omega_{ij} - \omega_{ji} = 0, \tag{2.8}$$

with

$$\omega_{ij} \equiv \omega_i^l g_{lj}. \tag{2.9}$$

Notice that the ω_i^j's are no longer independent since they must satisfy (2.3). There are only $n(n-1)/2$ independent ω_i^j's in the affine extension of the Euclidean frame bundle.

Given that, on sections, the ω_i^j's are of the form

$$\omega_i^j = \Gamma_{i\,l}^{\ j}dx^l, \tag{2.10}$$

and that dg_{ij} is $g_{ij,m}dx^m$, Eq. (2.8) can be written as

$$g_{ij,m} - \Gamma_{i\,m}^{\ l}g_{lj} - \Gamma_{j\,m}^{\ l}g_{li} = 0. \tag{2.11}$$

6.3   Meanings of covariance

This section might have gone in the previous chapter, except that Euclidean spaces provide us the concept of reciprocal vector basis fields, which affine geometry does not. These fields will enter our discussion at some point in this section.

The term covariant has a variety of meanings. One does speak of contravariant and covariant vectors (and vector fields). The contravariant vectors are the ones of which we have spoken in the previous and present chapters, except when we brought up the ϕ^i, precisely the covariant vectors or linear functions of contravariant vectors. This terminology is not very fortunate since "contra" means against. But, against what? Without entering details, let us say that this terminology has its roots in the early times of the tensor calculus, when a factor of great relevance was how a set of components transformed under a change of coordinates. We have in mind specifically equations such as (5.1)-(5.2) of chapter 3. In the following, when we say transform read "change of basis".

In affine space, things are simple. The components of contravariant (covariant) vectors are said to transform contravariantly (respectively covariantly). The elements of the bases of contravariant (respectively covariant) vectors then transform (watch it!) covariantly (respectively contravariantly). In that way, contravariant (respectively covariant) components are components with respect to contravariant (respectively covariant) bases of contravariant (respectively covariant) vectors. But why the redundancy? Is there a need to specify contravariant (covariant) bases of contravariant (vectors)? Not in affine space, where the bases are general vector bases. But we are preparing the way for the next paragraph.

We have said that contravariant bases transform covariantly, i.e. oppositely to the contravariant components so that the contraction of the components with the basis will yield an invariant. But, in Euclidean spaces, a tangent vector can be referred also to dual bases, which transform contravariantly, not covariantly. The components with those bases then transform covariantly. In other words, vectors of Euclidean spaces have both, contravariant and covariant components. Hence the behavior of the components of a vector (or of a vector field) no longer speaks of the nature of the object to which the components belong, since it depends on the type of basis that one is changing. For this reason, it is best to use the terms tangent and cotangent vectors for what one often calls contravariant and covariant vectors. Tangent vector can thus have contravariant and covariant components.

Similar considerations can be made for cotangent vectors, but we shall not get into that since we do not feel a need for it, specially in view of what we said about replacing bases ϕ^i with reciprocal bases $\mathbf{e}^i$. The idea of using $\mathbf{e}^i$ is to unify the conceptual spaces. Introducing also ϕ_i would be legitimate but would not help with conceptual unification.

Next consider functions of curves, i.e. what we have referred to as differential 1−forms. Their components transform like components of cotangent vector fields. In fact, they are defined as cotangent vector fields in many publications, but not here. Ours is a different concept: a differential 1−form is a function of curves, not of vectors. This is also the case in Rudin [65], and also, though less explicitly so, in Cartan's and Kähler's publications.

There is a second meaning of covariant. One often says that an object is covariant when one wants to say that its components behave tensorially. This may refer to contravariant, strictly covariant, or mixed contravariant-covariant behavior. For example, the covariant derivative of a vector field combines contravariant transformation properties relative to one of its two indices, and covariant transformations for the other one.

Now comes the most important remark of this section. Unlike the $v^i_{;j}$, which involves the connection, the $v^i_{,j}$ do not transform tensorially, unless the v^i are the components of a vector field with respect to a constant field of bases. The exterior derivative of a differential 1−form, on the other hand, does not depend on connection but still yields an object that has the appropriate "tensorial or covariant" transformation properties. Hence it is unfortunate to speak of the covariant derivative, as if the exterior derivative did not give rise to covariant quantities.

The second most important remark is as follows. As used by Cartan, Kähler and Flanders [41], d is an operator which, acting on respectively scalar-valued differential forms, vector-valued differential 0−forms and vector-valued differential r−forms ($r > 0$) yields what are commonly called the exterior, covariant and exterior-covariant derivatives. One might be tempted to say that the effects of the operator d are different depending on what object they are acting on. Not so, since scalar-valued differential forms and vector-valued differential 0−forms are particular cases of vector-valued differential r−forms. Once it is recognized that one only needs the encompassing concept of exterior covariant derivative in order to deal with those derivative"s", one may use instead simply the term exterior derivative, which is what Cartan and Kähler do. The display of bases (as in, say, $v_i \mathbf{e}^i$ or $v_i \omega^i$ or $v_i \phi^i$), helps avoid confusion.

We proceed to discuss differentiation in terms of reciprocal bases $\mathbf{e}^i$. We define the basis $(\mathbf{e}^i)$ by the relation

$$\mathbf{e}^i \cdot \mathbf{e}_j = \delta^i_j, \tag{3.1}$$

regardless of whether the basis $(\mathbf{e}_j)$ is orthonormal or not. We obtain

$$d\mathbf{e}^i \cdot \mathbf{e}_j = -\mathbf{e}^i \cdot \omega^k_j \mathbf{e}_k = -\delta^i_k \omega^k_j = -\omega^i_j, \tag{3.2}$$

which in turn implies

$$d\mathbf{e}^i = -\omega^i_j \mathbf{e}^j. \tag{3.3}$$

Equation (3.3) permits us to differentiate a vector field $\mathbf{v}$ when written as $v_i \mathbf{e}^i$. Indeed,

$$d\mathbf{v} = v_{i/j} \omega^j \mathbf{e}^i - v_k \omega^k_j \mathbf{e}^j = (v_{i/j} - v_k \Gamma^k_{ij}) \omega^j \mathbf{e}^i. \tag{3.4}$$

Let us compare (3.4) with the exterior derivative of $\alpha = v_i \omega^i$. For the purpose of differentiation, it is best to use coordinate bases of 1−forms. It is then clear that

$$d\alpha = d(v'_i dx^i) = (v'_{i,j} - v'_{j,i})(dx^j \wedge dx^i) \tag{3.5}$$

where the parenthesis around $dx^j \wedge dx^i$ makes reference to summation only over elements of a basis (If $dx^1 \wedge dx^2$ is in the expansion, $dx^2 \wedge dx^1$ is not). This is a covariant expression, in the sense that the exterior derivative of the same differential $1-$form given in terms of another coordinate system, $v_i'' dy^i$, yields the same differential $2-$form $d\alpha$:

$$(v_{i,j}' - v_{j,i}')(dx^j \wedge dx^i) = (v_{i,j}'' - v_{j,i}'')(dy^j \wedge dy^i). \tag{3.6}$$

In other words, $d\alpha$ is coordinate and frame field independent. Its components (i.e. the contents of the first parenthesis on each side of (3.6)), transform linearly because so do the elements of the basis of differential $2-$forms. (3.4) is connection dependent; (3.5) is not.

One should be aware of the fact that sometimes expressions show the connection explicitly, and yet they do not depend on it. For example, in later chapters we shall deal with the so called Levi-Civita connection, for which the following is correct:

$$d\alpha = v_{i/j}\omega^j \wedge \omega^i + v_k d\omega^k = (v_{i/j} + v_k \Gamma^{\ k}_{j\ i})\omega^j \wedge \omega^i. \tag{3.7}$$

The presence of the gammas is often an indication of dependence of connection. But not always, (3.7) being an example. The first equation of structure, $d\omega^k = \omega^j \wedge \omega^k_j$ does not imply that $d\omega^k$ is connection dependent since it is defined without resort to it. This equation simply states that connections that satisfy it have zero torsion. The spurious dependence on connection propagates to equations where it has been used to replace $d\omega^k$ with $\omega^j \wedge \omega^k_j$.

6.4   Hodge duality and star operator

The topic of Hodge duality is best treated using Clifford algebra. That will be dealt with in another book being prepared by this author. In this section, we simply present some basic results of that algebra, specifically Hodge duality, without resorting to a formal presentation of the same.

Given a differential $r-$form, we define its Hodge dual (or simply dual, when there is no ambiguity) as the differential $(n - r)$-form that one obtains by its Clifford product with the unit differential $n-$form. In the interest of expediency, we proceed to explain this concept with examples and the easy computations that result when one expresses differential forms as products of the differential $1-$forms $dx, dy, dz, dr, rd\theta, r\sin\theta d\phi, d\rho, \rho d\phi, dz$ associated with orthonormal tangent vector basis fields, specifically $(\mathbf{i}, \mathbf{j}, \mathbf{k})$, $(\widehat{\mathbf{e}}_r, \widehat{\mathbf{e}}_\theta, \widehat{\mathbf{e}}_\phi)$ and $(\widehat{\mathbf{e}}_\rho, \widehat{\mathbf{e}}_\phi, \widehat{\mathbf{e}}_z)$. in ξ^3. We shall refer to those bases of differential $1-$forms themselves as orthonormal. The notation will be (ω^i) in three dimensions, and (ω^μ) in other dimensions.

In ξ^3, we use the symbol w to refer to the unit differential $3-$form, i.e.

$$w = \omega^1 \wedge \omega^2 \wedge \omega^3 = dx \wedge dy \wedge dz = dr \wedge rd\theta \wedge r\sin\theta d\phi = \dots \tag{4.1}$$

A simple rearrangement of factors yields

$$w = r^2 \sin\theta \, dr \wedge d\theta \wedge d\phi = g^{1/2} dr \wedge d\theta \wedge d\phi, \tag{4.2}$$

where g is the determinant of the matrix made with the coefficients of the metric. We shall show at the end of this section (Eq. (4.25)) the generality of this type of result.

Consider, for instance, $(dx \wedge dy) \vee w$, i.e. $(dx \wedge dy) \vee (dx \wedge dy \wedge dz)$. When there is orthonormality, $dx \wedge dy$ equals $dx \vee dy$, and $dx \wedge dy \wedge dz$ equals $dx \vee dy \vee dz$. To minimize clutter, we use juxtaposition instead of the symbol $\vee$, and shall also represent $\alpha \vee w$ as $^*\alpha$, which is common in the literature. The notation $^*dx \wedge dy$ will be viewed as signifying $^*(dx \wedge dy)$ rather that $(*dx) \wedge dy$. We then have

$$^*dx \wedge dy = (dxdy)(dxdydz) = -dxdydydxdz = -dz, \tag{4.3}$$

where we have used the associative property and that

$$dxdy = -dydx, \qquad dxdx = dydy = dzdz = 1. \tag{4.4}$$

Let us observe that

$$w \vee w = dxdydzdxdydz = dxdxdydzdydz = dydzdydz = -dydydzdz = -1. \tag{4.5}$$

Also, using (4.3) and (4.5), we get

$$^*dz =^* (-^*dx \wedge dy) = -^*(^*dx \wedge dy) = -dxdyww = dx \vee dy, \tag{4.6}$$

and similarly, by cyclic permutation of (4.6),

$$^*dx = dy \wedge dz, \quad ^*dy = dz \wedge dx \tag{4.7}$$

and

$$^*dy \wedge dz = -dx, \quad ^*dz \wedge dx = -dy. \tag{4.8}$$

Finally,

$$^*1 = dx \wedge dy \wedge dz, \tag{4.9}$$

$$^*dx \wedge dy \wedge dz = -1, \tag{4.10}$$

and

$$^*fdx \wedge dy \wedge dz = -f, \tag{4.11}$$

by virtue of

$$(f\alpha)w = f(\alpha w). \tag{4.12}$$

Clearly $^{**}\alpha = -\alpha$.

Let $\omega^1, \omega^2, \omega^3$ be a basis of differential $1-$forms dual to an orthonormal basis field. Momentarily, and in order to emphasize that the equations about to follow are only valid for ω^i's dual to orthonormal bases, we shall use the circumflex, $\hat{\omega}^i$. We thus have

$$^*\hat{\omega}^1 \wedge \hat{\omega}^2 \wedge \hat{\omega}^3 = -1. \tag{4.13}$$

In parallel to equations (4.3)-(4.10), we have

$$^*\widehat{\omega}^2 \wedge \widehat{\omega}^3 = -\widehat{\omega}^1, \quad ^*\widehat{\omega}^3 \wedge \widehat{\omega}^1 = -\widehat{\omega}^2, \quad ^*\widehat{\omega}^1 \wedge \widehat{\omega}^2 = -\widehat{\omega}^3, \qquad (4.14)$$

$$^*\widehat{\omega}^1 = \widehat{\omega}^2 \wedge \widehat{\omega}^3, \quad ^*\widehat{\omega}^2 = \widehat{\omega}^3 \wedge \widehat{\omega}^1, \quad ^*\widehat{\omega}^3 = \widehat{\omega}^1 \wedge \widehat{\omega}^2, \qquad (4.15)$$

$$^*1 = \widehat{\omega}^1 \wedge \widehat{\omega}^2 \wedge \widehat{\omega}^3, \qquad (4.16)$$

$$^*\widehat{\omega}^1 \wedge \widehat{\omega}^2 \wedge \widehat{\omega}^3 = -1. \qquad (4.17)$$

We define the unit n-volume differential form as

$$\widehat{\omega}^1 \wedge \widehat{\omega}^2 \wedge \ldots \wedge \widehat{\omega}^n. \qquad (4.18)$$

We wish to express it in terms of coordinate bases of differential forms. These are not orthonormal in general. We replace the $\widehat{\omega}^i$ with their expressions in terms of arbitrary coordinate bases of differential 1−forms, i.e.

$$\widehat{\omega}^\mu = A_\nu{}^\mu dx^\nu, \qquad (4.19)$$

thus obtaining

$$\widehat{\omega}^1 \wedge \widehat{\omega}^2 \wedge \ldots \wedge \widehat{\omega}^n = A_{\lambda_1}{}^1 A_{\lambda_2}{}^2 \ldots A_{\lambda_n}{}^n dx^{\lambda_1} dx^{\lambda_2} \wedge \ldots \wedge dx^{\lambda_n}. \qquad (4.20)$$

All the indices λ_i are different and all products are equal to $dx^1 \wedge dx^2 \wedge \ldots \wedge dx^n$ up to sign. The coefficient of this factor is the determinant A of the matrix of the $A_\mu{}^\nu$. We thus have

$$\widehat{\omega}^1 \wedge \widehat{\omega}^2 \wedge \ldots \wedge \widehat{\omega}^n = |A| dx^1 \wedge dx^2 \wedge \ldots \wedge dx^n. \qquad (4.21)$$

where $|A|$ is the absolute value of A (in case the orientation of the bases $\widehat{\omega}^i$ and dx^i were different).

It is not customary to work with A. One rather uses its relation to the metric, which we proceed to derive. Let $\widehat{\mathbf{e}}_\mu$ and $\mathbf{e}_\mu$ be dual to $\widehat{\omega}^\mu$ and dx^μ respectively. We when have

$$g_{\mu\nu} = \mathbf{e}_\mu \cdot \mathbf{e}_\nu = A_\mu{}^\lambda A_\nu{}^\pi \widehat{\mathbf{e}}_\lambda \cdot \widehat{\mathbf{e}}_\pi = A_\mu{}^\lambda \widehat{g}_{\lambda\pi} A_\nu{}^\pi. \qquad (4.22)$$

In terms of matrices, this is written as

$$g = A\widehat{g}A^\top = AA^\top \qquad (4.23)$$

where $A^\top$ is the transpose of A. Taking determinants, we have

$$|A| = |g|^{1/2}, \qquad (4.24)$$

and, therefore,

$$\widehat{\omega}^1 \wedge \widehat{\omega}^2 \wedge \ldots \wedge \widehat{\omega}^n = |g|^{1/2} dx^1 \wedge dx^2 \wedge \ldots \wedge dx^n, \qquad (4.25)$$

which is the sought result.

When the metric is not positive definite, the most interesting metric is spacetime's. We have the option of choosing the signatures (-1,1,1,1) and (1,-1,-1,-1). In both cases, $(dt\, dx\, dy\, dz)^2 = -1$, and we can choose to define the dual through multiplication with $dt \vee dx \vee dy \vee dz$, or with $dx \vee dy \vee dz \vee dt$ $(= -dt \vee dx \vee dy \vee dx)$.

6.5 The Laplacian

In this author's opinion, Kähler's calculus is the best tool to deal with Laplacians. It involves an operator, called the Dirac operator (not exclusive to that calculus), that encompasses the curl and the divergence and is applicable to differential forms of arbitrary grade and arbitrary valuedness. Let us denote it as ∂.

A differential form, α, is said to be strictly harmonic if and only if

$$\partial\alpha = 0. \tag{5.1}$$

In the particular case of scalar-valued differential $1-$form α, we obtain what, in terms of components, are known as the curl and divergence of a vector field with the same components as the differential $1-$form.

The Laplacian of α is defined as $\partial\partial\alpha$. For $0-$forms, $\partial\partial$ is the usual Laplacian operator. A differential form is called harmonic if

$$\partial\partial\alpha = 0. \tag{5.2}$$

Kähler's most elegant approach to ∂ and $\partial\partial$ will be considered only in the next book, presented in the last chapter of the present book. Consequently, we shall not approach here the Laplacian operator as $\partial\partial$. Luckily, $\partial\partial$ becomes simply "$-^*d^*d\alpha$" in Euclidean spaces and on manifolds with Euclidean connections with zero torsion (called Levi-Civita connections).

We shall again consider bases $\widehat{\omega}^i$ dual to orthonormal basis fields, i.e. such that

$$d\mathbf{P} = dA^i\mathbf{a}_i = \widehat{\omega}^i\widehat{\mathbf{e}}_i. \tag{5.3}$$

Let us examine from this simplified perspective the Laplacian of a scalar function f in 3D in terms of orthonormal coordinate systems. The Hodge dual of

$$df = f_{/i}\widehat{\omega}^i. \tag{5.4}$$

is

$$^*df = f_{/i}\widehat{\omega}^j \wedge \widehat{\omega}^k, \tag{5.5}$$

with summation over cyclic permutations, writing then (5.5) as in Eq. (6.4) of chapter 4, differentiating it, obtaining the dual and changing sign, we get

$$\frac{1}{h^1h^2h^2}\left[\frac{\partial(f_{/1}h^2h^3)}{\partial x^1} + \frac{\partial(f_{/2}h^3h^1)}{\partial x^2} + \frac{\partial(f_{/3}h^1h^2)}{\partial x^3}\right]. \tag{5.6}$$

The expression (5.6) is known in the vector calculus as the Laplacian of the function f in terms of ortogonal coordinate systems. In terms of Cartesian coordinates ($h^1 = h^2 = h^3 = 1$), (5.6) reduces to

$$\frac{\partial^2 f}{\partial x^2} + \frac{\partial^2 f}{\partial y^2} + \frac{\partial^2 f}{\partial z^2}. \tag{5.7}$$

Exercise. Use (5.6) to compute the Laplacian in spherical and cylindrical coordinates. In spherical coordinates, we have

$$ds^2 = dr^2 + r^2 d\theta^2 + r^2 \sin^2\theta d\phi^2, \tag{5.8}$$

and, therefore,

$$\widehat{\omega}^1 = dr, \quad \widehat{\omega}^2 = rd\theta, \quad \widehat{\omega}^3 = r\sin\theta d\phi. \tag{5.9}$$

We similarly have, for cylindrical coordinates,

$$\widehat{\omega}^1 = d\rho, \quad \widehat{\omega}^2 = \rho d\phi, \quad \widehat{\omega}^3 = dz. \tag{5.10}$$

The definition $-{}^*d^*df$ does not require the use of bases ω^i dual to orthonormal fields of tangent vector bases. We have used them in order to connect with formulas used by electrical engineers. We shall now obtain the Laplacian in less specialized cases for the exclusive purpose of connecting with practitioners of tensor calculus, whose tangent basis fields are not orthogonal. Again, readers are informed that it would be much simpler to deal with $\partial\partial$ in each particular case when in possession of some basic knowledge of the Kähler calculus.

The Hodge dual of the "gradient" of a scalar field f is ${}^*f_{,l}dx^l$. is

$$ {}^*df = f_{,l}|g|^{1/2}dx^l(dx^1 \wedge \ldots \wedge dx^n) \tag{5.11}$$

where we have used (4.25). In the next book, we shall show that

$$dx^l(dx^1 \wedge \ldots \wedge dx^n) = (-1)^{l-1}\sum_{l=1}^{n} g^{lm}dx^1 \wedge \ldots \wedge \overline{dx^m} \wedge \ldots \wedge dx^n, \tag{5.12}$$

where the overbar means that the factor dx^m is missing. Thus (5.11) can be written further as

$$(-1)^{l-1}f_{,l}|g|^{1/2}g^{lm}dx^1 \wedge \ldots \wedge \overline{dx^m} \wedge \ldots \wedge dx^n. \tag{5.13}$$

When we exterior differentiate this expression, the mth term only contributes with differentiation with respect to x^m. Thus

$$d^*df = (f_{,l}|g|^{1/2}g^{lm})_{,m}dx^1 \wedge \ldots \wedge dx^n, \tag{5.14}$$

where the factor $(-1)^{l-1}$ disappears because it has been used to move dx^m from the front to its natural position. Using again (4.25), we have

$$d^*df = |g|^{-1/2}(f_{,l}|g|^{1/2}g^{ln})_{,m}\omega^1 \wedge \ldots \wedge \omega^n \tag{5.15}$$

and, therefore,

$$-{}^*d^*df = |g|^{-1/2}(f_{,l}|g|^{1/2}g^{lm})_{,m}. \tag{5.16}$$

This is the Laplacian that one finds in books on the tensor calculus, which we have obtained without the use of tensors.

Exercise. In $\mathcal{E}^3$, relate this to (5.6) when the system of coordinates is orthogonal. Hint g^{lm} then is a diagonal matrix.

6.6 Euclidean structure and integrability

From now on, we drop the circumflex. Since we know what the affine equations of structure mean in affine space, and since we know what the relation between the affine and Euclidean spaces is, it is clear that the equations of structure of these spaces are the same

$$d\omega^\mu - \omega^\nu \wedge \omega_\nu^\mu = 0, \tag{6.1}$$

$$d\omega_\mu^\nu - \omega_\mu^\lambda \wedge \omega_\lambda^\nu = 0, \tag{6.2}$$

but augmented by the restriction

$$\omega_{\mu\nu} + \omega_{\nu\mu} = 0, \tag{6.3}$$

in the Euclidean case. Equation (6.3) is the differential form of the statement that the bases of Euclidean frame bundles, thus to be called Euclidean bases, are orthonormal.

The system of equations

$$d\mathbf{P} = \omega^\mu \mathbf{e}_\mu, \quad d\mathbf{e}_\mu = \omega_\mu^\nu \mathbf{e}_\nu, \tag{6.4}$$

for some given set of differential forms is integrable if and only if the equations (6.1)-(6.2) are satisfied. In that case, the integration yields the $\mathbf{P}$ and $\mathbf{e}_\mu$ of affine space, and of Euclidean space in particular (i.e. when (6.3) is added to the system to be integrated).

We may be given a set of differential forms of the type (6.3) in terms of some unknown system of coordinates and we want to know whether they belong or not to a Euclidean space (meaning to its frame bundle and sections thereof). The test lies in checking whether they satisfy or not the (6.1)-(6.2) system. If the answer is positive and, in addition, Eq. (6.3) is satisfied, the differential forms $(\omega^\mu, \omega_\nu^\lambda)$ will belong to the frame bundle of a Euclidean space.

In general, these forms may look very complicated. When using Cartesian coordinates, X^μ they take a rather simple form,

$$\omega^\mu = dX^\mu (R^{-1})_\mu^\nu, \quad \omega_\nu^\kappa = dR_\nu^\mu (R^{-1})_\mu^\lambda, \tag{6.5}$$

as specializations or restrictions that these equations are of the invariant forms of affine space. Here R is the general matrix of the orthogonal group for the given number of dimensions. Given that even the pull-backs to sections of the invariant forms may take very cumbersome forms, one needs an integrability test to ascertain their belonging to affine space (Euclidean space if (6.3) is satisfied). As one could expect (we shall not give a proof, the pull-back of the exterior derivative equals the exterior derivative of the pull-back. This implies that the aforementioned test of integrability can be used also in the sections, which is where one works most of the time. We have already seen since the previous chapter what those integrability conditions are for affine space. Let us now say a little bit more about integrability conditions in general.

Consider the very simple question of whether the equation

$$dF(x,y) = f(x,y)dx + g(x,y)dy \qquad (6.6)$$

is integrable or not. All readers know the answer, but we want to expose them with a simple example to Frobenius test of integrability. We first write (6.6) as

$$dF - fdx - gdy = 0, \qquad (6.7)$$

and then see whether the exterior derivative of the left hand side vanishes identically or not. For this purpose, one may use, if needed, the equations(s) itself (themselves) whose integrability is being considered. In applying the theorem, one assumes that the solution (F in this case) exists, and reaches a contradiction (not getting identically zero) if it does not exist. We would thus write $ddF = 0$, and the differentiation of the left hand side of (6.6) would simply yield

$$0 - f_{,y}dy \wedge dx - g_{,x}dx \wedge dy, \qquad (6.8)$$

which does not vanish identically unless $f_{,y} = g_{,x}$.

Let us look at the same problem in a slightly different way. Consider the system

$$dF = fdx + gdy, \quad f_{,y} = g_{,x}. \qquad (6.9)$$

Now (6.8) is identically zero using the equation $f_{,y} = g_{,x}$ of the system, which is, therefore, integrable. The (in general) not integrable equation (6.7) yields the integrable system (6.9) by adding to it its integrability conditions.

What has just been said applies equally well to systems of differential equations involving vector-valued differential forms. The differentiation of the left hand side of the equations

$$d\mathbf{P} - \omega^i \mathbf{e}_i = 0, \quad d\mathbf{e}_j - \omega^i_j \mathbf{e}_i = 0, \qquad (6.10)$$

yields

$$(-d\omega^i + \omega^j \wedge \omega^i_j)\mathbf{e}_i, \qquad (6.11)$$

$$(-d\omega^i_j + \omega^k_j \wedge \omega^i_k)\mathbf{e}_i, \qquad (6.12)$$

after using the equations $d\mathbf{e}_j = \omega^i_j \mathbf{e}_i$, which makes part of the system (6.10). Frobenius theorem thus states that the integrability conditions for (6.10) are given by the equations obtained by equating to zero the expressions (6.11) and (6.12).

In the Euclidean case, we also have to differentiate $\omega_{ij} + \omega_{ji}$, which yields, using 6.3 itself,

$$d\omega_{ij} + d\omega_{ji} = d\omega_{ij} + d(-\omega_{ij}), \qquad (6.13)$$

which is identically zero. Hence the integrability conditions are then simply the same ones as in the affine case. This implication would have been the same if we took into account the so called affine extension of the Euclidean connection. In this case, Eq. (6.4) is replaced with

$$dg_{ij} + \omega^k_i g_{kj} + \omega^k_j g_{ki} = dg_{ij} + \omega_{ij} + \omega_{ji} = 0 \qquad (6.14)$$

for the affine extension of the Euclidean frame bundle. As if it were needed, one easily verifies that this equation is integrable.

Exercise. Given $\omega^1 = dx^1$, $\omega^2 = x^1 dx^2$ and $\omega_2^2 = dx^2 = -\omega_1^2$, find if these differential forms belong or not to a section of the frame bundle of Euclidean space (They do!).

Exercise. Do the same for $\omega^1 = dx^1$, $\omega^2 = \sin x^1 dx^2$, $\omega_i^j = 0$ (They do not!).

If the system of connection equations is not integrable, one can still integrate them on curves between any two given points. The result will depend on path. The "P's" resulting from the integration of $\omega^i e_i$ along two different curves between the same two points will be different in the case of $\omega^1 = dx^1$, $\omega^2 = \sin x^1 dx^2$, $\omega_i^j = 0$, even through the subsystem $de_i = \omega_i^j e_j = 0$ is obviously integrable. We would no longer be in $\mathcal{E}^2$.

6.7   The Lie algebra of the Euclidean group

Except for the fact that the role of the affine and linear groups is now taken by the Euclidean and orthogonal groups, what was said in section 5.9 applies also here.

As was the case then, we have

$$\begin{bmatrix} 0 & \omega^i \\ 0 & \omega_j^i \end{bmatrix}, \tag{7.1}$$

with $\omega_i^j = -\omega_j^i$ in the positive definite case. Instead of (9.6) of the previous chapter, we now have

$$\omega^k \alpha_k + \omega_j^i \beta_i^j, \tag{7.2}$$

where $\beta_j^i = \alpha_i^j - \alpha_j^i$, and with summation over, say, $i < j$, rather than over all pairs of indices (if $n = 3$, it is best to sum over cyclic permutations, namely $(12, 23, 31)$). Recall that the matrices represented by the bold faced characters in (7.2) are $(n + 1) \times (n + 1)$ matrices. To illustrate a point as to what has to be the product in the Lie algebra, let us provide some simplifications. We shall represent with primes the blocks that constitute the linear part of the transformations, both for the α_i^j and β_j^i, and shall consider $n = 3$. We have

$$\beta_i'^j = \alpha_j'^i - \alpha_i'^j. \tag{7.3}$$

Since the $\alpha_i'^j$ are given by

$$\alpha_1'^2 = \begin{bmatrix} 0 & 1 & 0 \\ 0 & 0 & 0 \\ 0 & 0 & 0 \end{bmatrix}, \quad \alpha_2'^3 = \begin{bmatrix} 0 & 0 & 0 \\ 0 & 0 & 1 \\ 0 & 0 & 0 \end{bmatrix}, \quad \alpha_3'^1 = \begin{bmatrix} 0 & 0 & 0 \\ 0 & 0 & 0 \\ 1 & 0 & 0 \end{bmatrix}, \tag{7.4}$$

we have

$$\beta_1'^2 = \begin{bmatrix} 0 & -1 & 0 \\ 1 & 0 & 0 \\ 0 & 0 & 0 \end{bmatrix}, \quad \beta_2'^3 = \begin{bmatrix} 0 & 0 & 0 \\ 0 & 0 & -1 \\ 0 & 1 & 0 \end{bmatrix}, \quad \beta_3'^1 = \begin{bmatrix} 0 & 0 & 1 \\ 0 & 0 & 0 \\ -1 & 0 & 0 \end{bmatrix}, \tag{7.5}$$

One readily checks that the β' matrices constitute an algebra under skew-symmetrized matrix multiplication, though not under simply matrix multiplication. Thus:

$$\beta'^2_1\beta'^3_2 - \beta'^3_2\beta'^2_1 = \beta'^1_3, \tag{7.6}$$

and cyclic permutations thereof. The same relations apply to the unprimed betas.

6.8 Scalar-valued clifforms: Kähler calculus

A neglected area in the study of Euclidean spaces is its Kähler calculus. Start by defining its underlying algebra, called Kähler algebra. It is like the tangent Clifford algebras of section 8 of chapter 3, except that the sigmas or the gammas are now replaced by the differentials dx^μ.

Let α be the differential form

$$\alpha = a_{\nu\lambda}\ldots dx^\nu \wedge dx^\lambda \wedge \ldots \tag{8.1}$$

Define its covariant differential $d_\mu\alpha$ [46], [48] as

$$d_\mu\alpha = \frac{\partial a_{\nu\lambda}\cdots}{dx^\mu}dx^\nu \wedge dx^\lambda \wedge \ldots \tag{8.2}$$

in Cartesian coordinates. If α is a differential $r-$form, so is $d_\mu\alpha$. The components of $d_\mu\alpha$ in other coordinate systems are no longer the partial derivatives of the components of α. Readers need only recall their knowledge of the divergence of a vector field in terms of arbitrary coordinate systems to realize that it is not simply a matter of replacing the partial derivatives with respect to Cartesian coordinates with partial derivatives with respect to arbitrary coordinates.

Kähler defines a "derivative" $\partial\alpha$ as

$$\partial\alpha = dx^\mu \vee d_\mu\alpha = d\alpha + \delta\alpha, \tag{8.3}$$

having defined $d\alpha$ and $\delta\alpha$ as

$$d\alpha \equiv dx^\mu \wedge d_\mu\alpha, \qquad \delta\alpha \equiv dx^\mu \cdot d_\mu\alpha. \tag{8.4}$$

These two differential forms play the respective roles of curl and divergence. These concepts are now being made to apply to differential forms, rather than to vector fields.

Kähler created and used this calculus to obtain the fine structure of the hydrogen atom with scalar-valued clifforms [47]. More importantly, he used these to show how the concept of spinor emerges in solving equations for differential forms, even if the equations are not restricted to the ideals to which the spinors belong [47]. In 1962 [48], he presented the general theory more comprehensively than in 1960 [46], and used it to get the fine structure more expeditiously than in 1961 [47]. A very important and yet overlooked result that he obtained is that antiparticles (at least in his handling of the electromagnetic interaction) emerge with the same sign of energy as particles [48] .

In [83], the author of this book has shown that computations in relativistic quantum mechanics with Kähler's calculus of scalar-valued clifforms are much easier and less contrived than with Dirac's calculus. The Kähler calculus of scalar-valued differential forms is to quantum mechanics what the calculus of vector-valued differential forms is to differential geometry and, thus, to general relativity. Missing is the use in physics of a Kähler's calculus of differential forms to arbitrary valuedness, specially Clifford valuedness.

6.9 Relation between algebra and geometry

We explained in section 7 of chapter 1 that a point separates algebra from geometry. Let us say it again. In algebra there is a zero; in geometry, there is not. We now proceed with additional differences.

In Euclidean vector spaces, there is the concept of orthonormal bases. But the non-orthonormal bases are as legitimate as those which are. On the other hand, in the geometry of Euclidean spaces, orthonormal bases have a legitimacy that arbitrary bases do not. Why? Because in geometry the role of groups becomes paramount. Euclidean geometry is the study of figures and properties that are invariant under the Euclidean group, the basic figures being the orthonormal bases at all points. Affine geometry is the study of figures and properties invariant under the affine group, the basic figures being all vector bases at all points.

Related to what has just been said, groups G_0 in the pair (G, G_0) are not given in algebra the relevance they have in geometry. On the same grounds, one speaks of fibers and bundles in geometry; much less so, or even not at all, in algebra.

Part IV

CARTAN CONNECTIONS

Chapter 7

GENERALIZED GEOMETRY MADE SIMPLE

7.1 Of connections and topology

This is a transitional chapter in the sense that we study simple examples of Cartan's generalization of affine and Euclidean Klein geometries. For further perspective, we also devote a section to the original Riemannian geometry, where Riemann's curvature had to do with a so-called problem and method of equivalence. "Equivalence" is not a topic of differential geometry proper, but may be used there nevertheless.

We shall study three common surfaces under two different rules to compare tangent vectors at different points on those surfaces. One of these rules is the so called Levi-Civita connection (LCC). Formulated in 1917, it was the first known affine connection (we shall later call it Euclidean rather than affine). It was adopted, may be unnecessarily, by general relativity. We say "may be unnecessarily" because eventually there were other connections compatible with the original Riemannian geometry and with general relativity, but with better properties.

The most obvious alternative to the LCC is the Columbus connection (see Preface), the reason for this choice of name later to become clear. It is of the type called teleparallel, which means that one can then establish a relation of geometric equality of tangent vectors at different points of a region of a manifold. If there are exceptional points (even as few as just one) where we cannot define the equality of its tangent vectors to the tangent vectors at other points of M, we say that the teleparallelism is local or limited to regions of the manifold. The topology of the manifold may impede defining the connection everywhere, while allowing it on regions.

Consider the following easy-to-understand example of a local property. The direction East (or any other direction for that matter) is not defined at the poles where, correspondingly, the Columbus connection is not defined either. One cannot put connections other than the LCC on the full 2-sphere, i.e. the sphere in 3-dimensional Euclidean space. We then say that they are not globally defined. It is a topological issue.

The Columbus connection is globally defined on manifolds that are regions of a 2-sphere, among them the one resulting from puncturing it at the poles. Puncturing changes the topology of the sphere. This difference is largely academic for the physicist who integrates his equation to find the world in which he lives. Cartan indeed spoke of systems of equations whose solutions are the manifolds in which the connection lives [29]. Thus the 2-sphere would be the solution to a system of equations of structure with the LCC, but not with any other connection.

Assume that, making abstraction of dimensionality and of signature of the metric, the spacetime solution of some hypothetical system of equations were like a sphere but with Columbus connection. It would not be the whole sphere but just a region of it. The energy required to create the region would be greater and greater as one approached the full sphere. Only infinite energy would close the surface, since it is to be expected that the singularity of the torsion at the poles would entail a singularity of required energy, as well as singularity of other physical quantities related to affine (said better, Euclidean) structure. Hence, we shall take the concept of teleparallelism and zero affine curvature as synonymous, at least from a physical perspective.

The LCC, on the other hand, is globally defined. But the LCC does not include a concept of geometric equality. Most practitioners of differential geometry would consider it to be the natural connection because it is canonically defined by the metric. But there is, for example, the torus, where, by reason of symmetry, the LCC connection is less natural than the always richer Columbus connection.

The decision as to what is the affine connection of spacetime was taken when there was not a general theory of connections, and of teleparallel connections in particular. The adoption might one day be viewed as a mistake induced by historical circumstance. For the moment, let us understand the mathematics.

7.2   Planes

The term plane means different things in different contexts. Under a first meaning, plane is any linear space, not necessarily 2-dimensional. This was the meaning of plane when this term was used in the definition of a Lie algebra. We might have used the term hyperplane, but we have followed custom in using the term plane in that case.

Other times plane means the two dimensional affine space. Or one may be referring to an $n-$dimensional structure where the square of the distance between two points may be defined by an expression of the form

$$(x^1)^2 + (x^1)^2 + ... + (x^n)^2, \tag{2.1}$$

even if a comparison of vectors at different points has not been defined. In particular, there is the 2−dimensional metric plane. The Euclidean plane is, as we have seen, the affine plane endowed additionally with a square distance as in (2.1) with $n = 2$.

7.2.1   The Euclidean 2-plane

We return to the Euclidean plane for later easy comparison of its LLC with the Columbus connection. Recall the important point that the constant basis $(\mathbf{i}, \mathbf{j})$ may be viewed as a constant basis field, i.e.

$$\mathbf{i}(A) = \mathbf{i}(B), \quad \mathbf{j}(A) = \mathbf{j}(B), \tag{2.2}$$

for any two points A and B in the space. Let us advance that the connection defined in this way is the LCC, which we shall later define. It is zero in this particular section,

$$d\mathbf{i} = 0, \qquad d\mathbf{j} = 0, \tag{2.3}$$

but not in the bundle, as we have already seen.

Consider now the Euclidean basis field in terms of which the elementary displacement $d\mathbf{P}$ may be written as:

$$d\mathbf{P} = d\rho\,\mathbf{e}_\rho + \rho\,d\phi\,\mathbf{e}_\phi. \tag{2.4}$$

If we perform the dot product of $d\mathbf{P}$ with itself, we get

$$d\mathbf{P} \cdot d\mathbf{P} = d\rho^2 + \rho^2\,d\phi^2, \tag{2.5}$$

where we have taken into account that $\mathbf{e}_\rho \cdot \mathbf{e}_\rho = 1$, $\mathbf{e}_\rho \cdot \mathbf{e}_\phi = 0$, $\mathbf{e}_\phi \cdot \mathbf{e}_\phi = 1$. It is clear that

$$\mathbf{e}_\rho = \cos\phi\,\mathbf{i} + \sin\phi\,\mathbf{j}, \tag{2.6a}$$
$$\mathbf{e}_\phi = -\sin\phi\,\mathbf{i} + \cos\phi\,\mathbf{j}. \tag{2.6b}$$

We differentiate (2.6) and obtain:

$$d\mathbf{e}_\rho = -\sin\phi\,d\phi\,\mathbf{i} + \cos\phi\,d\phi\,\mathbf{j}, \tag{2.7a}$$
$$d\mathbf{e}_\phi = -\cos\phi\,d\phi\,\mathbf{i} - \sin\phi\,d\phi\,\mathbf{j}. \tag{2.7b}$$

We refer $(d\mathbf{e}_\rho, d\mathbf{e}_\phi)$ to the bases $(\mathbf{e}_\rho, \mathbf{e}_\phi)$ themselves and invert (2.6) to obtain

$$\left\{\begin{matrix} \mathbf{i} \\ \mathbf{j} \end{matrix}\right\} = \begin{pmatrix} \cos\phi & -\sin\phi \\ \sin\phi & \cos\phi \end{pmatrix} \left\{\begin{matrix} \mathbf{e}_\rho \\ \mathbf{e}_\phi \end{matrix}\right\}, \tag{2.8}$$

which permits us to then get $d\mathbf{e}_\rho$ and $d\mathbf{e}_\phi$ as

$$\begin{Bmatrix} d\mathbf{e}_\rho \\ d\mathbf{e}_\phi \end{Bmatrix} = d\phi \begin{pmatrix} -\sin\phi & \cos\phi \\ -\cos\phi & -\sin\phi \end{pmatrix} \begin{pmatrix} \cos\phi & -\sin\phi \\ \sin\phi & \cos\phi \end{pmatrix} \begin{Bmatrix} \mathbf{e}_\rho \\ \mathbf{e}_\phi \end{Bmatrix} \qquad (2.9)$$

from (2.7) and, finally,

$$\begin{Bmatrix} d\mathbf{e}_\rho \\ d\mathbf{e}_\phi \end{Bmatrix} = d\phi \begin{pmatrix} 0 & 1 \\ -1 & 0 \end{pmatrix} \begin{Bmatrix} \mathbf{e}_\rho \\ \mathbf{e}_\phi \end{Bmatrix}, \qquad (2.10)$$

or

$$d\mathbf{e}_\rho = d\phi\,\mathbf{e}_\phi, \qquad d\mathbf{e}_\phi = -d\phi\,\mathbf{e}_\rho. \qquad (2.11)$$

We thus read for the LCC of the Euclidean plane in the section under consideration:

$$\omega_1^1 = 0, \qquad \omega_1^2 = d\phi, \qquad \omega_2^1 = -d\phi, \qquad \omega_2^2 = 0. \qquad (2.12)$$

Readers may want to test their understanding by building from (2.12) the differential forms $\omega_i{}^j$ in the bundle of orthonormal bases. We do not need to compute the torsion and the Euclidean curvature since we have not abandoned at any point the Euclidean plane and they are, therefore, zero.

7.2.2   Post-Klein 2-plane with Euclidean metric

We now introduce the Columbus connection on the metric plane (punctured at one point), which generates a Cartan geometry. Specifically, this will be a post-Kleinean Cartan geometry (space, manifold) that is flat in the metric sense. This type of flatness means that the metric can be written as a sum of squares of the differential of the coordinates because we keep the $ds^2 = dx^2 + dy^2$ of the Euclidean space. But we shall only change the relation of equality of vectors attached to different points of our manifold. We do not yet need to understand everything, as all of it will become obvious little by little.

Suppose we had a flat world in which there were an advanced society of 2-dimensional beings. Suppose further that there were a point O on the plane playing the role of our sun. Barring other causes for differences in climate, unlike between different points with the same latitude on the earth, all points at the same distance from O would be equivalent from a climatological perspective. There would be some temperate region in a circular band with radii close to some ideal climate radius, $r = r_0$. We would put those who have to sweat to make a living at distances $r \gg r_0$, where it is cold. The very hot region where $r \ll r_0$, we would try to populate with trial lawyers, chief executive officers of large corporations, preachers, thieves and many politicians. Unfortunately, those same guys would already have bought all the land in the temperate region.

In the world that we have just outlined circles centered at O and radial lines would constitute the lines of constant direction (autoparallels). Sailors, cartographers and applied geometers would naturally use the Columbus connection

$$d\mathbf{e}_\rho = 0, \qquad d\mathbf{e}_\phi = 0, \qquad (2.13)$$

rather than (2.11). Also, since we are in the metric plane, we would still have

$$ds^2 = d\rho^2 + \rho^2\, d\phi^2. \tag{2.14}$$

Said better, we would still have (2.14) as a consequence of defining ds^2 as $d\mathbf{P}\cdot d\mathbf{P}$ with $d\mathbf{P}$ given by

$$d\mathbf{P} = d\rho\, \mathbf{e}_\rho + \rho\, d\phi\, \mathbf{e}_\phi, \tag{2.15}$$

and, again, with $\mathbf{e}_\rho \cdot \mathbf{e}_\rho = 1$, $\mathbf{e}_\rho \cdot \mathbf{e}_\phi = 0$, $\mathbf{e}_\phi \cdot \mathbf{e}_\phi = 1$.

For completeness, we express this connection in terms of the $(\mathbf{i}, \mathbf{j})$ basis field. Differentiating Eqs. (2.8), we would have, using (2.13):

$$\left\{ \begin{matrix} d\mathbf{i} \\ d\mathbf{j} \end{matrix} \right\} = d\phi \begin{pmatrix} -\sin\phi & -\cos\phi \\ \cos\phi & -\sin\phi \end{pmatrix} \left\{ \begin{matrix} \mathbf{e}_\rho \\ \mathbf{e}_\phi \end{matrix} \right\}. \tag{2.16}$$

The inversion of (2.8) yields

$$\left\{ \begin{matrix} \mathbf{e}_\rho \\ \mathbf{e}_\phi \end{matrix} \right\} = \begin{pmatrix} \cos\phi & \sin\phi \\ -\sin\phi & \cos\phi \end{pmatrix} \left\{ \begin{matrix} \mathbf{i} \\ \mathbf{j} \end{matrix} \right\}. \tag{2.17}$$

Substituting (2.17) in (2.16), we get

$$\left\{ \begin{matrix} d\mathbf{i} \\ d\mathbf{j} \end{matrix} \right\} = d\phi \begin{pmatrix} 0 & -1 \\ 1 & 0 \end{pmatrix} \left\{ \begin{matrix} \mathbf{i} \\ \mathbf{j} \end{matrix} \right\}, \tag{2.18}$$

or, equivalently,

$$d\mathbf{i} = -d\phi\, \mathbf{j}, \qquad d\mathbf{j} = d\phi\, \mathbf{i}. \tag{2.19}$$

We provide some remarks for readers to think about (A systematic presentation of the theory that addresses the issues in question will be given in coming chapters). Define the Euclidean curvature as we did the affine curvature:

$$\Omega_i^j \equiv d\omega_i^j - \omega_i^k \wedge \omega_k^j. \tag{2.20}$$

We call it Euclidean rather affine because it lives in a different bundle and takes values in a different algebra. It pertains to Euclidean connections, which are called affine connections (specifically metric compatible affine connections) in the literature, but only very exceptionally in Cartan's writings. We shall return to this later on. The Euclidean curvature still is zero in the punctured plane endowed with the Columbus connection since there is equality of vectors at a distance, though not in the same way as in the previous subsection. Define the torsion also as in the affine case:

$$\Omega^i \equiv d\omega^i - \omega^j \wedge \omega_j^i. \tag{2.21}$$

Readers may verify that the torsion is not zero but

$$\Omega^1 \equiv d(d\rho) - 0 = 0, \quad \Omega^2 \equiv d(\rho d\phi) - 0 = d\rho \wedge d\phi. \tag{2.22}$$

Consider next the development in the Euclidean plane of a closed curve of this universe. The radial lines and concentric circles centered at the origin of the radial lines have constant direction. Since the circles are perpendicular to the radii, a curvilinear quadrilateral constituted by the intersection of two radial lines with two of those circles develops into an open curve in Euclidean space when the four segments of the curvilinear quadrilateral go into rectilinear segments in Euclidean space, these being orthogonal to the adjacent ones. This curve does not close because the radial lines intercept arcs of different size on the two circles. In this case, the failure to close is the manifestation of non-vanishing torsion. Warning: the development of a curve may not close also if the torsion were zero, but for a different reason. We shall see more on this later.

There is a different way of looking at this failure to close. Let the curvilinear quadrilateral have consecutive vertices A, B, C, D. Assume we represent on the plane not just one of the four segments in succession, but first ABC and then ADC. We cannot assign a vector to the pair of points (A, C) since the developments of ABC and ADC end at different points. Equation (2.15) remains valid, but a vector $\mathbf{P}$ does not exist in such a manifold consistently with the Columbus rule that we have imposed on it. Coordinates (x, y) may still be chosen as

$$x = \rho \cos \phi, \qquad y = \rho \sin \phi, \tag{2.23}$$

and the metric (2.14) again becomes

$$ds^2 = dx^2 + dy^2. \tag{2.24}$$

Non-experienced readers are not expected to understand, much less remember and/or be able to justify, everything that has been said here. We are just trying to provide the flavor of the intrincacies of differential geometry, intricacies that are not accessible either by the treatments for physicists or in the very formal treatments for mathematicians. As proof of this, witness the contrived examples of differential manifolds with torsion that are given in the literature.

7.3   The 2-sphere

The Columbus connection of the plane may be a little bit confusing because it admits coordinates whose squared differentials diagonalize the metric (as we have just seen) and yet it does not have a displacement vector defined on it (of which the Cartesian coordinates would be its components). The Columbus connection is easier to understand on the punctured sphere. It made history in 1492.

7.3.1   The Columbus connection on the punctured 2-sphere

In 1924, Cartan gave the example of the earth endowed with the Columbus connection [13]. He did not use this name. Here is why we do. Starting his first mayor voyage Westward into the Atlantic, Christopher Columbus ordered the captains of the other two ships that accompanied his to maintain the same direction, West! If we stretch things a little bit and put things in modern terms,

Columbus said this: the unit vector in the West direction is equal to itself at any other point. The same comment applies to any other direction (where defined, thus not at the poles). Under the Columbus connection, the autoparallels or lines of constant direction are the meridians, the parallels and the rhumb lines (i.e. curves that make a constant angle with the parallels they intersect, while spiraling towards the poles). The parallels and the meridians are only particular cases of rhumb lines.

Tangent vectors are vectors on the tangent planes to the surface of the sphere. The Columbus connection is one where there is a relation of equality of such vectors at any two different points of the punctured sphere independently of any curve used to compare them. In contrast, the LCC on the sphere and in any manifold other than Euclidean spaces is not a relation of equality. We shall see this in the next subsection. Let us rush to add that, from a local perspective, the cylinder and the cone are Euclidean spaces: just cut them, open them up and extend them on the Euclidean plane.

The LCC and the Columbus connections are called Euclidean connections. They pertain to direct Cartanian generalizations of Euclidean space, not of affine space. The transformations in the fibers of their bundles of frames are rotations. Thus they preserve angles. We shall use this fact here, but shall leave the discussion of these issues for future chapters.

The representation in Euclidean space of curvilinear quadrilaterals formed by the intersection of pairs of meridians and of parallels fails to close in general, and thus fails to give what would otherwise be a rectangle. In the previous subsection, this was due to the fact that the segments of parallels will in general be of different length.

We are now ready to write down the Columbus connection on the sphere. If $(\mathbf{e}_\theta, \mathbf{e}_\phi)$ are unit vectors in the direction of the (θ, ϕ) coordinate lines, we have

$$d\mathbf{e}_\theta = 0, \qquad d\mathbf{e}_\phi = 0. \tag{3.1}$$

Another alternative field would be one $(\mathbf{e}_\theta, \mathbf{e}_\phi)$ associated with the spherical coordinates (θ', ϕ') based on a different pair of poles. The equation of the connection (3.1) in terms of the new basis field would be rather cumbersome without better mathematical machinery than the one used in the book. Of course there is a connection that takes a simple form in terms of (θ', ϕ'), namely

$$d\mathbf{e}'_\theta = 0, \qquad d\mathbf{e}'_\phi = 0. \tag{3.2}$$

The Columbus rule to compare directions on the 2-sphere becomes a connection when it is put together with a conservation of distances (a stick was considered equal to itself wherever it went). It was not thought to be a Euclidean connection because the concept did not exist in the mathematical literature until Cartan published it in 1924 [13]. There are other types of connections on the sphere which also are teleparallel, i.e where a path-independent equality of tangent vectors at different points is defined. Before attempting to build them on the sphere, readers should attempt to do so on the plane. Hint: Think of a different system of coordinate lines and define them as lines of constant direction. They do not need to be orthogonal coordinate systems in order to define a connection; the unit tangent vectors to the coordinate lines will not then constitute Euclidean (i.e. orthonormal) bases.

Exercise. Compute the torsion of the Columbus connection on the punctured sphere. Do not worry if you cannot. All this will be done systematically in coming chapters.

7.3.2   The Levi-Civita connection on the 2-sphere

We are not going to be distracted in this subsection with how the LCC is defined in general. Suffice to say at this point that, according to the LCC, the maximum circles and only they are the lines of constant direction on the sphere.

Madrid and New York are approximately on the same parallel of 40.5^0 North latitude. They have West longitudes of 4^0 and 74^0 respectively. When commercial planes fly from Madrid to New York, they follow a maximum circle. They thus follow a line of constant direction of the LCC. For that purpose and after leaving the proximity of Madrid's airport, they fly in direction t, which is, say, D degrees North of West, and arrive in the proximity of NY in a direction t' which is D degrees South of West and which the LCC considers to be the same as t from the perspective of that maximum circle. Since West in Madrid is D degrees counterclockwise from t, the corresponding (i.e. equal) direction in NY is D degrees counterclockwise from t' and, therefore, $2D$ degrees counterclockwise from NY's West direction. West in Madrid and NY are not equal by the LCC, from the perspective of the maximum circle that they jointly determine.

Let us consider now an example that is even more extreme. Consider two antipodal points on the equator. We start going North on the meridian from one of those points, pass the pole and continue on the same meridian. We reach the antipode going South. The North direction has become the South direction on its voyage.

Except in the plane, where it is trivial, the LCC has to be computed. One computes it by solving the system constituted by the statement that its torsion is zero,

$$d\omega^i - \omega^j \wedge \omega^i_j = 0, \tag{3.3}$$

and that the connection is Euclidean, i.e. $\omega_{ij} + \omega_{ji} = 0$, which in turn implies

$$\omega^i_j = -\omega^j_i. \tag{3.4}$$

The system of equations (3-3)-(3-4) can be solved for ω^j_i. We shall enter the general method of solution in a future chapter. For 2-dimensional Euclidean connections, the solving is very easy, actually by inspection. In the case of the sphere, we give the solution in the form

$$d\mathbf{e}_\theta = \cos\theta \, d\phi \mathbf{e}_\phi, \quad d\mathbf{e}_\phi = -\cos\theta \, d\phi \mathbf{e}_\theta, \tag{3.5}$$

so as to not forget its geometric significance.

Exercise. Show that $\Omega_1^2 = -\sin\theta\, d\theta \wedge d\phi = -\Omega_2^1$. Again, do not worry if you cannot. All this will be done systematically in coming chapters.

7.3.3 Comparison of connections on the 2-sphere

In the figure, the curvilinear triangle PJN is constituted by 90^0 arcs of maximum circles. They have been chosen so that those arcs are lines of constant direction both according to the LCC and Columbus connection. Using the LCC,

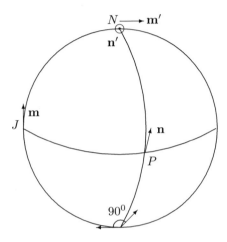

Figure 5: Transport of **n** from P to N along two different paths

let the unit vector **n** at P be "transported" to N, where it becomes the unit vector $\mathbf{n}'$ perpendicular to the plane of the paper and going in. We would then be tempted to write $\mathbf{n} = \mathbf{n}'$, and similarly $\mathbf{m} = \mathbf{m}'$. If we write $\mathbf{n} = \mathbf{m}$ as a matter of definition, all these equalities would then imply $\mathbf{n}' = \mathbf{m}'$, which is obviously wrong since they make an angle of 90^0. For reasons such as this, Levi-Civita could not write all these equalities. But he could say that $\mathbf{n}'$ is the result of parallel-transporting with his rule **n** from P to N along the meridian and that $\mathbf{m}'$ is the result of parallel transporting **n** from P to N along the line PJN. The result of the transport is a function of the path followed when one uses this connection.

Assume now the Columbus connection. We are no longer transporting anything. Since we cannot define directions at N without resort to basis fields associated with the sphere punctured at the poles, we consider a very small parallel near the North pole under the assumption that the earth is a perfect sphere. Let $\mathbf{m}''$ and $\mathbf{n}''$ be the unit vectors pointing North at the intersection of that parallel with the meridians through J and P. We have

$$\mathbf{m}'' = \mathbf{m} = \mathbf{n} = \mathbf{n}'', \tag{3.6}$$

if the rule to compare vectors is the Columbus connection. We have introduced that small parallel for comparison purposes with the LCC as to what happens when we get arbitrarily close to the pole, which we do next.

Consider the representation of the spherical triangle NJP that the LCC yields on the Euclidean plane. Since NJ and NP are perpendicular to JP, the representation does not close. It takes the form ⊔. In the case of the Columbus connection, we do not have a spherical triangle, but a curvilinear quadrilateral with *J* and *P* as vertices of one side, the opposite side being arbitrarily small. The result is also the same, except for an additional tiny horizontal contribution at the top pertaining to that very small parallel.

It is clear that if the torsion does not vanish, the failure to close of the representation on a plane of a closed curve on the sphere need not sound strange. This is clear since torsion is directly related to the translation element. But it is less clear why this would also happen when the torsion is zero and the curvature is not, since curvature is directly related only to the comparison of vectors. The reason is not other than the different vectors that we associate with the sides of curvilinear segments belong to different vector spaces when the affine (respectively Euclidean) curvature is not zero. The LCC does not permit one to identify the Euclidean tangent vector spaces at different points.

7.4   The 2-torus

In this book, the torus will always mean the 2-dimensional or ordinary torus. Its relevance vis a vis the theory of connections lies in that the Columbus connection on the torus is more natural than on the sphere, and also more natural than the LCC. We shall call it the canonical connection of the torus.

Any plane through the axis of the torus intersects the surface of this figure along two circles of equal radius. These circles we shall call meridians. A perpendicular plane to the axis intersect the torus along two circles that we shall call parallels. These two parallels approach each other as the perpendicular plane moves away from the equatorial plane, until the two become just one. Further away, the planes no longer intersect the torus.

7.4.1   Canonical connection of the 2-torus

By reason of symmetry, we should consider the parallels and meridians of the torus as lines of constant direction. Also lines of constant direction would be the rhumb lines, meaning lines that intersect the parallels at a constant angle. A unit vector along a rhumb line would be considered to be equal to itself. Actually, all unit vectors making the same angle with the parallels would be considered to be equal. This would be the Columbus connection on the torus. This a relation of vector equality on this figure. We do not need to remove any point since, unlike the case of the sphere, this equality of vectors is defined everywhere.

Let us compute the torsion of this connection. It will be given by just $d\omega^i$. Let r be the radius of the meridians, and let R be the distance from the center of the torus (i.e. the center to its equatorial parallels) to the centers of the meridians. Let θ and ϕ be angular coordinates on meridians and parallels respectively. As was the case with the Columbus connection on the sphere, we have

$$d\mathbf{e}_\theta = 0, \qquad d\mathbf{e}_\phi = 0. \tag{4.1}$$

The differential forms on this section of the frame bundle of the torus are

$$\omega^1 = (R + r\cos\theta)d\phi, \qquad \omega^2 = rd\theta. \tag{4.2}$$

The torsion of the Columbus connection is

$$\Omega^1 \equiv d\omega^1 = -r\sin\theta d\theta \wedge d\phi, \qquad \Omega^2 \equiv 0. \tag{4.3}$$

The Euclidean curvature is of course zero.

Consider the curvilinear quadrilateral determined by the intersection of two meridians and two parallels of different radii. Its representation in Euclidean space again is a quasi-rectangle. It fails to close. It is a reflection of the fact that the torsion of the Columbus connection is not zero.

We shall learn later that the autoparallels or lines of constant direction and the stationary curves (in actual practice, curves such that their length is smaller than the length of neighboring curves between the same two points) do not coincide in 2-dimensional surfaces under different connections. This is also the case in higher dimensions in general, but not always.

The Columbus connection is teleparallel since an equality of vectors at a distance is defined. It cannot be equal to the LCC of the torus since the LCC satisfies that its lines of constant direction coincide with its stationary curves. These curves are not in general the parallels and the meridians. We shall now show this without resort to calculation.

In one of the parallels of the "upper" semi-torus, stick two needles separated say, 60 or 90 degrees, or some other number in that general interval (for visual purposes). Run a length of chain between the two needles, loose enough to achieve that the chain sits exactly on the parallel. One can pull the chain from one end while keeping the chain fixed at the other end. The chain moves up. This shows that there are curves in the neighborhood of a segment of parallel between the same two points that have a smaller distance, and thus are not stationary at the parallel. To make the point even more evident by comparison, consider the same mental exercise with two needles on the sphere and the chain lying on the maximum circle through the two needles. One cannot pull the chain, since other curves in the neighborhood have greater length between the same two points.

We have described the Columbus connection on the torus. It satisfies that it is zero in the constant basis field just mentioned. What is the LCC of the torus? It is highly non trivial, in contrast to the Columbus connection. This is the reason why we said of the Columbus connection, which is trivial and respects the symmetry of the torus, that it is its canonical connection.

7.4.2   Canonical connection of the metric of the 2-torus

Consider now the Levi-Civita connection. Using again the system (3.3)-(3.4), one readily finds

$$\omega_1^2 = \sin\theta d\phi = -\omega_2^1. \tag{4.4}$$

Notice the similarity of this connection with the connection on the sphere (make some changes in notation to make the similarity more obvious). Think what would happen if you made the hole of the torus smaller and smaller, until $r = R$. Continue this process, making $R < r$, until the torus becomes a sphere.

Exercise. Compute the curvature of the LCC on the torus. Once again, do not worry if you cannot at this point.

The development on the Euclidean plane of closed curves on the torus is more difficult since the lines of constant direction under the Levi-Civita connection are not as easily identifiable. One cannot escape having to perform the computations.

7.5   Abridged Riemann's equivalence problem

A more comprehensive treatment of this issue will be dealt with in section 2 of chapter 10.

The concept of Riemannian curvature was born to deal with the issue of whether a quadratic symmetric differential form in n coordinates could be reduced to the form

$$(dx^1)^2 + (dx^2)^2 + ... + (dx^n)^2, \tag{5.1}$$

by a coordinate transformation. This is a problem of a type called of equivalence. It was solved by Riemann, Christoffel and Lipshitz independently. At the time of the Christoffel and Lipshitz publications, the paper by Riemann had not yet been published (See our note in reference [62]). Cartan solved it more elegantly in 1922 using differential forms. All those authors reached the same conclusion, namely that a necessary and sufficient condition for the metric to become (5.1) by a coordinate transformation is that the so called Riemannian curvature vanishes.

To illustrate the issue with a simple example, consider the metric on the unit sphere

$$ds^2 = d\theta^2 + \sin^2\theta\, d\phi^2. \tag{5.2}$$

Can we reduce it to the form $(dx^1)^2 + (dx^2)^2$ by a coordinate transformation? For Cartan, the problem is equivalent to finding whether the system of equations

$$dx^i - \omega^i = 0 \tag{5.3}$$

is integrable. Here the ω^i are those in the bundle, i.e.

$$\omega^1 = \cos\lambda\, d\theta + \sin\lambda\sin\theta d\phi, \tag{5.4a}$$

$$\omega^2 = -\sin\lambda\, d\theta + \cos\lambda\sin\theta d\phi, \tag{5.4b}$$

for arbitrary λ, rather than $\omega^1 = d\theta$ and $\omega^2 = \sin\theta d\phi$. The ω^i's in (5.4) are such that the sum of their squares yields (5.2), which is independent of λ.

Let us now consider this problem of equivalence in arbitrary dimension. Cartan remarks that, the ω^i being what they are, the $d\omega^i$ must be exterior products of the form

$$d\omega^i = \omega^j \wedge \sigma^i_j, \qquad (5.5)$$

for some undefined differential forms, σ^i_j, which will depend on both the ω^l and the differentials of the coordinates in the fibers, like the previous $d\lambda$. He shows that they must satisfy

$$\sigma_{ij} + \sigma_{ij} = 0, \qquad (5.6)$$

as a consequence of the fact that the differentiated ds^2 must not involve the $d\lambda$. The differentiation of ds^2 must be compatible with the structure to which ds^2 belongs, which is not a exterior algebra structure.

The system (5.5)-(5.6) can be solved for σ_{ij}. We shall deal with this system at length in a later chapter. At this point, let us continue with Cartan's argument. He develops further the consequences of the independence of ds^2 on the coordinates λ and finds the equation

$$d\sigma^j_i = \sigma^l_i \wedge \sigma^j_l + R^j_{ilm}\omega^l \wedge \omega^m \qquad (5.7)$$

This means that all the dependence of $d\sigma^j_i$ on the σ^m_k's and thus on differentials like the $d\lambda$'s is contained in the term $\sigma^l_i \wedge \sigma^j_l$. The $R^j_{ilm}\omega^l \wedge \omega^m$ constitute the Riemannian curvature. There are no $\omega^i \wedge \sigma^j_l$ terms, and no additional $\sigma^l_i \wedge \sigma^j_l$ terms.

Cartan finally considers the integrability of $dx^i - \omega^i = 0$ when accompanied by the integrability condition $\sigma^i_j = 0$. One has to apply the Frobenius test of integrability to the system

$$dx^i - \omega^i = 0, \qquad \sigma^i_j = 0. \qquad (5.8)$$

If σ^i_j is zero, $d\sigma^i_j$ also is zero. But this is not the way in which the theorem is to be used. One has to check whether $d\sigma^j_i$ is zero in (5.7) when using $\sigma^i_j = 0$. It is clear that it is not in general, unless the Riemannian curvature is zero. As for using the Frobenius test on (5.8), we have

$$d(dx^i - \omega^i) = -d\omega^i = -\omega^j \wedge \sigma^i_j = 0, \qquad (5.9)$$

by virtue of the (partially) defining relation (5.5) of σ^i_j and the last equation in the system (5.8). It thus follows that a necessary and sufficient condition for a change of coordinates to exist that reduce the metric to the form (5.1) is that the Riemannian curvature be zero.

7.6 Use and misuse of Levi-Civita

We now wish to discuss the significance of the specific form of the system (5.5)-(5.6) that defines σ^i_j, system which is instrumental in obtaining the Riemannian curvature through (5.7). That system is the same one as (3.3)-(3.4) for ω^i_j. The σ^j_i must thus be identified with the ω^i_j, but only in the form that the solutions to two different problems take. They are representations of the solution of two different problems. The σ^j_i solution was not a connection (of bases) before 1917, since manifolds other than the affine and Euclidean spaces themselves had not been endowed with some sort of comparison of tangent vectors at different points.

During the last third of the nineteenth century, the components of σ^j_i were known under the name of Christoffel symbols, combinations of the metric and its first order derivatives. In 1917, Levi-Civita (LC) conceived the possibility of their use for such comparisons [52](not a true concept of equality but a succedaneum for it). This was equivalent to using σ^i_j for the role that ω^i_j has played in this and the previous chapter. The concept of ω^i_j on arbitrary differential manifolds had not existed until then; the concepts of affine and Euclidean space existed, but much of what has been stated in this book about those spaces was not yet developed.

Many who should know better still fail to understand what exactly happened in 1917, partly because they are not aware of how Riemannian geometry emerged and grew. The Levi-Civita connection emerged in the absence of a general theory of connections. If such a theory had existed in 1917, one would have said simply that the differential 1−forms that one derives canonically from the metric and are instrumental in obtaining the Riemannian curvature can be chosen for one of many ways of transporting vectors. Due to ignorance, two concepts being represented by the same set of components became a set of quantities with two different roles; this set of quantities was treated as if it were a concept. It was only with the arrival of Cartan's general theory of connections that it became clear to many experts that one is actually dealing with two different concepts of curvature, which are represented in general by different differential forms. Everybody understands nowadays that the connection need not be LC's. What still is not clear to many is that σ^j_i has not ceased to exist on manifolds endowed with a metric but with a different affine connection. It simply happens that it only plays its pre-1917 role. Correspondingly, we have to consider Euclidean and metric curvatures, respectively derived from ω^i_j and σ^j_i. Unfortunately, many experts seem to have forgotten that the original role of Riemann's curvature had to do with the solution of the problem of whether a given metric can be reduced to the form (5.1) by a coordinate transformation, and with no other issue.

In view of these considerations, readers should by now understand that the Euclidean plane has zero torsion and zero Euclidean and metric curvatures. The punctured metric plane endowed with the Columbus connection has zero Euclidean and metric curvatures but non-zero torsion. The sphere endowed with the LC connection has zero torsion, and its Euclidean and metric curvatures are given by the same differential forms. The punctured sphere endowed with the Columbus connection has zero Euclidean curvature, non-zero metric curvature and non-zero torsion. Those comments for the sphere also apply to the torus, except that it need not be punctured.

Chapter 8

AFFINE CONNECTIONS

8.1 Lie differentiation, INVARIANTS and vector fields

For a change, readers may consider skipping paragraphs of their choice in the middle of this section and then try to capture the gist of the last paragraphs.

We would not care about Lie differentiation in this book were it not for the fact that, after reading the great classics (Cartan and Kaehler in this case), one can only be puzzled when reading modern authors in connection with the concepts of tangent vectors and Lie differentiation.

Starting in page 81 of his extraordinary 1922 book on Integral Invariants [9], Cartan dealt with what he called an infinitesimal transformation, nowadays viewed as the action of a Lie operator. He extended the action of such operators on ordinary functions to differential forms. A decade later, this operator was baptized as Lie derivative. It is under this name that we find it again in papers by Kähler, [46], [48], as well as Slebodzinski [69] (Both acknowledge Cartan for that extension). Reference [46] is the most interesting one in connection with the present discussion.

Cartan did not define a Lie derivative operator for differential forms. He did not see a need for it. It is clear in his work that the action of an infinitesimal transformation of a function on $\mathbb{R}^n$,

$$Af \equiv \xi^1 \frac{\partial f}{\partial x^1} + \xi^2 \frac{\partial f}{\partial x^2} + \ldots + \xi^n \frac{\partial f}{\partial x^n}, \tag{1.1}$$

implies the form of its action on a differential form u. In a stroke of genius, Cartan *derived* in a few simple steps the formula

$$Au = (du)A + d[u(A)], \tag{1.2}$$

which allows one to easily compute Au since $(du)A$ and $u(A)$ are what in modern terms is called the evaluation of du and u on a vector field **A** with the same

143

components as Cartan's infinitesimal transformation. This is a well known theorem in the theory of Lie derivatives, though this term was not yet in use. Notice (a) the absence of a symbol for partial differentiation, (b) that apart from A and u, there is in (1.2) only the invariant operator d (which makes clear that A is a coordinate independent concept), and (c) A comes at the end on both terms on the right. This does not impede the validity of (1.2) when A is a function of surfaces and hypersurfaces; the exterior calculus is the same for skew-symmetric multilinear functions of vectors as for integrands (i.e. functions of hypersurfaces).

With A given as

$$A = \xi^i \frac{\partial}{\partial x^i}, \tag{1.3}$$

Kähler *derives* the formula [46]

$$Au = \xi^i \frac{\partial u}{\partial x^i} + d\xi^i \wedge e_i u. \tag{1.4}$$

The operator e_i removes the factor dx^i from u, with sign $(+, -, +, -...)$ depending on the position of dx^i in the product).

Once again, the action of the operator (1.3) on differential forms has been derived from its action on scalar-valued $0-$forms. Direct proof of the equivalence of (1.2) and (1.3) is simple [82].

Let us be more precise regarding what Kähler did. A coordinate transformation allowed him to write the pull-back of the operator (1.3) as

$$A = \frac{\partial}{\partial y^n}, \tag{1.5}$$

since $\partial/\partial y^i$ $(i = 1, ..., n - 1)$ are always zero in his new coordinate system. After performing the differentiation (1.5) of u, Kähler undoes the pull-back, i.e. returns to x coordinates, thus obtaining (1.4). Again, the Lie derivative of functions on $\mathbb{R}^n$ implies what the Lie derivative of differential forms is. One more treatment, namely by Slebodzinski [69], has the same implication.

In the modern literature on differential geometry, the dominant definition of tangent vector $\mathbf{A}$ at a point of a differentiable manifold is (1.3). One then gives the name of Lie derivative to a new operator on differential forms, based on the concept of flow of a vector field. The action of this operator is not different from what we have just reported, though one may wonder why introduce the concept in the modern way, which looks ad hoc and causes a lot of confusion? (just look for the term Lie derivative in the web). Cartan, Kähler and Slebodzinski obtained (1.4) in respective different ways without resorting to a new definition. For practical purposes, you can then make a definition of Lie derivative by (1.3) of a differential form, like Kähler did. In the end, it is a matter of extending the concept of partial derivative to differential forms, starting with the elements of a basis of differential 1-forms.

Assume that we had been given the operator (1.5) rather than (1.3) itself. To be more relevant, assume that y^n is the rotation angle ϕ around an axis of cylindrical symmetry. $\partial u/\partial \phi$ is the partial derivative with respect to the parameter of the 1-parameter symmetry group. A change of $\partial/\partial \phi$ to Cartesian coordinates yields $x\partial_y - y\partial_x$. This corresponds to just the first term on the right of (1.4). But that is so because that concept of partial derivative is peculiar to functions on $\mathbb{R}^n$, not to functions of hypersurfaces. If u is given in terms of a system of coordinates (x) instead of a system of coordinates (y), where y^n equals ϕ, one should differentiate with respect to ϕ not only the coefficients $a_{i_1 i_2 \ldots}$ of u, but also the basis $(dx^{i_1} \wedge dx^{i_2} \wedge \ldots)$. One should obtain the partials $\partial(dx^{i_1} \wedge dx^{i_2} \wedge \ldots)/\partial \phi$, which is what Kähler explicitly did. We refer readers to the source [46] (It is in German but, for this specific purpose, readers need only read the formulas).

Let us look at the problem from the perspective of invariant operators. Let y^n be the specific coordinate with respect to which we wish to differentiate while everything else remains constant. We write the invariant d as

$$d = \sum_{i=1}^{n-1} dy^i \frac{\partial}{\partial y^i} + dy^n \frac{\partial}{\partial y^n}. \tag{1.6}$$

The differentials dy are invariants since we have

$$y^i(B) - y^i(A) = \int_A^B dy^i, \tag{1.7}$$

regardless of whether we pull dy^i to another coordinate system to perform the integration. We still obtain $y^i(B) - y^i(A)$ because we are dealing with functions on the manifold. If we managed to choose coordinates such that

$$\sum_{i=1}^{n-1} dy^i \frac{\partial}{\partial y^i} \tag{1.8}$$

were an invariant, so would $\partial/\partial y^n$, as a consequence of the invariant character of d and of the difference

$$d - dy^n \frac{\partial}{\partial y^n}. \tag{1.9}$$

Kähler devised a coordinate system such that (1.8) would be zero [46].

What the modern literature calls a tangent vector field is, for Cartan as for Kähler, a Lie operator or Lie derivative. One needs a concept of vector field different from the modern one if one wants to stick to the letter and spirit of Cartan and Kähler. Incidentally, $d\xi^i \wedge e_i u$ is spin in the case of rotations, as Kähler showed [48]. It is a sophisticated theoretical development that involves, in addition, metric structure (in this case, but not required to simply obtain (1.4)). Be as it may, readers should ask themselves what is for rotations the last term of (1.4).

Cartan and Kähler never referred to (1.1) as vector field. For them, vector fields are passive objects, not differential operators; they do not act on anything. We are interested in their view because of its significance for relativistic quantum mechanics [48], [83] and presumably beyond (See last part of this book). Our problem then is to define a tangent vector as a passive operator. Such a concept is not new, but largely overlooked.

Because differentiations are of the essence in dealing with tangent vectors, we shall have to deal with parametrizations, while, at the same time, defining a parameter-independent concept. One has the intuitive notion of common tangency of curves at a point, through their tangency to the same straight line through the point. We now wish to formulate a concept of equivalence of curves that go in the same direction at a point and at the same rate with respect to a common parameter.

In order to have the same parametrization on all curves, we use one of the coordinates as parameter, the same one for all curves through the point. Any coordinate in a coordinate system is in principle a good parameter except occasionally. Thus, when dealing with circles on the plane, centered at the origin, we cannot use as parameter the radial coordinate ρ, since it does not change as we move on the circle.

Let x^i, for given i and coordinate patch U, be the chosen parameter. Call it y. Let ϕ_1 and ϕ_2 be smooth functions (from intervals I_1, I_2 on the real line to a manifold M) such that $\phi_1(y) = \phi_2(y)$ for the value of the coordinate y. We say that ϕ_1 and ϕ_2 have the same tangent at P if $(d/dy)(f_U \circ \phi_1)$ and $(d/dy)(f_U \circ \phi_2)$ take the same value at that point (Review the first paragraph of Chapter 2, section 8). Using the concept of differentiable manifold one shows that if this condition is satisfied in one coordinate patch, it also holds in any other coordinate patch.

Readers who may have had problems with the previous paragraph need simply think of the *tangent to a curve* at a point as a set of quantities $\partial x^i / \partial \lambda$, where λ is the parameter on the curve. The digression in a previous paragraph about the choice of a coordinate as the common parameter deals with the need to compare rates of change on different curves, and also on different parametrizations of the same curve.

One now defines a tangent vector at P as an equivalence of curves that have there the same tangent. This is closely related to the intuitive idea of identifying small pieces of the manifold with small pieces of affine space, except that we now do it through curves. A small piece of curve between its points $P(y)$ and $P(y + \Delta y)$ is closely represented by a set of quantities $\left[\dfrac{dx^i}{dy} \right]_P \Delta y$. There is, therefore, an obvious and thus natural physical interpretation of tangent vectors as such equivalences (read equivalence classes).

We next endow tangent vectors at any given point with a structure of vector space and build bases thereof. And, using the group $GL(n)$, we create a fiber at each point of the manifold. We have thus got a frame bundle very much like the frame bundle F^n. The difference is, however, that we do not have projection maps like we had in affine space because we cannot assign vectors to arbitrary pairs of points of the base manifold. Correspondingly, we shall not have, in general, rectilinear coordinates A^i.

To summarize: on an arbitrary $n-$dimensional differentiable manifold we introduce a bundle of tangent vector spaces, i.e. a tangent vector space at each point of the manifold. Pictorially, we divide the latter into small overlapping pieces, which we match with small pieces of affine spaces through the tangent vector spaces, the point of tangency playing the role of zero.

8.2 Affine connections and equations of structure

A more formal approach to the subject of this section will be given in the last one of this chapter. We shall explicitly indicate whether a subscript or a superscript comes first in order to later deal with issues of skew-symmetry.

By way of the bundle, we give in simple terms the integrated concepts of connections and equations of structure, formally introduced at the end of the chapter.

A differentiable manifold M of dimension n is said to be affinely connected if we associate with it another manifold endowed with $n + n^2$ linearly independent differential forms ω^i and $\omega_k^{\ j}$ with properties that guarantee that the system of "connection equations"

$$d\mathbf{P} = \omega^i \mathbf{e}_i, \qquad d\mathbf{e}_i = \omega_i^{\ j} \mathbf{e}_j \qquad (2.1)$$

will make small pieces of M look like small pieces of affine space of the same dimension. In the table at the end of the chapter we use notation that reflects that d in (2.1) does not represent differentiations of anything, as explained in its footnote.

The ω^i must be linearly independent combinations of the differential of the coordinates on M and only they. In other words, they must be horizontal. The coefficients in the linear combination will depend not only on the x's but also on the n^2 additional coordinates defining the fibers. As a consequence, the $d\omega^i$ will not be horizontal. The $\omega_i^{\ j}$, though not horizontal, must be such that, when pulled to the fibers, they become the $\omega_i^{\ j}$ of the linear group. "Pulling to the fibers" is the technical way of saying that we set the dx^i, equivalently the ω^i, to zero.

We further require that, when we compute $d\omega^i - \omega^j \wedge \omega_j^{\ i}$ and $d\omega_i^{\ j} - \omega_i^{\ k} \wedge \omega_k^{\ j}$, we obtain expressions which are quadratic expressions in the ω^j exclusively. We may then write

$$\Omega^i = d\omega^i - \omega^j \wedge \omega_j^{\ i}, \qquad \Omega_i^{\ j} = d\omega_i^{\ j} - \omega_i^{\ k} \wedge \omega_k^{\ j}, \qquad (2.2)$$

where

$$\Omega^i \equiv R^i_{\ kl}(\omega^k \wedge \omega^l) = \frac{1}{2} R^i_{\ kl} \omega^k \wedge \omega^l, \qquad (2.3a)$$

$$\Omega_i^{\ j} \equiv R_i^{\ j}_{\ kl}(\omega^k \wedge \omega^l) = \frac{1}{2} R_i^{\ j}_{\ kl} \omega^k \wedge \omega^l. \qquad (2.3b)$$

Notice that $d\omega^i$ and $d\omega_i{}^j$ must be non-horizontal, so that their non-horizontal contributions cancel out with the non-horizontal contributions of the last terms in Eqs. (2.2).

For simplicity, we take either $k < l$ or $l < k$ in $R^i{}_{kl}(\omega^k \wedge \omega^l)$ and $R_i{}^j{}_{kl}(\omega^k \wedge \omega^l)$. But we can mix these options, like $\omega^1 \wedge \omega^2$, $\omega^2 \wedge \omega^3$, $\omega^3 \wedge \omega^1$, provided that we do not use $\omega^k \wedge \omega^l$ and $\omega^l \wedge \omega^k$ at the same time. The $R^i{}_{kl}(\omega^k \wedge \omega^l)$ and $R_i{}^j{}_{kl}(\omega^k \wedge \omega^l)$ do not determine what the $R^i{}_{kl}$ and $R_i{}^j{}_{kl}$ in $\frac{1}{2}R^i{}_{kl}\omega^k \wedge \omega^l$ and $\frac{1}{2}R_i{}^j{}_{kl}\omega^k \wedge \omega^l$ are, since the factors $\omega^k \wedge \omega^l$ only pick their skew-symmetric part. Equivalently, one can write $\omega^k \wedge \omega^l$ in an infinite number of ways as a linear combination of itself with $\omega^l \wedge \omega^k$. It is standard to choose them so that

$$R^i{}_{kl} = -R^i{}_{lk}, \qquad R_i{}^j{}_{kl} = -R_i{}^j{}_{lk}. \tag{2.4}$$

Without the notational simplification (2.3), we have

$$d\omega^i - \omega^j \wedge \omega_j{}^i = \frac{1}{2}R^i{}_{kl}\omega^k \wedge \omega^l, \tag{2.5a}$$

$$d\omega_i{}^j - \omega_i{}^m \wedge \omega_m{}^j = \frac{1}{2}R_i{}^j{}_{kl}\omega^k \wedge \omega^l, \tag{2.5b}$$

which are known as equations of structure of affinely connected manifolds.

Differential forms are cumbersome, even in simple bundles. Thanks to the fact that the pull-back of the exterior differential and of products of differential forms are equal to the exterior differential and products of their pull-backs, we can work with pull-backs of ω^i and $\omega_k{}^j$ to the sections. However, the foregoing introduction of the concept of affine connection cannot be translated into sections since, for instance, the concept of horizontality does not make sense in them.

Assume that, computing the left hand sides of (2.5), we obtained zeroes:

$$0 = d\omega^i - \omega^j \wedge \omega_j{}^i, \qquad 0 = d\omega_i{}^j - \omega_i{}^k \wedge \omega_k{}^j. \tag{2.6}$$

A theorem called Frobenius theorem of integrability of exterior systems implies that (2.1) is then integrable. The result of the integration can of course be given the form

$$\mathbf{P} = \mathbf{Q} + A^i \mathbf{a}_i, \qquad \mathbf{e}_i = A_i{}^j \mathbf{a}_j, \tag{2.7}$$

which is one way to look at the affine frame bundle of Af^n, and to the action of the affine group. A^i and $A_l{}^m$ constitute a set of coordinates in this bundle.

The details for the integrability of the first equation (2.1) is not the same as for the second one, which is independent of the former. The $\mathbf{e}_i$ certainly exist in the tangent vector spaces. But the integrability issue concerns whether the different vector spaces can be identified. It is only then that it makes sense to ask oneself for the existence of $\mathbf{P}$, which, if existing, would live in the vector space resulting from that identification. Hence, the integrability of the first of (2.1) is conditional to the integrability of the second, but not the other way around.

We formally differentiate the system (2.1) to obtain

$$d(d\mathbf{P}) = (d\omega^i - \omega^j \wedge \omega_j{}^i)\mathbf{e}_i \tag{2.8a}$$

$$d(d\mathbf{e}_i) = (d\omega_i{}^j - \omega_i{}^k \wedge \omega_k{}^j)\mathbf{e}_j. \tag{2.8b}$$

The left hand sides of (2.2) are short hand expressions for the right hand sides. Equations (2.5) can now be given the compact form

$$d(d\mathbf{P}) = \Omega^i \mathbf{e}_i \tag{2.9a}$$

$$d(d\mathbf{e})_i = \Omega_i{}^j \mathbf{e}_j. \tag{2.9b}$$

Once this is understood, one may use the term torsion to refer to any of the left and right hand sides of (2.8a) and (2.9a). Similarly, one may use the term affine curvature to refer to any of the left and right hand sides of (2.8b) and (2.9b).

We proceed to repeat the formulas of section 5.6 for the components of the curvature, since they will guide us when making similar important considerations for the less well understood concept of torsion. We momentarily introduce a small change of notational convention to make the point that, exceptionally, there will not be duality between the different bases of tangent tensors, on the one hand, and bases of differential forms, on the other.

On sections, we express $\omega_i{}^j$ as $\Gamma_i{}^j{}_{k'} dx^k$ rather than as $\Gamma_i{}^j{}_k \omega^k$ in order to simplify differentiations, since ddx^k is zero but $d\omega^k$ is not. We primed the index k rather than the main line character Γ to make explicit that $_i{}^j$ makes reference to a, in general, non-holonomic field of vector bases, and $_{k'}$ belongs to a holonomic or coordinate basis of differential $1-$forms. We then have:

$$d\omega_i{}^j = d(\Gamma_i{}^j{}_{k'} dx^k) = (\Gamma_i{}^j{}_{k',l'} - \Gamma_i{}^j{}_{l',k'})(dx^l \wedge dx^k), \tag{2.10a}$$

$$-\omega_i{}^m \wedge \omega_m{}^j = (\Gamma_i{}^m{}_{k'} \Gamma_m{}^j{}_{l'} - \Gamma_i{}^m{}_{l'} \Gamma_m{}^j{}_{k'})(dx^l \wedge dx^k), \tag{2.10b}$$

$$\Omega_i{}^j = (\Gamma_i{}^j{}_{k',l'} - \Gamma_i{}^j{}_{l',k'} + \Gamma_i{}^m{}_{k'} \Gamma_m{}^j{}_{l'} - \Gamma_i{}^m{}_{l'} \Gamma_m{}^j{}_{k'})(dx^l \wedge dx^k). \tag{2.10c}$$

Defining $R'^j_{i\,lk}$ as

$$R_i{}^j{}_{l'k'} \equiv \frac{1}{2}(\Gamma_i{}^j{}_{k',l'} - \Gamma_i{}^j{}_{l',k'} + \Gamma_i{}^m{}_{k'} \Gamma_m{}^j{}_{l'} - \Gamma_i{}^m{}_{l'} \Gamma_m{}^j{}_{k'}), \tag{2.11}$$

we can write

$$\Omega_j{}^i = R_j{}^i{}_{l'm'} dx^l \wedge dx^m. \tag{2.12}$$

It is also common to define

$$R'^j_i{}_{lk} = \Gamma_i{}^j{}_{k',l'} - \Gamma_i{}^j{}_{l',k'} + \Gamma_i{}^m{}_{k'} \Gamma_m{}^j{}_{l'} - \Gamma_i{}^m{}_{l'} \Gamma_m{}^j{}_{k'}. \tag{2.13}$$

Then

$$\Omega_j{}^i = \frac{1}{2} R_j{}^i{}_{l'm'} dx^l \wedge dx^m. \tag{2.14}$$

At this point, it is advisable to return to an expansion in terms of $\omega^l \wedge \omega^m$, so that there will be correspondence (i.e. duality) between bases in the algebras of differential forms and of valuedness. Substitution of the dx^i in terms of the ω^j in

$$\frac{1}{2}R_j{}^i{}_{l'm'}dx^l \wedge dx^m = \frac{1}{2}R_j{}^i{}_{lm}\omega^l \wedge \omega^m \qquad (2.15)$$

allows one to relate coefficients. Here the change of the components of the curvature is bilinear (concerns only the indices l and m) rather than quadrilinear. The reason is that, when in order to simplify differentiations we express $\omega_i{}^j$ of (2.9) as $\Gamma_i{}^j{}_k dx^k$ rather than as $\Gamma_i{}^j{}_k \omega^k$, we change bases of differential forms but not bases of tangent vectors and tensors. From this point on, the bases of valuedness and the bases of differential forms will correspond to each other.

Consider now the torsion. In terms of coordinate basis fields, Ω^i equals $-dx^j \wedge \omega_j{}^i$. The equation

$$R^i_{jk} = \Gamma_k{}^i{}_j - \Gamma_j{}^i{}_k \qquad (2.16)$$

follows. All indices correspond here to holonomic bases. This has given rise to the statement that the torsion is the antisymmetric (what we have called skew-symmetric) part of the connection. That is a generally incorrect statement since it does not concern arbitrary bases of differential forms and frame fields. When they are arbitrary, one has

$$R^i_{jk}(\omega^j \wedge \omega^k) = (M^i_{jk} + \Gamma_k{}^i{}_j - \Gamma_j{}^i{}_k)(\omega^j \wedge \omega^k), \qquad (2.17)$$

where $M^i_{jk}(\omega^j \wedge \omega^k)$ stands for $d\omega^i$. Equation (2.16) does not follow in general.

Notice the different positioning of the indices j and k in R relative to Γ. The skew-symmetry of R^i_{jk} in (2.16), in addition to pertaining to coordinate basis fields, concerns the first and third indices in Γ. On the other hand, the skew-symmetry $\omega_{ij} + \omega_{ji} = 0$ discussed in chapter 6 (Eq. (1.5)) pertains to orthonormal basis fields and concerns the first and second indices in the gammas. Only on Euclidean spaces we have frame fields where the skew-symmetry $0 = \Gamma_{kij} - \Gamma_{jik}$ (zero torsion) and the symmetry $\Gamma_{ijk} + \Gamma_{ikj} = 0$ (metric compatibility) coexist. In generalizations of Euclidean spaces known as Riemannian spaces, both equations can be the case, but not at the same time. The first one requires holonomic bases. We must then use Eq. (2.11) of that chapter,

$$g_{ij,m} - \Gamma_i{}^l{}_m g_{lj} - \Gamma_j{}^l{}_m g_{li} = 0, \qquad (2.18)$$

which implies

$$\Gamma_{ijk} + \Gamma_{ikj} = g_{ij,k} \qquad (2.19)$$

rather than $\Gamma_{ijk} + \Gamma_{ikj} = 0$. In later chapters, readers will be in a better position to understand all this.

8.3   Tensoriality issues and second differentiations

In this section, we prepare the ground for the study of second differentiations. We first consider tensoriality issues related to $d(d\mathbf{e}_j)$. Unlike $d\mathbf{e}_j$, it transforms linearly under a change of frame field. It does so in such a way that its contraction with the components v^j of a vector field is invariant.

One sometimes finds in the literature the statement that a tensor transforms in such and such way under a coordinate transformation. Tensors do not transform. They are invariants. Their components do, under changes of section of the frame bundle. The changes of section are elements of the group G_0, different elements in general at different points in each section. G_0 acts p+q+r times in the change of basis of (p, q)−valued differential r−forms, unless we change only the basis of valuedness or only the basis of differential forms. When, in some of the literature, scalar functions γ_i^j are said to be tensor-valued, authors have in mind $\gamma_i^j \phi^i \mathbf{e}_j$. They fail to exhibit the basis $\phi^i \mathbf{e}_j$ for (1,1)-tensor valuedness.

In an equation such as (5.1) of chapter 3 for the transformation of components of a vector field, only coordinates *on the base manifold* show up. This does not conflict with what we have just said about transformations which are members of G_0, which thus depend on the coordinates *in the fibers*. At any given point, the parameters $A_{i'}^j$ of the element G_0 is the matrix whose components are the $\partial x'^j / \partial x^i$ evaluated at that point. They bring up the coordinates in the section by pullback.

We saw in the previous section —though it must have been obvious even before— that the tensorial transformation of components on pull-backs to sections reflects horizontality in the bundle. But horizontality concerns the scalar-valued differential form that multiplies a tensor to form a tensor-valued differential form. The question is: how does one reconcile the fact that, whereas tensoriality involves all indices in the components, horizontality only involves the differential form indices among them?

An invariant of the type $\gamma_i^j \phi^i \mathbf{e}_j$ is not so due to "its looks", since $\omega_i^j \phi^i \mathbf{e}_j$ has the same looks but is not invariant from section to section. Assume that the γ_i^j and ω_i^j were differential 1−forms. In the bundle, where we cannot set the $A_{i'}^j$ to be functions of x, it is the absence of the differentials of the $A_{i'}^j$ that defines horizontality. The difference between the tensorial transformation of γ_i^j and the non-tensorial one of ω_i^j lies exclusively in that the first one is horizontal and the second one is not. It has noting to do with the i and j indices. The linear transformation properties for valuedness indices is guaranteed by construction, namely by the action of the group on the fibers for both γ_i^j and ω_i^j. That is why horizontality guarantees tensoriality, i.e. linear transformation in connection with all explicit (i.e. valuedness) and implicit (i.e. differential form) indices.

Let $\mathbf{e}_i$ and $\mathbf{e}_i'$ be vector bases at the intersection of two given sections with the fiber at x. We have

$$\mathbf{e}_i' = A_{i'}^j(x)\mathbf{e}_j. \qquad (3.1)$$

Formal differentiation yields

$$d\mathbf{e}_i' = dA_{i'}^j \mathbf{e}_j + A_{i'}^j d\mathbf{e}_j. \qquad (3.2)$$

The first term on the right hand side of (3.2) is not in general a linear combination of the $d\mathbf{e}_j$. Since the second term is, their sum and, therefore, the left hand side of the equation are not. Thus the de_i's do not transform linearly. But, if we differentiate (3.2), we get

$$d(de_i') = A_{i'}^j d(de_j), \qquad (3.3)$$

since the first of the four terms resulting from differentiating the right hand side of (3.2) is zero and the next two cancel each other out. This equation shows that $d(de_i)$ transforms *in the fibers* like the basis itself, so that contraction with the components v^i is the same in all sections. Once again, this is not the case with $v^j de_j$. Comparison of Eqs. (3.2) and (3.3) exhibits the contrast between the respective non-tensorial and tensorial character of the subscript i in the two cases.

When one defines de_i —which is what we did in the previous section— we still have, as in affine-Klein geometry,

$$\begin{aligned} d\mathbf{v} &= \mathbf{v}(P + dP) - \mathbf{v}(P) \\ &= v^i(P + dP)\mathbf{e}_i(P + dP) - v^i(P)\mathbf{e}_i(P) \\ &= dv^i\mathbf{e}_i + v^i d\mathbf{e}_i. \end{aligned} \qquad (3.4)$$

With the notation $P + dP$ we mean that, when evaluating $d\mathbf{v}$, the coordinate functions are $x^i + dx^i$. All the formulas of chapter 5 apply here. We have in particular that

$$dv^i\mathbf{e}_i + v^i d\mathbf{e}_i = dv'^i\mathbf{e}_i' + v'^i d\mathbf{e}_i'. \qquad (3.5)$$

But we did not have equalities $dv^i\mathbf{e}_i = dv'^i\mathbf{e}_i'$ and $v^i d\mathbf{e}_i = v'^i d\mathbf{e}_i'$ then, and we do not have them now.

In search for deep understanding, we proceed to formally differentiate twice the field of bases of vector-valued $1-$forms, $\phi^i\mathbf{e}_j$, so as to then differentiate $(1,1)$-valued differential $0-$forms, $\gamma_i^j\phi^i\mathbf{e}_j$. As a first step, we find $d(d\phi^i)$:

$$d(d\phi^i) = d(-\omega_j^i\phi^j) = -(d\omega_j^i)\phi^j + \omega_k^i \wedge d\phi^k = -\Omega_j^i\phi^j. \qquad (3.6)$$

When we differentiate $\phi^i\mathbf{e}_j$ twice, two of the terms resulting from the second differentiation cancel each other out and we obtain

$$d[d(\phi^i\mathbf{e}_j)] = [d(d\phi^i)]\mathbf{e}_j + \phi^i dd\mathbf{e}_j. \qquad (3.7)$$

Therefore, using (2.10b) and (3.6) in (3.7), we further get

$$d[d(\phi^i\mathbf{e}_j)] = -\Omega_k^i\phi^k\mathbf{e}_j + \Omega_j^k\phi^i\mathbf{e}_k. \qquad (3.8)$$

The second differential of $\gamma_j^i\phi^j\mathbf{e}_i$ is now obvious. We abbreviate γ_j^i as γ and $\phi^j\mathbf{e}_i$ as $\mathbf{f}$. Thus:

$$d(\gamma\mathbf{f}) = (d\gamma)\mathbf{f} + \gamma d\mathbf{f}. \qquad (3.9)$$

Further differentiation yields

$$dd(\gamma\mathbf{f}) = -d\gamma \wedge d\mathbf{f} + d\gamma \wedge d\mathbf{f} + \gamma d[d(\mathbf{f})], \qquad (3.10)$$

where we have used $dd\gamma = 0$. Using now (3.8), we finally get

$$dd[\gamma_i^j(\phi^i \mathbf{e}_j)] = \gamma_i^j(\Omega_j^k \phi^i \mathbf{e}_k - \Omega_k^i \phi^k \mathbf{e}_j). \tag{3.11}$$

We leave it for interested readers the minor modifications required in the last three equations when the γ_i^j's are not $0-$forms. No new terms appear, but changes in sign may emerge in the process of exterior differentiation.

8.4 Tangent developments and annulment of connection at a point

The connection equation for $\mathbf{e}_i$ can be written as

$$\mathbf{e}_i(P + dP) = \mathbf{e}_i(P) + \omega_i^j(P)\mathbf{e}_j(P), \tag{4.1}$$

to be integrated on tiny curves. A small vector $\mathbf{\Delta P}$ is thus assigned to the end points of the curve. We reverse the roles of the initial and end points of the curve and denote $\mathbf{e}_i$ as $\mathbf{a}$. We get

$$\mathbf{a}(P) = \mathbf{e}_i(P + dP) - \omega_i^j(P + dP)\mathbf{e}_j(P + dP). \tag{4.2}$$

The minus sign reflects change of orientation when evaluating (4.1) (i.e. integrating ω_i^j) on parametrized tiny curves. We thus rewrite (4.2) as

$$\mathbf{a}(P) = a^j(P + \Delta P)\mathbf{e}_j(P + \Delta P), \tag{4.3}$$

where

$$a^i = 1 - \omega_i^i, \qquad \text{(no sum)} \tag{4.4a}$$

$$a^j = -\omega_i^j, \qquad j \neq i, \tag{4.4b}$$

ω_i^j exceptionally representing here the actual evaluation of ω_i^j in the tiny curve. We have thus expressed the vector $\mathbf{a}$ at P in terms of a basis at $P + dP$.

We shall now remove the restriction of the curve being tiny. On parametrizations of curves, all partial derivatives become derivatives with respect to the parameter, λ. We take a basis $(\mathbf{e}_i)$ at $P(\lambda = 0)$ as "initial condition" for the integration along a curve of the system of differential equations

$$d\mathbf{e}_i(\lambda) = \omega_i^j(\lambda)\mathbf{e}_j(\lambda). \tag{4.5}$$

For clarity, we shall denote these initial values of $(\mathbf{e}_i)$ as $(\mathbf{a}_i)$. The result of integrating (4.5) is

$$\mathbf{e}_i(\lambda) = C_i^j(\lambda)\mathbf{a}_j. \tag{4.6}$$

If we denote the matrix $[C_i^j]$ as C and solve for $\mathbf{a}_j$, we get, with some abuse of notation,

$$\{\mathbf{a}_j\} = \{\mathbf{e}_j(0)\} = C^{-1}\{\mathbf{e}(\lambda)\}. \tag{4.7}$$

Equation (4.7) shows how the basis at $P(0)$ is expressed in terms of the basis at $P(\lambda)$. The matrix C^{-1} depends on connection and curve between points $P(0)$ and $P(\lambda)$. The dependence on curve means that we do not have here a relation of equivalence; the transitive property is not satisfied.

We shall say that a vector $v^i(\lambda)\mathbf{e}_i(\lambda)$ at $P(\lambda)$ is the parallel transported vector $v^i\mathbf{a}_i$ along a curve γ from $P(0)$ to $P(\lambda)$ if

$$v^i(\lambda)\mathbf{e}_i(\lambda) = v^i\mathbf{a}_i. \tag{4.8}$$

The components $v^i(\lambda)$ can be obtained by substitution of (4.7) in (4.8) and equating coefficients on both sides. $v^i(\lambda)\mathbf{e}_i(\lambda)$ is a constant vector field on the curve.

We now explicitly show that ω_i^j can be made to vanish at any given point by appropriate choice of basis field. We rewrite (4.1) as

$$\mathbf{e}_i(x + dx) = \mathbf{e}_i(x) + \omega_i^j(x)\mathbf{e}_j(x). \tag{4.9}$$

On the other hand, rewriting x as $(x + dx) - dx$, we have

$$\mathbf{e}_i(x) = \mathbf{e}_i(x + dx) - \omega_i^j(x + dx)\mathbf{e}_j(x + dx). \tag{4.10}$$

From these equations, we get

$$\omega_i^j(x)\mathbf{e}_j(x) = \omega_i^j(x + dx)\mathbf{e}_j(x + dx), \tag{4.11}$$

and, therefore, (4.9) can further be written as

$$\mathbf{e}_i(x + dx) - \omega_i^j(x + dx)\mathbf{e}_j(x + dx) = \mathbf{e}_j(x). \tag{4.12}$$

Defining the left and right hand sides of (4.12) respectively as $\mathbf{e}_j'(x + dx)$ and $\mathbf{e}_j'(x)$, (4.12) implies

$$\mathbf{e}_j'(x + dx) = \mathbf{e}_j'(x), \tag{4.13}$$

or, in other words,

$$\omega_i'^j(x) = 0. \tag{4.14}$$

8.5   Interpretation of the affine curvature

We proceed to give a geometric interpretation of the curvature and of $d(d\mathbf{e}_i)$. Let P, M, M' and P' be the four corners of a small curvilinear quadrilateral on the manifold (Figure 6). Since the result to be obtained is tensorial, it will be valid in any section.

Figure 6 has a mnemonic purpose. It is not a development in the tangent plane, which does not close in general. Nor does it belong to the manifold since d_1P and d_2P, belong to tangent planes (Some would find more appropriate to write d_1P and d_1P as if to signify that we are not dealing with changes of a vector but with vector associated with pairs of points in the manifold). We have closed the rectangle in order to make clear that we are getting to the same point of the manifold, P', in two different ways.

We write the parallel-transported $\mathbf{e}_i(P)$ from P to M as $(\mathbf{e}_i)_{P \to M}$. The equations that follow are then obvious.

$$(\mathbf{e}_i)_{P \to M} = \mathbf{e}_i(P) + d_1 \mathbf{e}_i(P), \tag{5.1}$$

$$(\mathbf{e}_i)_{P \to M \to P'} = (\mathbf{e}_i)_{P \to M} + d_2[(\mathbf{e}_i)_{P \to M}], \tag{5.2}$$

$$(\mathbf{e}_i)_{P \to M \to P'} = \mathbf{e}_i(P) + d_1 \mathbf{e}_i(P) + d_2 \mathbf{e}_i(P) + d_2 d_1 \mathbf{e}_i(P). \tag{5.3}$$

where $d_1 \mathbf{e}_i$ and $d_2 \mathbf{e}_i$ are not differentials but evaluations on two tiny curves.

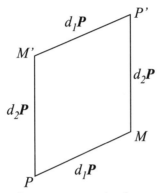

Figure 6: Assignment of segments, not development in Euclidean space

Parallel transport along the other path, $P \to M' \to P'$, yields

$$(\mathbf{e}_i)_{P \to M' \to P'} = \mathbf{e}_i(P) + d_1 \mathbf{e}_i(P) + d_2 \mathbf{e}_i(P) + d_1 d_2 \mathbf{e}_i(P), \tag{5.4}$$

We subtract (5.4) from (5.3) and obtain

$$(\mathbf{e}_i)_{P \to M \to P'} - (\mathbf{e}_i)_{P \to M' \to P'} = d_2 d_1 \mathbf{e}_i(P) - d_1 d_2 \mathbf{e}_i(P). \tag{5.5}$$

The expansion of $d_2 d_1 \mathbf{e}_i$, which belongs to the path $P \to M \to P'$, is

$$\begin{aligned}
d_2 d_1 \mathbf{e}_i &= d_2[\omega_i^j(1)\mathbf{e}_j] = dx^k(1) d_2(\Gamma_{ik}^j \mathbf{e}_j) \\
&= dx^k(1) dx^l(2) \Gamma_{ik,l}^j \mathbf{e}_j + dx^k(1) dx^l(2) \Gamma_{ik}^m \Gamma_{ml}^j \mathbf{e}_j.
\end{aligned} \tag{5.6}$$

Clearly the symbols $dx^k(1)$ and $dx^k(2)$ are increments in the coordinates. For the other path, we interchange the numbers 1 and 2 in (5.6), exchange the dummy indices k and l and get

$$d_1 d_2 \mathbf{e}_i = dx^k(1) dx^l(2) \Gamma_{il,k}^j \mathbf{e}_j + dx^k(1) dx^l(2) \Gamma_{il}^m \Gamma_{mk}^j \mathbf{e}_j, \tag{5.7}$$

We subtract (5.7) from (5.6) and, to facilitate later developments, we write $dx^k(1)$ and $dx^k(2)$ as $dx^k(g)$ and $dx^k(n)$ respectively. We thus get

$$d_2 d_1 \mathbf{e}_i - d_1 d_2 \mathbf{e}_i = dx^k(g) dx^l(n) [\Gamma^j_{ik,l} - \Gamma^j_{il,k} + \Gamma^m_{ik} \Gamma^j_{ml} - \Gamma^m_{il} \Gamma^j_{mk}] \mathbf{e}_j. \qquad (5.8)$$

The contents of the square bracket is skew-symmetric in k and l. If the product $dx^k(g) dx^l(n)$ were symmetric, the right hand side of (5.8) would vanish. But it is not. Thus, for instance, $dx^1(g) dx^2(n)$ is not the same as $dx^2(g) dx^1(n)$.

Recall that dx, dy, ... are not small increments but functions of curves. On the other hand, $dx^k(g)$ and $dx^l(n)$ represent the result of evaluations (read integrations) of dx^k and dx^l on small curves. Readers may think of $dx^k(g)$ and $dx^l(n)$ as, for example, Δx and Δy. And they may prefer to use the notation $\Delta_2 \Delta_1 \mathbf{e}_i$ instead of $d_2 d_1 \mathbf{e}_i$. All depends on what one wishes to emphasize. We replace small quantities with differentials for obtaining a differential equation. We thus state the result of our analysis so far as

$$(\mathbf{e}_i)_{P-M-P'} - (\mathbf{e}_i)_{P-M'-P'} \rightarrow dd\mathbf{e}_i = \Omega^j_i \mathbf{e}_j. \qquad (5.9)$$

Notice the ordering on the left hand side of (5.9). Putting together $P - M - P'$ and, in reverse order, $P - M' - P'$, we get $P - M - P' - M - P$, which amounts to returning to the same point following a closed curve with anti-clockwise orientation. Equation (5.9) yields the geometric interpretation of the affine curvature, which can also be read from

$$d_2 d_1 \mathbf{v} - d_1 d_2 \mathbf{v} \rightarrow v^i d(d\mathbf{e}_i) = v^i \Omega^j_i \mathbf{e}_j = dd\mathbf{v}. \qquad (5.10)$$

The difference between a vector and the same vector after we take it along a closed curve is obtained by integration of $v^i \Omega^j_i \mathbf{e}_j$ on the small surface enclosed by a tiny closed curve on the manifold. The final step in the interpretation of the curvature takes us back to section 7 of chapter 5, where one connects Eq. (7.7), which is another way of writing (5.10), with $\mho$ defined as

$$\mho \equiv \Omega^j_i \phi^i \mathbf{e}_j. \qquad (5.11)$$

(We lack a bold faced $\mho$ symbol).

8.6 The curvature tensor field associated with the curvature differential form

When we transport a vector from one point to a nearby point along two different paths, we get in general two different vectors at the end point. The affine curvature allows us to expediently compute the difference between the two. For a future application, we shall specialize this problem to when the two paths make a curvilinear quadrilateral constituted by the intersection of curves of two different continuous families of curves. We can approximate the integrals by

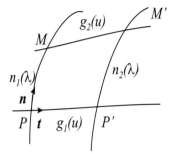

Figure 7: Notation for pairs of intersecting curves

replacing the sides of small curvilinear quadrilaterals with segments tangent to
the curves involved. In doing so, the component of the tensor-valued differential
2−form curvature become the components of a (1,3)-tensor.

In Figure 7, we denote as g and n the members of the two intersecting families
of curves. In the next section, one of the families will be made of autoparallels
(lines of constant direction). In section 3 of chapter 10, they will be geodesic
curves (i.e. curves of stationary distance). We move a vector from one point
to another along a g curve and continue by moving it on an n curve to a "final
point". We also move between the same points, first on an n curve and then
on a g curve (Notice the slight change of the labelling of the points in figure
7 relative to what it was in figure 6). For simplicity we use coordinate bases.
Evaluations on the curves of Fig. 7 under the assumption that they are very
small yields

$$d_2 d_1 \mathbf{v} - d_1 d_2 \mathbf{v} = v^i R_i{}^j{}_{kl} dx^k(g) dx^l(n) \mathbf{e}_j, \qquad (6.1)$$

where $dx^k(g)$ and $dx^l(n)$ are increments, not differentials.

Notice that, at this point, we are not dealing with the curvature differential
form, since we have already performed its "integration" on the surface delimited
by the small pieces of curves. The $R_i{}^j{}_{kl}$ are the components of a (1,3) tensor field
$\phi^i R_i{}^j{}_{kl} \phi^k \phi^l \mathbf{e}_j$ that has just been evaluated on the triple of vectors ($\mathbf{v}$, $dx^k(g)\mathbf{e}_k$,
$dx^l(n)\mathbf{e}_l$) tangent to the manifold at P.

The $dx^k(g)$ and $dx^l(n)$ are $t^m \Delta u$ and $n^m \Delta \lambda$ on the curves g and n respectively.
The tangent vectors $\mathbf{t}$ and $\mathbf{n}$ are indicated in the figure. We thus have

$$d_u d_\lambda \mathbf{v} - d_\lambda d_u \mathbf{v} = v^i R_i{}^j{}_{kl} t^k n^l \Delta u \, \Delta \lambda \, \mathbf{e}_j. \qquad (6.2)$$

Equation (6.2) lends itself to be rewritten as

$$\lim_{\Delta u, \, \Delta \lambda \to 0} \frac{d_u d_\lambda \mathbf{v} - d_\lambda d_u \mathbf{v}}{\Delta u \Delta \lambda} = v^i R_i{}^j{}_{kl} t^k n^l \mathbf{e}_j, \qquad (6.3)$$

Again, the $R_i{}^j{}_{kl}$ have to be viewed as the components of a (1,3)-tensor $R_i{}^j{}_{kl} \phi^i \otimes$
$\mathbf{e}_j \otimes \phi^k \otimes \phi^l$. The right hand side of (6.3) represents the evaluation of this tensor
on the triple of vectors ($\mathbf{v}$, $\mathbf{t}$, $\mathbf{n}$).

8.7 Autoparallels

An *autoparallel* is a curve where the tangent vector $d\mathbf{P}/d\lambda$ defined by a given parametrization remains equal to itself up to a factor, which may vary from point to point and from one parametrization to another.

For a given autoparallel curve and parameter u', we define $\alpha(u')$ by

$$d\mathbf{t}' \equiv \alpha(u')\mathbf{t}'du'. \tag{7.1}$$

We reserve the symbol u for parametrizations yielding

$$d\mathbf{t} = 0. \tag{7.2}$$

Since $d\mathbf{P}$ on the curve can be written as $\mathbf{t}'du'$, but also as $\mathbf{t}du$, we have

$$\mathbf{t}'du' = \mathbf{t}du. \tag{7.3}$$

Equation (7.3) implies that

$$\beta(u')\mathbf{t} = \mathbf{t}', \tag{7.4}$$

where $\beta = du/du'$. Differentiating (7.4) and taking into account (7.1)-(7.3), one obtains

$$d\beta\mathbf{t} + \mathbf{0} = \alpha\mathbf{t}'du' = \alpha\beta du'\mathbf{t}. \tag{7.5}$$

A first integration yields

$$\beta = C_1 e^{\int \alpha du'} = \frac{du}{du'}, \tag{7.6}$$

and, integrating again,

$$u = \int C_1 e^{\int \alpha du'}du' + C_2, \tag{7.7}$$

which shows how to obtain affine parameters (meaning those for which $dt/du = 0$) from non-affine ones.

It is clear that all affine parameters are given in terms of one of them by

$$u' = c_1 u + c_2, \tag{7.8}$$

as follows from (7.7) with α equal to zero.

We shall change parameters as needed so that autoparallels satisfy $d\mathbf{t} = 0$. Let $\mathbf{v}$ be the tangent vector $\mathbf{t}$ to a family of autoparallels. We then have

$$\lim_{du,d\lambda\to 0} \frac{d_u d_\lambda \mathbf{t}}{du\,d\lambda} = t^i R^j_{ikl} t^k n^l \mathbf{e}_j, \tag{7.9}$$

having set $d_u \mathbf{t} = 0$ because $d\mathbf{t} = 0$. We then have

$$\frac{\partial^2 \mathbf{t}}{\partial u \partial \lambda} = t^k R^j_{kil} t^i n^l \mathbf{e}_j. \tag{7.10}$$

In a later application of this equation, we shall get a better feeling for expressions such as the one on the left hand side of (7.10).

8.8 Bianchi identities

Exterior differentiation of the equations of structure yields

$$d\Omega^i = -d\omega^j \wedge \omega^i + \omega^j \wedge d\omega^i_j, \tag{8.1a}$$

$$d\Omega^j_i = -d\omega^k_i \wedge \omega^j_k + \omega^k_i \wedge d\omega^j_k. \tag{8.1b}$$

The differentials of the ω's are obtained from Eqs. (2.3) and replaced in (8.1) to yield

$$d\Omega^i = -\Omega^j \wedge \omega^i_j + \omega^j \wedge \Omega^i_j, \tag{8.2a}$$

$$d\Omega^j_i = -\Omega^k_i \wedge \omega^j_k + \omega^k_i \wedge \Omega^j_k. \tag{8.2b}$$

If you did some algebra to get from (8.1) to (8.2), you made an unnecessary effort. Just replace $d\omega^j$ and $d\omega^k_i$ with Ω^j and Ω^i_k in Eqs. (8.1). This is more than just a mnemonic rule: we can ignore all terms that do not involve torsion and curvature since they cancel out among themselves when both are zero. This is so because the Bianchi identities of affine space read $0 = 0$.

Equations (8.2) are known as Bianchi identities. More often than not, one deals with their pull-back to sections of the frame bundle, where they receive the same name. They are integrability conditions for the system of equations of structure, so that invariant connection forms on the manifold exist.

Cartan refers to (8.2a) and (8.2b) as the laws of conservation of torsion and curvature respectively. Recall from the last section of chapter 4 how he dealt with conservation. The starting point is a differential form whose exterior derivative is zero. Different daughter conservation laws then follow depending on grade r of the differential form, type of r-surface that one chooses as domain of integration, dimension of the manifold and signature of the metric. The question then is: if $d\alpha = 0$ for, say, a vector-valued differential form α, what is a corresponding statement of conservation? If there is a general rule for such differential forms, the first Bianchi identity, (8.2a), should be a particular case of conservation of vector-valued differential forms.

We introduce the compact notation

$$\Omega \equiv \Omega^i \mathbf{e}_i, \qquad \mho \equiv \Omega^j_i \boldsymbol{\phi}^i \mathbf{e}_j. \tag{8.3}$$

The Bianchi identities can then be written as

$$d\Omega = \omega^j \wedge \Omega^i_j \mathbf{e}_i, \qquad d\mho \equiv \mathbf{0}. \tag{8.4}$$

These are equations rather than identities if Ω and $\mho$ are explicitly given differential forms. And they are identities if we instead replace them with their definitions in terms of the fundamental invariants of the differentiable manifold and their derivatives.

8.9 Integrability and interpretation of the torsion

Torsion is a vector-valued differential 2−form of components:

$$\Omega^j \equiv \frac{1}{2} R^i_{kj} \omega^j \wedge \omega^k = R^i_{kj} (\omega^j \wedge \omega^k). \tag{9.1}$$

$\Omega^i \mathbf{e}_i$ is the vector-valued differential form $d(d\mathbf{P})$, i.e. $d(\omega^i \mathbf{e}_i)$. Remember that $d\mathbf{P}$ is not the result of differentiating some vector valued function in general; both $d\mathbf{P}$ and $d\mathbf{e}_i$ are either given explicitly or a solution of some system of differential equations, say, the equations of structure. The option used in affine space of differentiating $\mathbf{e}_i = A^j_i(x)\mathbf{a}_j$ and replacing $\mathbf{a}_j$ in terms of $\mathbf{e}_j$ in order to obtain $d\mathbf{e}_i = \omega^j_i \mathbf{e}_j$ is not available on general differentiable manifolds; one such expression for $\mathbf{e}_i$ in terms of $\mathbf{a}_j$ at some point does not exist unless the manifold is endowed with equality of vectors at a distance, condition called teleparallelism (TP). The affine curvature would have to be zero over regions of M for that option to apply.

Recall that affine curvature can be viewed as being about transporting a vector around a closed curve and returning to the point of departure with a different vector. Torsion, on the other hand, is about representing in affine space small curves of general affinely connected differentiable manifolds. That the paths have to be very small for integration of vector-valued integrals to mean anything at all is seldom if ever considered in books on differential geometry, more interested in formal aspects than on geometric interpretations. Or they have it wrong for failing to refer to very small paths.

Élie Cartan, author of the theory of affine connections, makes emphasis time and again on smallness of curves if the affine curvature is not zero. The results obtained depend on what path is chosen to bring to the same vector space the contributions by small pieces of surface, contributions whose sum constitutes (in the limit) the vector-valued surface integral. Recall that those "tiny terms" live in different tangent spaces.

Because of its relevance, we quote from a seminal paper by Cartan of 1923 [11] where he presents his theory of affine connections. He wrote (minor changes to more modern notation have been introduced): "One finally arrives, *for an infinitely small contour*, to the formula

$$\int d\mathbf{P} = \iint (d\omega^i - \omega^j \wedge \omega^i_j)\mathbf{e}_i, \tag{9.2}$$

identical to the formula found for a proper affine space. The vectors $\mathbf{e}_i$ on the right hand side are here relative to any point, provided that it be infinitely close to the contour ..." (emphasis in original). The very last statement is obviously making reference to the need to be using just one vector space. If and when that is the case, the vector basis can be taken out of the integral, which justifies the use of Stokes theorem when dealing with vector-valued differential forms.

Equation (9.2) assigns a vector to a path, vector which is zero in affine space but not in general. In the case of $PP'MM'$ of Figure 7, its representation on affine space is a set of four vectors, one after another, to make a "quasi-quadrilateral". The integral then represents the tiny vector joining the two ends of the quasi quadrilateral.

Cartan makes the same consideration about *infinitely small contours* when, in the same section, he reaches an equation that parallels (9.2) but involves the curvature.

Finally, when in the same spirit Cartan explains the geometric significance of the Bianchi identities, he writes: "*The geometric sum of the infinitely small displacements associated with the different elements of a closed surface is zero when the closed surface is* **infinitely small**." ([11], full test in italics in original, but bold face has been added). Notice Cartan's insistence on smallness of integration domains.

To summarize, affine torsion is about representing on affine space itself (i.e. on the corresponding Klein space) tiny closed curves of an affinely connected manifold and "quantifying" its failure to close in the form of a vector, which one obtains through integration of the torsion. Let it not be forgotten that the representation of closed curves fails to close even if the torsion is zero but the affine curvature is not. But the reason for the failure to close is not the same one when the torsion is zero as when the affine curvature is zero. In general, both torsion and curvature will contribute to the failure to close.

8.10 Tensor-valuedness and the conservation law

In this section we speak of the limitations of the conservation law in dealing with tensor-valued differential forms if the curvature of the affine connection does not vanish.

In tensor calculus, the covariant derivative is the all pervading derivative. In calculus of differential forms, it emerges only in the differentiation of tensor-valued differential 0−forms with what Cartan and Kähler call exterior derivative. In recognition of the practice in most of the literature (not necessary here) of the term exterior covariant, we shall take the intermediate course of writing down "exterior(-covariant)" where Cartan and Kähler would use the term exterior. The nature of the object being differentiated determines whether one gets what goes in the literature by the names of exterior, covariant or exterior covariant derivatives.

The conservation law of vector fields is of particular interest because of its relative simplicity. It is the next step up in valuedness after scalar valuedness. In chapter 4, section 7, we learned that, if α is scalar-valued and $d\alpha = 0$ in some simply connected domain R, we have

$$\int_{\partial R} \alpha = 0. \tag{10.1}$$

If, in addition, α is a 0−form, f, we have for any two points in R:

$$f(B) - f(A) = 0. \tag{10.2}$$

Let us now replace scalar-valuedness with vector-valuedness. The exterior derivative of a vector field (in Cartan and Kähler's terminology) is a vector-valued differential $1-$form,

$$d\mathbf{v} = d(v^i \mathbf{e}_i) = v^i_{;j} \omega^j \mathbf{e}_i, \qquad (10.3)$$

where

$$v^i_{;j} \equiv v^i_{/j} + v^k \Gamma_k{}^i{}_j, \qquad (10.4)$$

which is known in the literature as the covariant derivative.

In a similar way to the conservation of α, one would be tempted to say (temptation that we must resist) that, if $d\mathbf{v} = 0$, Stokes theorem,

$$\int_R d\mathbf{v} = \int_{\partial R} \mathbf{v}, \qquad \text{hold it!} \qquad (10.5)$$

yields

$$\int_{\partial R} \mathbf{v} = 0, \qquad \text{hold it!} \qquad (10.6)$$

where ∂R is the pair of end points A and B of the curve R. We would then conclude that $\mathbf{v}_B = \mathbf{v}_A$ in simply connected domains. That is wrong. $\mathbf{v}_B$ and $\mathbf{v}_A$ belong to two different spaces and its equality is not even defined in general. This wrong argument emphasizes something about integration of anything that is tensor-valued, and vector valued in particular: in general, those integrations are not defined.

Those integrations are, however, defined if there are constant basis fields in our manifold. Such is the case in affine and Euclidean spaces, and in their generalizations endowed with teleparallel connections. Using constant basis fields, the relation between vector-valued quantites becomes a relation between their components with respect to one and the same vector basis resulting from the identification of tangent vector bases at different points. We thus have in those constant fields

$$\int_R dv^i = v^i \big|_{\partial R}. \qquad (10.7)$$

It then follows that, if $d\mathbf{v} = \mathbf{0}$,

$$0 = v^i(B) - v^i(A). \qquad (10.8)$$

This is the case for any pair of points. Thus v^i is a constant and so is $\mathbf{v}$. Once this result has been obtained, we may take arbitrary basis fields to express the constancy of $\mathbf{v}$, i.e.

$$v^i \mathbf{e}_i \big|_A = v^i \mathbf{e}_i \big|_B, \qquad (10.9)$$

though this equality does not hold for the components themselves. Since the equation $d\mathbf{v} = 0$ is equivalent to $v^i_{;j} = 0$, the conservation law, (10.6), may be said to be a consequence of $v^i_{;j} = 0$, but only if there is teleparallelism. If there are no constant frame fields, the equation $d\mathbf{v} = 0$ is deceitful, since it is meaningless.

Similarly, let $\mathbf{T}$ be a tensor field such that $d\mathbf{T} = 0$. If and only if the geometry has teleparallelism (Klein geometries in particular have it), we may conclude that

$$\mathbf{T}(A) = \mathbf{T}(B) \qquad (10.10)$$

for any pair of points (A, B). In other words, $\mathbf{T}$ is a constant tensor field. But, as was the case with vector fields, (10.10) does not even mean anything without teleparallelism.

One can try to save the conservation law when there is not teleparallelism by confining oneself to very small regions. In that case, we can follow Cartan in discussing the specific case of a vector-valued differential $3-$form $\boldsymbol{\tau}$ in spacetime [8] (he was discussing energy-momentum tensors, which are disguised forms of such differential forms). After defining its exterior(-covariant) derivative, $d\boldsymbol{\tau}$, he stated that $\boldsymbol{\tau}$ is conserved if $d\boldsymbol{\tau} = 0$, in which case

$$\int_{\partial R} \boldsymbol{\tau} = 0. \qquad (10.11)$$

But non-scalar-valued integration is not without problems even for small domains, as we now explain.

Cartan identified the first Bianchi identity with the statement of conservation of the torsion. But the first Bianchi identity does not state that the exterior(-covariant) derivative of the torsion is zero unless the affine curvature is zero. So, the conservation of the vector-valued differential form torsion provides a very telling example of how intrinsically limited is any statement of conservation of vector and tensor-valued forms when there is not teleparallelism.

The relevance that the curvature has in depriving the conservation of non-scalar-valued differential forms of true meaning manifests itself in particular in connection with the torsion, i.e. the formal exterior(-covariant) derivative of $d\mathbf{P}$. We would think that, when the torsion is zero, the integral of $d\mathbf{P}$ on two different paths between the same two points must be equal:

$$\int_{A;\ (path\ 1)}^{B} d\mathbf{P} = \int_{A;\ (path\ 2)}^{B} d\mathbf{P}. \qquad (10.12)$$

But we saw with simple examples in the previous chapters that this is not correct. Strictly speaking, the development of a closed curve does not close in general, independently of how small the curve is.

Returning to the point we previously made that the conservation law has to do with exterior derivatives in the sense of Cartan and Kähler and not with covariant derivatives (unless they coincide, as is the case with vector fields), the conservation of the Einstein tensor would seem to contradict it. We shall see in a later chapter that this is a very special case. It happens that the annulment of the components of the covariant derivative of the Einstein tensor is equivalent to the annulment of the components of the exterior(-covariant) derivative of the Einstein vector-valued differential $3-$form, which is what that famous mathematical object really is. But covariant derivatives are not relevant in general. For instance, the vanishing of the exterior(-covariant) derivative of the curvature,

$$d(\Omega_i^j \mathbf{e}^i \wedge \mathbf{e}_j) = 0, \tag{10.13}$$

does not contain the same information as the vanishing of its covariant derivatives

$$R_{ikl;s}^j = 0 \tag{10.14}$$

(See section 12 of this chapter).

Those trivial observations are necessary in order to identify misstatements about the conservation law in general relativity. If one performs such illegal integrations and one gets the right results (meaning that they appear to explain physical observations), it may happen that the connection of spacetime is a teleparallel one, rather than the specific one that is presently ascribed to it and which is known as Levi-Civita's. Teleparallelism may be assumed inadvertently when one makes a specific basis field play a special role in a computation. In chapters 9 and 10 we shall learn about the relation between teleparallel connections and Levi-Civita's.

8.11   The zero-torsion case

When the torsion is zero, the first equation of structure and the first Bianchi identities respectively become

$$d\omega^i = \omega^j \wedge \omega_j^i \tag{11.1}$$

$$0 = \omega^j \wedge \Omega_j^i. \tag{11.2}$$

Equation (11.2) implies

$$R_{jkl}^i + R_{klj}^i + R_{ljk}^i = 0, \tag{11.3}$$

which is valid in arbitrary bases. The use of coordinate bases makes the left hand side of (11.1) disappear and the first equation of structure become

$$\Gamma'_{k\ j}^{\ i} = \Gamma'_{j\ k}^{\ i} \tag{11.4}$$

in terms of those bases, indicated by primes, but not in general.

The property of zero torsion of a space takes a particularly interesting form in terms of intersecting families of curves. The representation $\mathbf{PP'} + \mathbf{P'M'} + \mathbf{M'M} + \mathbf{MP}$ of $PP'M'MP$ (Fig. 7) can be rewritten as

$$(\mathbf{P'M'} - \mathbf{PM}) - (\mathbf{MM'} - \mathbf{PP'}). \tag{11.5}$$

Let $\Delta\lambda$ and Δu be small quantities. Then

$$\mathbf{P'M'} - \mathbf{PM} = \frac{\partial(\mathbf{n}\Delta\lambda)}{\partial u}\Delta u = \frac{\partial\mathbf{n}}{\partial u}\Delta\lambda\Delta u \tag{11.6a}$$

$$\mathbf{MM'} - \mathbf{PP'} = \frac{\partial(\mathbf{t}\Delta u)}{\partial \lambda}\Delta \lambda = \frac{\partial \mathbf{t}}{\partial \lambda}\Delta \lambda \Delta u. \tag{11.6b}$$

These are connection independent equations. Both $\mathbf{MM'} - \mathbf{PP'}$ and $\mathbf{P'M'} - \mathbf{PM}$ go to zero as Δu and $\Delta \lambda$ do. However,

$$\frac{\mathbf{P'M'} - \mathbf{PM}}{\Delta u \Delta \lambda} \qquad \text{and} \qquad \frac{\mathbf{MM'} - \mathbf{PP'}}{\Delta u \Delta \lambda}$$

do not. The statement that $\mathbf{PP'} + \mathbf{P'M'} + \mathbf{M'M} + \mathbf{MP}$ is zero on infinitesimal curves means that $\frac{\partial \mathbf{n}}{\partial u} - \frac{\partial \mathbf{t}}{\partial k}$ goes to zero in the limit and, therefore,

$$\frac{\partial \mathbf{n}}{\partial u} = \frac{\partial \mathbf{t}}{\partial \lambda}. \tag{11.7}$$

Taking this to (7.10) makes it become

$$\frac{\partial^2 \mathbf{n}}{\partial u^2} = t^i R_{ikl}^j t^k n^l \mathbf{e}_j. \tag{11.8}$$

Notice that, as defined in section 6, $\mathbf{n}$ is the vector assigned to an "instantaneous" change of the parameter λ by one unit. The term instantaneous is used in the absence of a better word, meaning for the parameter λ what it literally means when the parameter is time.

We have derived equation (11.8) in a purely affine context. It already looks like the formula for geodesic deviation in Riemannian geometry. But we have not yet considered that geometry; metrics have not been used in obtaining (11.8).

8.12 Horrible covariant derivatives

We shall now express the Bianchi identities (i.e. the differentiated equations of structure) in terms of components of covariant derivatives as an example of how cumbersome such expressions become. Torsion and curvature are, of course, not tensors. Their covariant derivatives would be the components of the Cartan-Kähler exterior derivative if they were tensors of respective ranks 3 and 4. But they are rather (1,0)-valued and (1,1)-valued differential 2−forms.

The first Bianchi identity involves both curvature and torsion. We thus start with the second Bianchi identity,

$$d\Omega_i^j + \Omega_k^i \wedge \omega_k^j - \omega_i^k \wedge \Omega_k^j = 0, \tag{12.1}$$

since it involves only curvature. If we give Ω_i^j in terms of a coordinate basis of 2−forms, i.e. as $\frac{1}{2}R'^{\,j}_{i\,kl}dx^k \wedge dx^l$, there is in $d\Omega_i^j$ only the term resulting from the differentiation of $R'^{\,j}_{i\,kl}$. If, on the other hand, we represent it as $\frac{1}{2}R^{\,j}_{i\,kl}\omega^k \wedge \omega^l$, three terms result from $d\Omega_i^j$, for a total of five, on the left hand side of (12.1). Only the last two terms in this first development, which are the same last terms as in (12.1), involve the connection.

On the other hand, the covariant derivative $R_{i\,kl;m}^{j}$ has five terms of which four depend on connection since $R_{i\,kl;m}^{j}$ is the coefficient of the (1,3)-valued differential 1−form $d(R_{i\,kl}^{j}\phi^{i}\phi^{k}\phi^{l}\mathbf{e}_{j})$. In order to make $R_{i\,kl;m}^{j}$ appear in (12.1), we invoke the first equation of structure to substitute $d\omega^{k}$ and $d\omega^{l}$ in the differentiation of $\frac{1}{2}R_{i\,kl}^{j}\omega^{k}\wedge\omega^{l}$. We thus obtain seven terms on the left hand side of (12.1). Five of the terms yield $R_{i\,kl;m}^{j}\omega^{m}\wedge\omega^{k}\wedge\omega^{l}$. The two extra terms are

$$\frac{1}{2}R_{i\,hl}^{j}\Omega^{h}\wedge\omega^{l} - \frac{1}{2}R_{i\,kh}^{j}\omega^{k}\wedge\Omega^{h} = R_{i\,hl}^{j}\Omega^{h}\wedge\omega^{l} = \frac{1}{2}R_{i\,hl}^{j}R_{mk}^{h}\omega^{m}\wedge\omega^{k}\wedge\omega^{l}. \quad (12.2)$$

Equation (12.1) thus becomes

$$(R_{i\,kl;m}^{j} + R_{i\,hl}^{j}R_{mk}^{h})\omega^{m}\wedge\omega^{k}\wedge\omega^{l} = 0. \quad (12.3)$$

We complete the translation to the tensor calculus by expressing (12.3) in terms of components:

$$(R_{i\,kl;m}^{j} + R_{i\,hl}^{j}R_{mk}^{h}) + (R_{i\,lm;k}^{j} + R_{i\,hm}^{j}R_{kl}^{h}) + (R_{i\,mk;l}^{j} + R_{i\,hk}^{j}R_{lm}^{h}) = 0. \quad (12.4)$$

There are 36 indices in (12.4) versus only ten in (12.1). There may be a discrepancy by a factor of two with much of the literature in the second of the two terms in each parenthesis. The difference vanishes in one defines the components of the torsion by $R_{mk}^{h}\omega^{m}\wedge\omega^{k}$ rather than by $R_{mk}^{h}(\omega^{m}\wedge\omega^{k})$. The new R_{mk}^{h} would be half what the old one is, which would cause (12.4) to become

$$(R_{i\,kl;m}^{j}+2R_{i\,hl}^{j}R_{mk}^{h})+(R_{i\,lm;k}^{j}+2R_{i\,hm}^{j}R_{kl}^{h})+(R_{i\,mk;l}^{j}+2R_{i\,hk}^{j}R_{lm}^{h}) = 0. \quad (12.5)$$

Assume that the torsion were zero. Even then the conservation law of the curvature would not be given by the annulment of its covariant derivative, but by a less restrictive condition, as (12.5) immediately shows. All this speaks of how inimical covariant derivatives are to the issue of conservation.

Consider now the first Bianchi identity

$$d\Omega^{i} + \Omega^{j}\wedge\omega_{j}^{i} - \omega^{j}\wedge\Omega_{j}^{i} = 0. \quad (12.6)$$

We write the torsion as

$$\Omega^{i} = \frac{1}{2}R_{jk}^{i}\omega^{j}\wedge\omega^{k}. \quad (12.7)$$

When we differentiate (12.7), we shall have five terms on the right hand side if we use the first equation of structure to substitute for $d\omega^{j}$ and $d\omega^{k}$. Three of those terms together with $\Omega^{j}\wedge\omega_{j}^{i}$ form

$$\frac{1}{2}R_{jk;l}^{i}\omega^{l}\wedge\omega^{j}\wedge\omega^{k}. \quad (12.8)$$

The other two terms arising from substituting $d\omega^{j}$ and $d\omega^{k}$ on the differentiated Ω^{i} result in

$$\frac{1}{2}R_{hk}^{i}\Omega^{h}\wedge\omega^{k} - \frac{1}{2}R_{jh}^{i}\Omega^{h}\wedge\omega^{j} \quad (12.9)$$

and can be written further as

$$\ldots = R^i_{hk}\Omega^h \wedge \omega^k = \frac{1}{2}R^i_{hk}R^h_{lj}\omega^l \wedge \omega^j \wedge \omega^k. \tag{12.10}$$

The last term in (12.6) is

$$-\frac{1}{2}R^i_{ljk}\omega^l \wedge \omega^j \wedge \omega^k. \tag{12.11}$$

When we substitute (12.8), (12.10) and (12.11) in (12.6), we get

$$(R^i_{jk;l} + R^i_{hk}R^h_{lj} - R^i_{ljk})\omega^l \wedge \omega^j \wedge \omega^k = 0, \tag{12.12}$$

and, therefore,

$$(R^i_{jk;l} + R^i_{hk}R^h_{lj} - R^i_{ljk}) + (R^i_{kl;j} + R^i_{hl}R^h_{jk} - R^{\ i}_{j\ kl}) + (R^i_{lj;k} + R^i_{hj}R^h_{kl} - R^i_{klj}) = 0. \tag{12.13}$$

Defining again the components of the torsion as $R^h_{mk}\omega^m \wedge \omega^k$ rather than as $R^h_{mk}(\omega^m \wedge \omega^k)$, we would get, instead of (12.13),

$$(2R^i_{jk;l} + 4R^i_{hk}R^h_{lj} - R^i_{ljk}) + (2R^i_{kl;j} + 4R^i_{hl}R^h_{jk} - R^i_{jkl}) +$$

$$+ (2R^i_{lj;k} + 4R^i_{hj}R^h_{kl} - R^i_{klj}) = 0. \tag{12.14}$$

Notice that there are 42 indices here, to be compared with the seven indices in (12.6). Notice also that, even if the curvature were zero, the conservation of torsion would not be given by the annulment of its *covariant* derivative.

8.13 Affine connections: rigorous APPROACH

We start by recalling how Lie algebras of affine groups emerge in the context of the moving frame. Recall the affine group: $\mathbf{P} = \mathbf{Q} + A^i\mathbf{a}_i$, $\mathbf{e}_i = A^j_i\mathbf{a}_j$, where $(\mathbf{Q}, \mathbf{a}_1 \ldots \mathbf{a}_n)$ is a fixed frame. From it we obtained

$$d\mathbf{P} = \omega^i\mathbf{e}_i, \quad d\mathbf{e}_i = \omega^j_i\mathbf{e}_j, \tag{13.1}$$

with

$$\omega^i = dA^m B^i_m, \quad \omega^j_i = dA^m_i B^j_m. \tag{13.2}$$

In other words, $dg \cdot g^{-1}$ is the $1-$form of components (ω^i, ω^j_i). The differential forms $d\omega^i - \omega^j \wedge \omega^i_j$ and $d\omega^j_i - \omega^k_i \wedge \omega^j_k$ also are going to play a prominent role in the theory of affine connections, but we no longer have

$$0 = d\omega^i - \omega^j \wedge \omega^i_j, \quad 0 = d\omega^j_i - \omega^k_i \wedge \omega^j_k. \tag{13.3}$$

We exceptionally use the term "affine-linear" rather than simply affine as a means to exclude the affine connections of the more general Finsler type, which can be defined in a similar but slightly more complicated way [73].

Definition 1 *An affine-linear connection is a one-form (ω^i, ω_i^j) on an (n^2+n)-dimensional manifold $B(M)$, taking values in the Lie algebra of the affine group for dimension n and satisfying the conditions*

(1) *The $n^2 + n$ real-valued one-forms are linearly independent.*

(2) *The differential forms ω^i are the soldering forms, meaning that they are the coefficients of the translation differential form $d\mathbf{P}$ when pushed forward from a section to the frame bundle by the action of the general linear group $GL(n)$.*

(3) *The pullbacks of ω_i^j to the fiber of the fibration $\Pi : B(M) \to M$ are the left invariant forms of $GL(n)$.*

(4) *The forms $\Omega^i = d\omega^i - \omega^j \wedge \omega_j^i$, called torsion, and $\Omega_i^j = d\omega_i^j - \omega_i^k \wedge \omega_k^j$, called affine curvature, are quadratic exterior polynomials in the n forms ω^i*

$$\Omega^i = R_{jk}^i \omega^j \wedge \omega^k \quad (R_{jk}^i + R_{kj}^i = 0) \tag{13.4}$$

$$\Omega_j^i = R_{jkl}^i \omega^k \wedge \omega^l \quad (R_{jkl}^i + R_{jlk}^i = 0). \tag{13.5}$$

Theorem 8.1 *The integration of the system of differential equations (13.1) along a curve in $B(M)$ gives an affine transformation between tangent spaces to M.*

Proof. On parametrizations $f(u)$, $(a \le u \le b)$ of curves in $B(M)$ (we assume that f is sufficiently smooth), the (ω^i, ω_i^j) are now the pullbacks of (ω^i, ω_i^j) into $[a, b]$. Because the equations are linear, the solution is linear with respect to the initial conditions and, therefore, takes the form $(\mathbf{P} = \mathbf{A} + f^i(u)\mathbf{a}_i, \mathbf{e}_i = g_i^j \mathbf{a}_j)$, with $\mathbf{A} = \mathbf{P}(a)$ and $\mathbf{a}_i = \mathbf{e}_i(a)$ as initial conditions. But $\mathbf{e}_i(a)$ is the initial basis, i.e. the basis $f(a)$. $\mathbf{P}(a)$ is the origin of the tangent vector space at $\Pi(f(a))$. Similar interpretations apply to $\mathbf{P}(u)$ and $\mathbf{e}_i(u)$ at $\Pi(f(u))$. Hence we have obtained an affine transformation between the vector spaces tangent at the points $\Pi(f(a))$ and $\Pi(f(u))$. $\mathbf{P}(u)$ describes a curve in the tangent vector space at $\Pi(f(a))$ called the *development* of the curve f into that vector space. ∎

Theorem 8.2 *The integration of the connection is independent of the section used to perform it.*

Proof. Let $g(u)$, $(a \le u \le b)$ be a curve in M. Take another curve $G(u)$, $(c < u < d)$ with a, $b \in (c, d)$ and whose restriction to $[a, b]$ is the curve $g(u)$. Let $(\widetilde{\omega}^i, \widetilde{\omega}_i^j)$ be the push-forwards to $B(M)$ of the forms (ω^i, ω_i^j) on $G(u)$. The horizontality of the soldering forms implies that the $\widetilde{\omega}^i$'s are proportional to du, whereas $\widetilde{\omega}_i^j$ will be a linear combination of du and of the differentials of the n^2 coordinates on the fibers. Since torsion and curvature are quadratic and horizontal, they become zero when substituting $\widetilde{\omega}^i$ for ω^i. It follows that the forms $\widetilde{\omega}^i$ and $\widetilde{\omega}_i^j$ are in involution. By the Frobenius theorem they generate an (n^2+1)-dimensional integral submanifold where $\widetilde{\omega}^i \mathbf{e}_i$ and $\widetilde{\omega}_i^j \mathbf{e}_j$ are the differentials $d\mathbf{P}$ and $d\mathbf{e}_i$ of vector-valued functions $\mathbf{P}$ and $\mathbf{e}_i$ ∎

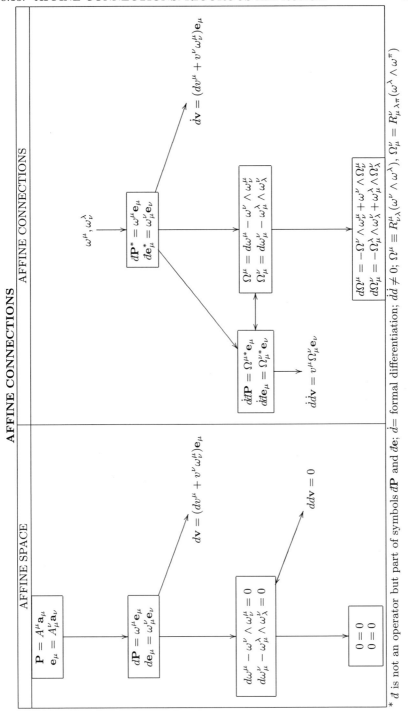

AFFINE CONNECTIONS

AFFINE CONNECTIONS

AFFINE SPACE

$\omega^\mu, \omega^\lambda_\nu$

$\dot{d}\mathbf{v} = (dv^\mu + v^\nu \omega^\mu_\nu)\mathbf{e}_\mu$

$$\dot{d}\mathbf{P}^* = \omega^\mu{}_{*}\mathbf{e}_\mu$$
$$\dot{d}\mathbf{e}^*_\mu = \omega^\nu{}_\mu{}_{*}\mathbf{e}_\nu$$

$$\Omega^\mu = d\omega^\mu - \omega^\nu \wedge \omega^\mu_\nu$$
$$\Omega^\nu_\mu = d\omega^\nu_\mu - \omega^\lambda_\mu \wedge \omega^\nu_\lambda$$

$$\dot{d}\dot{d}\mathbf{P} = \Omega^\mu{}_{*}\mathbf{e}_\mu$$
$$\dot{d}\dot{d}\mathbf{e}_\mu = \Omega^{\nu}_\mu{}_{*}\mathbf{e}_\nu$$

$\dot{d}\dot{d}\mathbf{v} = v^\mu \Omega^\nu_\mu \mathbf{e}_\nu$

$$d\Omega^\mu = -\Omega^\nu \wedge \omega^\mu_\nu + \omega^\nu \wedge \Omega^\mu_\nu$$
$$d\Omega^\nu_\mu = -\Omega^\lambda_\mu \wedge \omega^\nu_\lambda + \omega^\lambda_\mu \wedge \Omega^\nu_\lambda$$

$\dot{d}\mathbf{v} = (dv^\mu + v^\nu \omega^\mu_\nu)\mathbf{e}_\mu$

$$\mathbf{P} = A^\mu \mathbf{a}_\mu$$
$$\mathbf{e}_\mu = A^\nu_\mu \mathbf{a}_\nu$$

$$d\mathbf{P} = \omega^\mu \mathbf{e}_\mu$$
$$d\mathbf{e}_\mu = \omega^\nu_\mu \mathbf{e}_\nu$$

$ddv = 0$

$$d\omega^\mu - \omega^\nu \wedge \omega^\mu_\nu = 0$$
$$d\omega^\nu_\mu - \omega^\lambda_\mu \wedge \omega^\nu_\lambda = 0$$

$$0 = 0$$
$$0 = 0$$

* d is not an operator but part of symbols $d\mathbf{P}$ and $d\mathbf{e}$; $\dot{d}=$ formal differentiation; $\dot{d}\dot{d} \neq 0$; $\Omega^\mu \equiv R^\mu_{\nu\lambda}(\omega^\nu \wedge \omega^\lambda)$, $\Omega^\nu_\mu = R^\nu_{\mu\lambda\pi}(\omega^\lambda \wedge \omega^\pi)$

Chapter 9

EUCLIDEAN CONNECTIONS

9.1 Metrics and the Euclidean environment

The term *Euclidean connection* is due to Cartan and pertains to the comparison of *Euclidean* neighborhoods at different points of a differentiable manifold. It is a better term than the more usual one of *Riemann plus torsion* geometry, which reflects the historical accident that there was Riemannian geometry before the birth of the concept of torsion (other than torsion of curves, which is something different).

Torsion is a property of connections. Euclidean space has a connection, the one that makes **i**, **j** and **k** be the same three vectors everywhere in Euclidean space. It thus has a concept of torsion that happens to be zero. In contrast, the original Riemannian geometry did not have a concept of comparison of *Euclidean* neighborhoods, and did not, therefore, have a concept of torsion, not even implicitly.

What would a "non-Euclidean environment" mean when speaking of metrics? It would mean a concept of metric as in Riemannian geometry in the first epoch of its existence. Cartan referred to those manifolds with the term "false spaces of Riemann" [13]. As he explained, they became true spaces in 1917, with the advent in that year of the Levi-Civita connection (LCC) [52]. It must be said, however, that connections were present in a crude implicit form in the work of Ricci and Levi-Civita in 1901 [60]. We shall devote the first two sections of the next chapter to the false spaces of Riemann, i.e. without a connection and, therefore, without a Euclidean (nor pseudo-Euclidean) structure. In the meantime, let us spend a couple of paragraphs to provide a little bit of background on manifolds with exclusively metric structure.

Assume we had a differentiable manifold endowed with a metric

$$ds^2 = g_{\mu\nu}(x)dx^\mu dx^\nu \tag{1.1}$$

but without a connection. One could obtain sets of ω^μ's that orthonormalize it, i.e. such that

$$ds^2 = \sum_\mu \eta_\mu \omega^\mu \omega^\mu, \tag{1.2}$$

where $\eta_\mu = \pm 1$. Take one of those sets and push it to the Euclidean frame bundle by the action of G_0 (it is the group of rotations in n dimensions, if the signature is Euclidean; and it is the group constituted by boosts, rotations in the Euclidean subspace and the products of all those, if the signature is Lorentzian). Consider the system of differential equations

$$dy^\mu - \omega^\mu(x, u, dx) = 0, \tag{1.3}$$

where u represents the parameters in G_0. This system may admit or not solutions for y^μ. If it does, the metric (1.1) can be rewritten as

$$ds^2 = \sum_\mu \eta_\mu (dy^\mu)^2, \tag{1.4}$$

η_μ being ± 1. A necessary and sufficient condition for existence of a solution to (1.3) is that the set of quantities nowadays known as components of the Riemannian curvature be zero. They are thus an answer to a specific problem of integrability without reference to a comparison of vectors, which do not enter the argument. But there still is a concept of bundle (implying a specific dependence on coordinates absent from the metric) where the group in the fibers (rotations, or rotations together with hyperbolic rotations) is the same as in the Euclidean bundle of frames for the same dimension.

When one tries to add affine structure to that manifold, the issue of compatibility of the two structures arises. This is not, however, Cartan's approach to Euclidean structure, generalized or not. For him, Euclidean geometry, and post-1917 Riemannian geometry, is first and foremost about connections that are restrictions of affine connections. This means that vector bases are replaced with orthonormal or pseudo-orthonormal ones. End of considerations on Riemannian pseudo-spaces in this chapter.

As in the geometry of Riemannian pseudo-spaces, the group G_0 in the fibers is the group of isometries (rotations in Euclidean spaces, and Lorentz transformations in the spacetime of special relativity), both in Euclidean spaces and in their Cartan generalizations. Once a Euclidean frame (i.e. where the vector basis is orthonormal or pseudo orthonormal) has been chosen, the points in each fiber are in a one to one correspondence with the members of G_0. The role of the pair (G, G_0) in defining geometries advises against the use of the term "metric compatible affine connection", so that we do not overlook that role.

Later in this chapter, we shall start to deal with pseudo-Euclidean differentiable manifolds, i.e. where the signature of the metric is pseudo-Euclidean, and specifically Lorentzian. Correspondingly, we shall use Greek indices to go from 0 to $n - 1$. Latin indices will go from 1 to $n - 1$. The notation $\mu = (0, m)$ will then be obvious.

Let us review how the concepts of Chapter 6 have to be understood when the metric neither is of the type (1.4), nor can be reduced to it. Regardless of affine connection, one defines ds^2 as $d\mathbf{P} \cdot d\mathbf{P}$. But one should rather write

$$ds^2 = d\mathbf{P}(\otimes, \cdot)d\mathbf{P} = g_{\mu\nu}(x)dx^\mu \otimes dx^\nu, \qquad (1.5)$$

where the first symbol in $(\otimes, \cdot)$ is meant to refer to the tensor product in the algebra of differential forms, symbol which is usually overlooked. More generally:

$$ds^2 = \omega^\mu \mathbf{e}_\mu(\otimes, \cdot)\omega^\nu \mathbf{e}_\nu = g_{\mu\nu}\omega^\mu \otimes \omega^\nu. \qquad (1.6)$$

where

$$g_{\mu\nu} = \mathbf{e}_\mu \cdot \mathbf{e}_\nu. \qquad (1.7)$$

For orthonormal and pseudo-orthonormal bases, $g_{\mu\nu}$ is diagonal and $g_{\mu\mu}$ equals η_μ, meaning ± 1. The metric can then be written as

$$ds^2 = \sum_\mu \eta_\mu \omega^\mu \otimes \omega^\mu, \qquad (1.8)$$

which is a more informative way of writing (1.2).

ds^2 is not an exterior differential 2−form, or, for that matter, an exterior differential r−form for some r. One integrates

$$\sqrt{g_{\mu\nu}(x)\frac{\omega^\mu}{d\lambda} \otimes \frac{\omega^\nu}{d\lambda}}d\lambda \qquad (1.9)$$

on curves. The notation $\omega^\mu/d\lambda$ is justified by the proportionality of ω^ρ to $d\lambda$ in one dimension (i.e. on curves). Of course, the tensor product in expression (1.8) is unnecessary since the $\omega^\rho/d\lambda$ are scalar valued.

The proper way of writing the metric in the Euclidean (or pseudo-Euclidean) frame bundle of a differentiable manifold endowed with a metric is given by (1.2), equivalently (1.8); on the other hand, (1.1), equivalently (1.5), belongs to the affine extension (see next section) of the Euclidean bundle. Our starting point in this section should have been (1.8), were it not for our respecting tradition on this matter.

9.2   Euclidean structure and Bianchi IDENTITIES

A Euclidean frame bundle of a differentiable manifold is the restriction of its affine frame bundle from vector bases to orthonormal or pseudo-orthonormal bases; one is replacing a pair (G, G_0) with another. At this point, knowledgeable readers (and also others who are not so much so but are in a hurry) might which to jump to the equations of structure, (2.9) to (2.11), to see that we are just adding Eq. (2.11) to the equations of structure of the previous chapter.

Whereas the affine frame bundle is of dimension $n + n^2$, the Euclidean frame bundle is of dimension $n + [n(n-1)/2]$. Hence, if and when one says that a Euclidean connection is a restriction of an affine connection from an affine frame bundle to a Euclidean frame bundle, one means that a Euclidean connection is to a Euclidean frame bundle what an affine connection is to an affine frame bundle.

One uses the term "extension of an Euclidean connection to an affine connection" when, by the action of the linear group for the same dimension, one extends the formulas for a Euclidean connection from a Euclidean frame bundle to an affine frame bundle, i.e. where the frames are not orthonormal.

Differentiation of (1.7) yields

$$dg_{\mu\nu} - \omega^k_\mu g_{k\nu} - \omega^k_\nu g_{k\mu} = dg_{\mu\nu} - \omega_{\mu\nu} - \omega_{\nu\mu} = 0 \qquad (2.1)$$

for the affine extension of a Euclidean connection. Recalling already defined notation for the slash subscript, we have

$$dg_{\mu\nu} = g_{\mu\nu/\lambda}\omega^\lambda. \qquad (2.2)$$

We define $\Gamma_{\mu\nu\lambda}$'s and $\Gamma'_{\mu\nu\lambda}$'s as components, respectively in terms of ω^λ's and of dx^λ's, of pull-backs of $\omega_{\mu\nu}$ to sections of the bundle:

$$g_{\mu\nu,\lambda} - \Gamma'_{\mu\nu\lambda} - \Gamma'_{\nu\mu\lambda} = 0, \qquad (2.3)$$

and

$$g_{\mu\nu/\lambda} - \Gamma_{\mu\nu\lambda} - \Gamma_{\nu\mu\lambda} = 0. \qquad (2.4)$$

Hence, neither the $\omega_{\mu\nu}$ nor the $\Gamma_{\mu\nu\lambda}$'s and $\Gamma'_{\mu\nu\lambda}$ are skew-symmetric unless the bases are orthonormal or pseudo-orthonormal, in which case $dg_{\mu\nu}$ is zero. The equations

$$\omega_{\mu\nu} + \omega_{\nu\mu} = 0. \qquad (2.5)$$

and

$$\Gamma_{\mu\nu\lambda} + \Gamma_{\nu\mu\lambda} = 0, \qquad \Gamma'_{\mu\nu\lambda} + \Gamma'_{\nu\mu\lambda} = 0. \qquad (2.6)$$

then follow.

In the properly Euclidean case (positive definite metric), equation (2.5)uclidean implies

$$\omega^\nu_\mu = -\omega^\mu_\nu, \qquad (2.7)$$

and, in the Lorentzian case,

$$\omega^i_0 = \omega^0_i, \qquad \omega^j_i = -\omega^i_j. \qquad (2.8)$$

The $\Gamma'_{\mu\nu\lambda}$'s are called Christoffel symbols. There is no need to retain this name, except for being able to recognize what it refers to, if found in the literature.

Equations (2.1)-(2.4) belong to the extension of an Euclidean connection to an affine connection. Equations (2.5)-(2.8) belong to the Euclidean (or pseudo-Euclidean) connections themselves. We repeat that the extension is not an affine connection proper.

Before proceeding with the definition of Euclidean connection, we advise inexperienced readers to revisit the last section of chapter 6, where the Lie algebra of the Euclidean group is introduced.

Definition 2 *A differentiable manifold M endowed with a metric is said to be endowed with a Euclidean connection if its Euclidean frame bundle $B(M)$, which is of dimension $n + \frac{1}{2}n(n-1)$, is endowed with a differential one-form $(\omega^\mu, \omega^\nu_\mu)$ taking values in the (real) Lie algebra of the Euclidean group and satisfying the conditions:*

(1) *The $n + \frac{1}{2}n(n-1)$ real-valued one-forms are linearly independent.*

(2) *The forms ω^μ are the soldering forms, meaning that they are the coefficients of the translation differential form $d\mathbf{P}$ when pushed forward from any section of $B(M)$ to $B(M)$ itself.*

(3) *The pullbacks of ω^ν_μ to the fiber of $\Pi : B(M) \to M$ are the left invariant forms of $O(n)$, or of the Lorentz group if the signature is $(1, -1, ..., -1)$.*

(4) *The forms $\Omega^\mu = d\omega^\mu - \omega^\nu \wedge \omega^\mu_\nu$, called the torsion, and $\Omega^\nu_\mu = d\omega^\nu_\mu - \omega^k_\mu \wedge \omega^\nu_k$, called the Euclidean curvature, are quadratic exterior polynomials in the n forms ω^μ*

$$\Omega^\mu = R^\mu_{\nu\lambda}(\omega^\nu \wedge \omega^\lambda) \qquad (R^\mu_{\nu\lambda} + R^\mu_{\lambda\nu} = 0) \qquad (2.9)$$

and

$$\Omega^\mu_\nu = R^\mu_{\nu\lambda\pi}(\omega^\lambda \wedge \omega^\pi) \qquad (R^\mu_{\nu\lambda\pi} + R^\mu_{\nu\pi\lambda} = 0). \qquad (2.10)$$

To these equations of structure, we add

$$\omega_{\mu\nu} + \omega_{\nu\mu} = 0, \qquad (2.11)$$

which is the differential form of (1.7). This is consistent with condition (3). It is easy to show from $\Omega^\nu_\mu = d\omega^\nu_\mu - \omega^k_\mu \wedge \omega^\nu_k$ and (2.11) that

$$\Omega_{\mu\nu} = -\Omega_{\nu\mu}, \qquad (2.12)$$

which was not the case for affine connections in general. Although we have proved this equation in orthonormal basis fields, this is clearly valid also for the affine extension. We do not speak of Euclidean and affine torsions since, unlike the case of the curvature, the torsion retains the same number of independent components in both cases.

In the Euclidean environment, the concept of Euclidean-Klein geometry is first, followed by the concept of Euclidean connection, then followed by the concept of metric as a derived invariant

$$ds^2 = \sum_\mu \eta_\mu (\omega^\mu)^2, \quad \text{meaning} \quad \sum_\mu \eta_\mu \omega^\mu \otimes \omega^\mu, \qquad (2.13)$$

The metric allows us to replace a linear function of vectors with a vector, the action of the linear function being replaced with the dot product with that vector. We may thus write the Euclidean curvature as

$$\mho \equiv \Omega^{\nu}_{\mu} \mathbf{e}^{\mu} \wedge \mathbf{e}_{\nu}, \tag{2.14}$$

instead of (8.3) of chapter 8, and the Bianchi identities as

$$d\mathbf{\Omega} = \omega^{j} \wedge \Omega^{i}_{j} \mathbf{e}_{i}, \qquad d\mho \equiv \mathbf{0}. \tag{2.15}$$

When the torsion is zero, the Euclidean connection is called the Levi-Civita connection (LCC), denoted α^{ν}_{μ}. We shall see in the next section that it depends only on the metric and its first derivatives. The curvature $d\alpha^{\nu}_{\mu} - \alpha^{\lambda}_{\mu} \wedge \alpha^{\nu}_{\lambda}$ receives the name of Riemann's curvature. Hence Riemannian curvature means two things. On the one hand, it is the set of quantities whose being zero constitutes the integrability condition of (1.3); it concerns differentiable manifolds regardless of whether they are endowed with a connection or not. But, if the manifold is endowed with the Levi-Civita connection, it is at the same time the Euclidean curvature, playing the role that affine curvatures play in manifolds endowed with affine connection. We write

$$\underline{\mho} \equiv \underline{\Omega}^{\nu}_{\mu} \mathbf{e}^{\mu} \wedge \mathbf{e}_{\nu} = \underline{R}^{\mu}_{\nu\lambda\pi}(\omega^{\lambda} \wedge \omega^{\pi})\mathbf{e}^{\mu} \wedge \mathbf{e}_{\nu}, \tag{2.16}$$

$$\underline{\Omega}^{\nu}_{\mu} \equiv d\alpha^{\nu}_{\mu} - \alpha^{\lambda}_{\mu} \wedge \alpha^{\nu}_{\lambda}, \tag{2.17}$$

to make explicit the meaning of different symbols.

When the torsion is not zero, the Euclidean connection of the differentiable manifold no longer is the LCC. We still may define symbols α^{ν}_{μ} and $\underline{\Omega}^{\nu}_{\mu}$ that will take the same form as if the torsion were zero, except that we shall no longer use for them the terms connection and Euclidean curvature, because they no longer are so. They are concepts derived from the metric, equivalently from the fundamental invariants ω^{μ}'s that participate in defining the Euclidean structure. Invariants derived from the fundamental ones need not involve all of the last ones at the same time. The two curvatures, Euclidean and Riemannian, are related through the torsion and its derivatives, as we shall see in section 8.

Of special interest is the case when the Euclidean curvature satisfies

$$\omega^{\mu} \wedge \Omega^{\nu}_{\mu} = 0, \tag{2.18}$$

like the Riemannian curvature does. But also do all those connections, affine or Euclidean, whose exterior(-covariant) derivative is zero

$$d\mathbf{\Omega} = 0. \tag{2.19}$$

Equation (2.19) implies

$$R_{\mu}{}^{\nu}{}_{\lambda\pi} + R_{\lambda}{}^{\nu}{}_{\pi\mu} + R_{\pi}{}^{\nu}{}_{\mu\lambda} = 0. \tag{2.20}$$

A property specific to Euclidean connections is

$$R_{\mu\nu\lambda\pi} = R_{\lambda\pi\mu\nu}. \tag{2.21}$$

for Euclidean curvatures. In order to prove (2.21), we lower the index ν in Eq. (2.20). We rewrite the resulting equation as

$$R_{\mu(\nu)\lambda\pi} = 0, \tag{2.22}$$

which defines $R_{\mu(\nu)\lambda\pi}$. We add to (2.22) three more copies of this equation with exchanged indices to get

$$R_{\mu(\nu)\lambda\pi} + R_{\nu(\mu)\lambda\pi} + R_{\lambda(\mu)\pi\nu} + R_{\pi(\mu)\nu\lambda} = 0. \tag{2.23}$$

We develop each of the four terms on the left hand side as per their definition and use the skew-symmetry under exchange of the last two indices. Terms cancel out and we finally get (2.21).

Because of current usage in the literature, we shall sometimes speak of affine connections when referring to extensions of Euclidean connections, and even to the Euclidean connections themselves. In doing so, one still makes the implicit point that the metric structure is subordinated to the affine structure, then called Euclidean.

9.3 The two pieces of a Euclidean connection

Sometimes, we may not know the set of fundamental invariants $(\omega^\mu, \omega^\nu_\lambda)$ of a differentiable manifold, but we may know Ω^μ and ω^μ. The system

$$\Omega^\mu = d\omega^\mu - \omega^\nu \wedge \omega^\mu_\nu, \tag{3.1}$$

$$0 = \omega_{\mu\nu} + \omega_{\nu\mu} \tag{3.2}$$

then determines ω^μ_ν.

When the torsion is not zero, we may still define Levi-Civita symbols, α^μ_ν, as those which satisfy the system

$$0 = d\omega^\mu - \omega^\nu \wedge \alpha^\mu_\nu, \tag{3.3}$$

$$0 = \alpha_{\mu\nu} + \alpha_{\nu\mu}. \tag{3.4}$$

We do not refer to α^μ_ν as the LCC in this case. Defining

$$\beta^\nu_\mu \equiv \omega^\nu_\mu - \alpha^\nu_\mu, \tag{3.5}$$

allows us to view α and β as two complementary pieces of Euclidean connections. One readily shows that β^ν_μ satisfies the system

$$\Omega^\mu = -\omega^\nu \wedge \beta^\mu_\nu. \tag{3.6}$$

$$0 = \beta_{\mu\nu} + \beta_{\nu\mu}. \tag{3.7}$$

β_μ^ν and, better yet, $\boldsymbol{\beta}$ ($= \beta_\mu^\nu \mathbf{e}^\mu \wedge \mathbf{e}_\nu$) are called contorsion. We introduce new symbols $L_{\mu\ \lambda}^{\ \nu}$ by

$$\boldsymbol{\beta} \equiv \beta_\mu^\nu\, \mathbf{e}^\mu \wedge \mathbf{e}_\nu \equiv L_{\mu\ \lambda}^{\ \nu} \mathbf{e}^\mu \wedge \mathbf{e}_\nu \omega^\lambda. \tag{3.8}$$

We now review the issue of tensoriality, equivalently, horizontality. $d\mathbf{P}$ is the unit vector-valued differential $1-$form

$$d\mathbf{P} = \delta_\mu^\nu \omega^\mu \mathbf{e}_\nu. \tag{3.9}$$

It is an invariant; the ω^μ are tensorial. $\boldsymbol{\Omega}$, defined as $d(d\mathbf{P})$, is an invariant because so are d and $d\mathbf{P}$. The components $R_{\nu\lambda}^\mu$ of $\boldsymbol{\Omega}$, defined by

$$\boldsymbol{\Omega} = R_{\nu\lambda}^\mu(\omega^\nu \wedge \omega^\lambda)\mathbf{e}_\mu = \frac{1}{2}R_{\nu\lambda}^\mu \omega^\nu \wedge \omega^\lambda \mathbf{e}_\mu, \tag{3.10}$$

are tensorial. We also saw the non-tensorial transformation of ω_μ^ν and α_μ^ν. Their difference, β_ν^μ ($=\omega_\mu^\nu - \alpha_\mu^\nu$), however, is tensorial since the non tensorial terms cancel out. Clearly, we have

$$\beta_\lambda^{\prime s} = A_{\lambda'}^\mu A_\nu^{s'} \beta_\mu^\nu, \tag{3.11}$$

and, equivalently,

$$\beta_\mu^\nu \mathbf{e}^\mu \wedge \mathbf{e}_\nu = \beta_\mu^{\prime\nu} \mathbf{e}^{\prime\mu} \wedge \mathbf{e}_\nu' = \beta_\mu^{\prime\prime\nu} \mathbf{e}^{\prime\prime\mu} \wedge \mathbf{e}_\nu' = \ldots. \tag{3.12}$$

Be aware of the fact that $\alpha^{\mu\nu}\mathbf{e}_\mu \wedge \mathbf{e}_\nu$ are not similar to $\beta^{\mu\nu}\mathbf{e}_\mu \wedge \mathbf{e}_\nu$ on sections of the Euclidean frame bundle since the α_μ^ν and ω_μ^ν do not transform linearly under the group of isometries.

We proceed with a clarification of the meaning of $\mathbf{e}_\mu$. Whereas we have $d\mathbf{e}_1 = \omega_1^\mu(x, dx)\mathbf{e}_\mu$ on a section, we have $d\mathbf{e}_1 = \omega_1^\mu(x, u, dx, du)\mathbf{e}_\mu$ in the bundle. This emphasizes that, at each point x of the base space, $\mathbf{e}_1$ does not represent a vector, but the first element of all possible (orthonormal) bases at x.

9.4 Computation in coordinate bases of the affine extension of the Levi-Civita connection

The affine extension of the equation (3.4) satisfied by the LCC is

$$dg_{\mu\nu} = \alpha_{\mu\nu} + \alpha_{\nu\mu}, \tag{4.1}$$

which implies, using primes for the implicit bases of differential $1-$forms,

$$g_{\mu\nu,\lambda} = \underline{\Gamma}_{\mu\lambda}^{\prime\pi} g_{\pi\nu} + \underline{\Gamma}_{\nu\lambda}^{\prime\pi} g_{\pi\mu}. \tag{4.2a}$$

We shall use underlined quantities whenever we refer to the components of α or of its curvature. We interchange the subscripts μ and λ to obtain

$$g_{\lambda\nu,\mu} = \underline{\Gamma}_{\lambda\mu}^{\prime\pi} g_{\pi\nu} + \underline{\Gamma}_{\nu\mu}^{\prime\pi} g_{\pi\lambda}. \tag{4.2b}$$

and the subscripts ν and λ to get

$$g_{\mu\lambda,\nu} = \underline{\Gamma}'^{\pi}_{\mu\nu} g_{\pi\lambda} + \underline{\Gamma}'^{\pi}_{\lambda\nu} g_{\pi\mu}. \tag{4.2c}$$

We add equations (4.2a) and (4.2b), subtract (4.2c) and use the symmetry

$$\underline{\Gamma}'^{\pi}_{\mu\nu} = \underline{\Gamma}'^{\pi}_{\nu\mu}, \tag{4.3}$$

which is equation (11.4) of the previous chapter applied to the LCC. We thus get

$$g_{\mu\nu,\lambda} + g_{\lambda\nu,\mu} - g_{\mu\lambda,\nu} = 2\underline{\Gamma}'^{\pi}_{\mu\lambda} g_{\pi\nu}. \tag{4.4}$$

With the appropriate contraction, one arrives at the sought result, namely

$$\underline{\Gamma}'^{\pi}_{\mu\lambda} = \frac{1}{2} g^{\nu\pi}(g_{\mu\nu,\lambda} + g_{\lambda\nu,\mu} - g_{\mu\lambda,\nu}), \tag{4.5}$$

which shows by explicit calculation that a solution exists and is unique.

Exercise. Find the $\underline{\Gamma}'^{\pi}_{\mu\lambda}$'s for the line element $ds^2 = d\mathbf{P} \cdot d\mathbf{P} = d\rho^2 + \rho^2 d\theta^2$. Verify that the Euclidean curvature is zero.

Exercise. Same problem for the line elements

$$ds^2 = d\mathbf{P} \cdot d\mathbf{P} = d\rho^2 - \rho^2 d\theta^2 \tag{4.6}$$

and

$$ds^2 = dr^2 + r^2 d\theta^2 + r^2 \sin^2\theta d\phi^2. \tag{4.7}$$

Exercise. Find the Levi-Civita connection for

$$ds^2 = e^{2\nu(r)} dt^2 - e^{2\lambda(r)}[dr^2 + r^2(d\theta^2 + \sin\theta d\phi^2)], \tag{4.8}$$

where $\nu(r)$ and $\lambda(r)$ are undetermined functions that admit second derivatives. The solution to this exercise is tedious through the use of (4.5). We shall later see a faster way to solve for the Levi-Civita connection (by inspection, which is advantageous in most cases).

9.5 Computation of the contorsion

We proceed to compute the contorsion of Euclidean connections. We rewrite (3.6) as

$$\Omega_\mu = -\omega^\nu \wedge \beta_{\nu\mu}, \tag{5.1}$$

and recall the definition of the components of $\beta_{\nu\mu}$ through

$$\beta_{\nu\mu} = L_{\nu\mu\lambda}\omega^\lambda. \tag{5.2}$$

In this way, we have

$$\Omega_\mu = -L_{\nu\mu\lambda}\omega^\nu \wedge \omega^\lambda = (L_{\lambda\mu\nu} - L_{\nu\mu\lambda})(\omega^\nu \wedge \omega^\lambda), \qquad (5.3)$$

where, as usual, the parenthesis in $(\omega^\nu \wedge \omega^\lambda)$ is used to signify summation over a basis of 2−forms, rather than over all possible values of the indices ν and λ. Recall that we defined the components of the torsion by

$$\Omega_\mu = \frac{1}{2}R_{\mu\nu\lambda}\omega^\nu \wedge \omega^\lambda = R_{\mu\nu\lambda}(\omega^\nu \wedge \omega^\lambda). \qquad (5.4)$$

Notice that the coefficients of $(\omega^\nu \wedge \omega^\lambda)$ in (5.4) are not differences of R's, unlike the coefficients on the right hand side of (5.3).

We equate corresponding coefficients in (5.3) and (5.4) and get

$$R_{\mu\nu\lambda} = L_{\lambda\mu\nu} - L_{\nu\mu\lambda} = L_{\mu\nu\lambda} + L_{\lambda\mu\nu}, \qquad (5.5)$$

where we have used the skew-symmetry of β, and thus of L, with respect to the first pair of indices in $L_{\nu\mu\lambda}$. Two more copies of this equation with the indices changed are

$$R_{\nu\lambda\mu} = L_{\nu\lambda\mu} + L_{\mu\nu\lambda} \qquad (5.6)$$

$$R_{\lambda\nu\mu} = L_{\lambda\nu\mu} + L_{\mu\lambda\nu}. \qquad (5.7)$$

Adding the last three equations and taking into account that $L_{\pi\sigma\rho} = -L_{\sigma\pi\rho}$, we have

$$L_{\mu\nu\lambda} = \frac{1}{2}(R_{\mu\nu\lambda} + R_{\nu\lambda\mu} + R_{\lambda\nu\mu}). \qquad (5.8)$$

A mnemonic rule for this equation is: form the cyclic sum of R's starting with the same indices $\mu\nu\lambda$ as on the left hand side, and replace $R_{\lambda\mu\nu}$ with its negative, $R_{\lambda\nu\mu}$. We get β_μ^ν as

$$\beta_\mu^\nu = \eta_\nu\beta_{\mu\nu} = \eta_\nu L_{\mu\nu\lambda}\omega^\lambda. \qquad (5.9)$$

where η_ν is ± 1, depending on the signature of the metric.

In the previous section, we computed the LCC in terms of coordinate bases. If we wish to compute it in terms of orthonormal bases, suffice to notice that α satisfies a system of equations similar to β's. Since $-d\omega^\nu = -\omega^\nu \wedge \alpha_\nu^\mu$, the computation that we carried out in this section can be repeated with $-d\omega^\mu$ playing the same role as Ω^μ. Do not overlook to skew-symmetrize $-d\omega^\mu$, i.e. to express it in terms of basis $(\omega^\mu \wedge \omega^\nu)$.

9.6 Computation of the Levi-Civita connection by inspection

We now solve for the LCC by inspection, using orthonormal bases. It is a convenient method when the metric is not very complicated. Most interesting cases are like that. Readers are warned that, in order to avoid a lengthy title of the section, we have referred in it to the LCC. What we are about to compute is α_μ^ν, which may or may not be the Euclidean connection of the manifold, but which also is present when the metric is.

One cannot give a simple algorithm that will always lead to a solution by inspection in a manageable way. Readers may wish to consider a few rules of thumb (illustrated with examples) in order to start becoming familiar with such a solving.

RULE ♯1. We inspect one by one all the equations in (3.3)-(3.4), starting with those whose left-hand side $(d\omega^i)$ is not zero.

The polar metric in ξ^2: $ds^2 = d\rho^2 + \rho^2 d\theta^2$. We are given this metric and the condition of zero torsion and we are asked to compute the ω^i's and ω_i^ν's. We readily obtain

$$\omega^1 = d\rho, \qquad \omega^2 = \rho d\theta, \qquad \omega_2^1 = -\omega_1^2. \tag{6.1}$$

The first equations of structure become

$$d\omega^1 = 0 = \omega^2 \wedge \omega_2^1, \tag{6.2}$$

$$d\omega^2 = d\rho \wedge d\theta = \omega^1 \wedge \omega_1^2. \tag{6.3}$$

Because of rule ♯ 1, we first inspect (6.3), which leads us to provisionally assume that ω_1^2 is proportional to $d\theta$. We now insert $\omega_2^1 = -d\theta$ in (6.3) and find that, indeed, it is satisfied. So $\omega_1^2 = -\omega_2^1 = d\theta$ is *the solution*.

It may appear that Eq. (6.3) was not used for finding the solution and that, therefore, there is redundancy in the equations. This is not correct. The most general solution of Eq. (6.3) is $\omega_1^2 = d\theta + f d\rho$, where f is an arbitrary function of ρ. Of all these solutions, Eq. (6.3) picks the one with $f = 0$. However, when we solve by inspection, we first focus solely on the term $d\theta$ that must be present in ω_1^2 for Eq. (6.3) to be satisfied, not those which, like $f d\rho$, might or might not be present. We thus formulate the rule that follows.

RULE ♯ 2. In order to avoid as much as possible the carrying of unknown coefficients, we only extract from each equation the information that some ω_i^ν must contain a term proportional to some specific ω^λ. We do not consider the other terms whose existence may not be needed until later on. At each stage, everything that is not necessarily different from zero is provisionally zero.

The polar metric in 2-D Lorentz space. Consider the metric $d\rho^2 - \rho^2 d\theta^2$ and zero torsion. We now have:

$$\omega^0 = d\rho \quad \omega^1 = \rho d\theta \quad \omega_0^1 = \omega_1^0. \tag{6.4}$$

From $d\omega^1 = d\rho \wedge d\theta = \omega^0 \wedge \omega_0^1$, we find that, provisionally, $\omega_0^1 = d\theta$. Since $\omega_1^0 = d\theta$ satisfies the $d\omega^0$ equation of structure, we conclude $\omega_0^1 = \omega_1^0 = d\theta$.

The spherical metric in ξ^3: $ds^2 = dr^2 + r^2 d\theta^2 + r^2 \sin\theta d\phi^2$. We readily write

$$\omega^1 = dr, \quad \omega^2 = rd\theta, \quad \omega^3 = r\sin\theta d\phi, \quad \omega_i^j = -\omega_j^i. \tag{6.5}$$

Hence,

$$d\omega^2 = dr \wedge d\theta = r^{-1}\omega^1 \wedge \omega^2 = \omega^1 \wedge \omega_1^2 + \omega^3 \wedge \omega_3^2 \tag{6.6}$$

$$d\omega^3 = \sin\theta dr \wedge d\theta = r^{-1}\omega^1 \wedge \omega^3 = \omega^1 \wedge \omega_1^3 + \omega^2 \wedge \omega_2^3 \tag{6.7}$$

$$d\omega^1 = 0 = \omega^2 \wedge \omega_2^1 + \omega^3 \wedge \omega_3^1. \tag{6.8}$$

From (6.6), $\omega_1^2 = r^{-1}\omega^2$ (provisionally). If ω_1^2 contains a term proportional to ω^1, and ω_3^2 contains a term proportional to ω^3, these terms will show up when inspecting other equations. For the same reason, if ω_1^2 contains a term proportional to ω^3 and ω_3^2 contains a term proportional to ω^1 such that their respective contributions cancel with each other, we do not bother for the time being. We now examine (6.7). Necessarily, ω_1^3 must contain the term $r^{-1}\omega^3$. So, provisionally, $\omega_1^3 = r^{-1}\omega^3$. These provisional ω_i^ν's become the solution since they also satisfy the last equation to be checked, (6.8).

RULE ♯3. If we are able to match by inspection all the equations with provisional ω_μ^ν's, they become the solution due to the uniqueness that was proved in Section 3.

The Schwarzschild metric in isotropic coordinates. It is given by

$$ds^2 = (\omega^0)^2 - \sum (\omega^i)^2, \tag{6.9}$$

where

$$\omega^0 = e^{\phi(r)}dt, \quad \omega^1 = e^{\mu(r)}dr, \quad \omega^2 = e^{\nu(r)}rd\theta, \quad \omega^3 = e^{\lambda(r)}r\sin\theta d\phi. \tag{6.10}$$

We proceed to inspect the equation for $d\omega^0$, using primes to denote derivatives with respect to r:

$$d\omega^0 = \phi'e^{(\phi-\mu)}\omega^1 \wedge dt = \dots \tag{6.11}$$

We do not need to write out $\omega^\rho \wedge \omega_\rho^0$. We provisionally have

$$\omega_1^0 = \phi'e^{(\phi-\mu)}dt. \tag{6.12}$$

We skip for the moment the $d\omega^1$ equation and consider the equation for $d\omega^2$:

$$d\omega^2 = (1 + \mu'r)e^\mu dr \wedge d\theta = (1 + \mu'r)\omega^1 \wedge d\theta = \omega^\lambda \wedge \omega_\lambda^2. \tag{6.13}$$

$d\theta$ is proportional to ω^2, which cannot be the first factor in $\omega^\lambda \wedge \omega_\lambda^2$ since ω_2^2 is zero. We compare $dr \wedge d\theta$ with $\omega^1 \wedge \omega_1^2$ and obtain, provisionally,

$$\omega_1^2 = (1 + \mu'r)d\theta = -\omega_2^1. \tag{6.14}$$

This was the first appearance of the ω_1^2 (equivalently ω_2^1) forms. We now inspect $d\omega^3$, where we write all the dx^μ in terms of the ω^ρ's except for $d\phi$. Thus

$$d\omega^3 = \sin\theta(\mu'r + 1)\omega^1 \wedge d\phi + \cos\theta\omega^2 \wedge d\phi. \tag{6.15}$$

This requires that, necessarily, ω_1^3 and ω_2^3 contain at least the terms exhibited in:

$$\omega_1^3 = \sin\theta(r\mu' + 1) = -\omega_3^1, \tag{6.16}$$

$$\omega_2^3 = \cos\theta d\phi = -\omega_3^2. \tag{6.17}$$

We had not detected a non-null ω_2^3 term in $d\omega^2$. We thus return to $d\omega^2$:

$$d\omega^2 = \ldots \omega^3 \wedge \omega_3^2. \tag{6.18}$$

In (6.17), ω_3^2 is proportional to ω^3 so that $\omega^3 \wedge \omega_3^2$ is zero and the equation for $d\omega^2$ continues to be balanced. We now inspect the $d\omega^1$ equation. It involves the ω_ρ^1 terms with $\rho \neq 1$. The provisional results (6.12)-(6.14) satisfy the equation

$$d\omega^1 = 0 = \omega^\rho \wedge \omega_\rho^1. \tag{6.19}$$

One finally checks everything, just in case.

RULE ♯4. Consider a term for ω_μ^ν emerging in the equation for $d\omega^\nu$, and whose being different from zero was not apparent when previously considering the equation for $d\omega^\mu$. One reinspects the $d\omega^\mu$ equation to see if said new term —also present in ω_ν^μ— yields or not a contribution to $d\omega^\mu$. If it does not, we keep inspecting new equations. If it does, see the next example.

The van Stockum's metric. It is defined by

$$\omega^0 = dt - ar^2 d\phi, \quad \omega^1 = e^{-\alpha^2 r^2/2}dr, \quad \omega^2 = e^{-\alpha^2 r^2/2}dz, \quad \omega^3 = rd\phi, \quad (6.20)$$

with signature $(+1,-1,-1,-1)$, where α and a are constants. For present purposes, we shall set $a = \alpha = 1$.

The equation for $d\omega^0$ reads

$$d\omega^0 = -2rdr \wedge d\phi = -2e^{r^2/2}\omega^1 \wedge \omega^3 = \omega^1 \wedge \omega_1^0 + \omega^2 \wedge \omega_2^0 + \omega^3 \wedge \omega_3^0. \tag{6.21}$$

The exterior product $\omega^1 \wedge \omega^3$ may arise from either $\omega^1 \wedge \omega_1^0$ or $\omega^3 \wedge \omega_3^0$, or both at the same time. So we leave it for future examination.

We proceed to examine the equation for $d\omega^2$:

$$d\omega^2 = -re^{-r^2/2}dr \wedge dz = -re^{r^2/2}\omega^1 \wedge \omega^2 = \sum_\mu \omega^\mu \wedge \omega_\mu^2. \tag{6.22}$$

Since ω_2^2 is zero, $\omega^1 \wedge \omega^2$ can only arise from $\omega^1 \wedge \omega_1^2$. Thus, we provisionally have

$$\omega_1^2 = -re^{r^2/2}\omega^2 = -\omega_2^1. \tag{6.23}$$

From the equation for $d\omega^3$, we obtain

$$d\omega^3 = dr \wedge d\phi = e^{r^2/2}r^{-1}\omega^1 \wedge \omega^3 = \sum_\mu \omega^\mu \wedge \omega^3_\mu, \tag{6.24}$$

which provisionally yields

$$\omega^3_1 = e^{r^2/2}r^{-1}\omega^3 = -\omega^1_3. \tag{6.25}$$

We then go to the $d\omega^1$ equation. We substitute (6.23) and (6.25) in

$$d\omega^1 = 0 = \omega^0 \wedge \omega^1_0 + \omega^2 \wedge \omega^1_2 + \omega^3 \wedge \omega^1_3, \tag{6.26}$$

and obtain

$$d\omega^1 = 0 = \omega^0 \wedge \omega^1_0, \tag{6.27}$$

which implies that ω^1_0 is proportional to ω^0 and has no other components.

We have to return to the equation for $d\omega^0$, whose consideration had been postponed. If ω^1_0 is proportional to ω^0, as required by (6.27), it is not possible to match $\omega^1 \wedge \omega^0_1$ with anything else in (6.21). So, we enter a new phase in the computation.

Because it appears that we have a problem with ω^0_1, we return to the $d\omega^0$ equation, (6.21), and assume that the term $\omega^1 \wedge \omega^0_1$ will make part of matching the left hand side. This means that ω^0_1 will have a term where ω^3 is a factor. This in turn leads us to consider that ω^0_3 may have a term where ω^1 is a factor. We thus set

$$\omega^0_1 = A\omega^3, \quad \omega^0_3 = B\omega^1. \tag{6.28}$$

Substitution in the equation (6.21) for $d\omega^0$ yields

$$-2e^{r/2} = A - B. \tag{6.29}$$

Because $\omega^1_0 = \omega^0_1$, we must recheck the equation for $d\omega^1$, namely (6.26):

$$0 = \omega^0 \wedge A\omega^3 + \omega^2 \wedge \omega^1_2 + \omega^3 \wedge \omega^1_3, \tag{6.30}$$

which indicates that ω^1_3 must necessarily contain a term proportional to ω^0, in addition to the term exhibited in (6.25):

$$\omega^1_3 = A\omega^0 - e^{r^2/2}r^{-1}\omega^3. \tag{6.31}$$

In this new phase, we still have to consider the $d\omega^2$ and $d\omega^3$ equations. We thus return to (6.24). Both ω^0_3 and ω^1_3 enter the equation for $d\omega^3$:

$$d\omega^3 = e^{r^2/2}r^{-1}\omega^1 \wedge \omega^3 = -B\omega^0 \wedge \omega^1 - \omega^1 \wedge A\omega^0 + \omega^1 \wedge e^{r^2/2}r^{-1}\omega^3 + \omega^2 \wedge \omega^3_2. \tag{6.32}$$

The matching of the $\omega^0 \wedge \omega^1$ terms requires $A = B$. Hence, using (6.29),

$$A = B = -e^{r^2/2}. \tag{6.33}$$

Thus, tentatively,

$$\omega_3^1 = -e^{r^2/2}\omega^0 + e^{r^2/2}r^{-1}\omega^3 = -\omega_1^3, \tag{6.34}$$

$$\omega_0^1 = -e^{r^2/2}\omega^3 = \omega_1^0, \tag{6.35}$$

$$\omega_3^0 = -e^{r^2/2}\omega^1 = \omega_0^3. \tag{6.36}$$

Notice that, in principle, ω_2^3 need not be zero, since it could be proportional to ω^2. But it is not necessarily different from zero at this point.

One easily verifies that all the equations for the $d\omega^\mu$ are verified with those values together with

$$0 = \omega_2^0 = \omega_0^2 = \omega_3^2 = \omega_3^2, \tag{6.37}$$

which thus yields the solution

From the example just solved, we extract a final rule:

RULE ♯5. if we cannot match the coefficients, we introduce undetermined coefficients (A and B in the example).

In the unlikely event that the process becomes too messy, resort to coordinate bases and use formula (4.5). You can always return to the orthonormal bases knowing how the connection transforms.

9.7 Stationary curves and Euclidean AUTOPARALLELS

A differentiable manifold endowed with an Euclidean connection has a metric structure superimposed on an affine structure. The affine structure allows for a concept of autoparallels, or lines of constant direction. The metric structure gives rise to a concept of stationary length between two points, not necessarily maximum or minimum length.

We saw in section 8.7 how to turn into equations the concept of autoparallels. The metric structure now provides us with the length of curves as a most desirable parameter, since the autoparallels then satisfy the equation $d\mathbf{t} = 0$, as follows from differentiation with respect to s of

$$\mathbf{t} \cdot \mathbf{t} = \frac{d\mathbf{P}}{ds} \cdot \frac{d\mathbf{P}}{ds} = \frac{ds^2}{ds^2} = 1, \tag{7.1}$$

since $d\mathbf{t}/ds$ is perpendicular to $\mathbf{t}$ and their dot product is zero.

When we use s as a parameter, we choose to refer to $\mathbf{t}$ as $\mathbf{u}$. We then have

$$0 = \frac{d\mathbf{u}}{ds} = \frac{du^\nu + u^\mu \omega_\mu^\nu}{ds} \mathbf{e}_\nu = \left(\frac{du^\nu}{ds} + u^\mu \Gamma_{\mu\lambda}^\nu u^\lambda\right) \mathbf{e}_\nu, \tag{7.2}$$

where we have replaced ω_ν^μ with $\Gamma_{\nu\lambda}^\mu \omega^\lambda$, and where $\frac{\omega^\lambda}{ds}$ means precisely the factor u^λ of ds in the expression for ω^λ. Needless to say that, since $d\mathbf{P} = \omega^\lambda \mathbf{e}_\lambda$, the

u^λ are components of the 4-velocity (in the case of spacetime). Equation (7.2) applies in particular to coordinate basis fields, i.e.

$$\ddot{x}^\nu + \Gamma'^\nu_{\mu\lambda}\dot{x}^\mu\dot{x}^\lambda = 0, \tag{7.3}$$

where overdot means differentiation with respect to distance, s. It thus means differentiation with respect to proper time, if dealing with the spacetime manifold. Of course, the issue becomes whether autoparallels would mean anything if the torsion of spacetime were not zero.

Consider now curves of stationary distance. By definition, we have

$$0 = \delta \int_Q^P ds = \delta \int_Q^P \left(g_{\mu\nu}\frac{dx^\mu}{ds}\frac{dx^\nu}{ds} \right) ds, \tag{7.4}$$

where δ means variation and where (P, Q) are the common end points of a family of curves. We rewrite (7.4) as

$$0 = \delta \int_Q^P L ds, \tag{7.5}$$

where

$$L = g_{\mu\lambda}\dot{x}^\mu\dot{x}^\lambda. \tag{7.6}$$

The Euler-Lagrange equations,

$$\frac{d}{ds}\frac{\partial L}{\partial \dot{x}^\rho} = \frac{\partial L}{\partial x^\rho}, \tag{7.7}$$

follow. Straightforward development of (7.7) yields

$$g_{\sigma\lambda}\ddot{x}^\lambda + g_{\sigma\lambda,\mu}\dot{x}^\mu\dot{x}^\lambda - \frac{1}{2}g_{\mu\sigma,\lambda}\dot{x}^\mu\dot{x}^\lambda = 0. \tag{7.8}$$

Contraction with $g^{\nu\sigma}$ and a minor manipulation results in

$$\ddot{x}^\nu + \frac{1}{2}g^{\nu\pi}(g_{\mu\pi,\lambda} + g_{\pi\lambda,\mu} - g_{\mu\lambda,\pi})\dot{x}^\mu\dot{x}^\lambda = 0. \tag{7.9}$$

We use eq. (4.5) to further write (7.9) as

$$\ddot{x}^\nu + \underline{\Gamma}'^\nu_{\mu\lambda}\dot{x}^\mu\dot{x}^\lambda = 0, \tag{7.10}$$

which is the equation for the extremals. Comparison with (7.3) shows that this equation is the same as the equation of the autoparallels with α as connection. This says that the extremals coincide with the autoparallels when the Euclidean connection is the LCC, i.e. in Riemannian geometry.

Before we proceed with the relation between autoparallels and extremals in the general case of non-zero torsion, one must be aware of the following. In general relativity, a "particle" having a mass comparable to the mass of nearby bodies cannot be viewed as a test particle. It will follow geodesics determined by the metric that the particle itself helps to create, by virtue of its affecting how the other particles move. Hence it has an indirect effect on its own motion.

In a hypothetical world with torsion, autoparallels and geodesics would not coincide in general since

$$\omega^\mu_\nu = \Gamma^\nu_{\mu\lambda}\omega^\lambda = \underline{\Gamma}^\nu_{\mu\lambda}\omega^\lambda + L^\nu_{\mu\lambda}\omega^\lambda, \tag{7.11}$$

where the $L^\nu_{\mu\lambda}$'s are the components of the contorsion. If the gravitational equations of motion were still given by geodesics, the gravitational interaction would not be directly sensitive to the torsion, since $\underline{\Gamma}^\nu_{\mu\lambda}$ is independent of it. We said "would not be directly sensitive" because the torsion might still contribute to how charged particles interact thus affecting the gravitational field they create.

An interesting issue is: what type of contribution to the equations of motion would the torsion yield if the autoparallels were the equations of motion? For that we need only consider $L^\nu_{\mu\lambda}\omega^\lambda$, since the contribution of $\underline{\Gamma}^\nu_{\mu\lambda}\omega^\lambda$ is additive to it and has already been discussed and assigned to gravitation. The question that we are asking is too general. We shall ask the simpler question of contribution of torsions of the specific form

$$\Omega^\mu = -v^\mu F, \tag{7.12}$$

where v^μ is some vector field and F is a scalar-valued differential 2−form.

In view of the considerations made about the additivity of the effects of the metric and the torsion, we may assume that the metric is flat for present purposes. We then have

$$0 = \ddot{x}^\nu + L^\nu_{\mu\lambda}\dot{x}^\mu\dot{x}^\lambda. \tag{7.13}$$

We raise the index ν in Eq. (5.8):

$$L^\nu_{\mu\lambda} = \frac{1}{2}(R^\nu_{\mu\lambda} + R^\nu_{\lambda\mu} + R^\nu_{\lambda\mu}). \tag{7.14}$$

The contraction with $\dot{x}^\mu\dot{x}^\lambda$ annuls the middle term because of skew-symmetry of $R^\nu_{\lambda\mu}$ with respect to the same indices. We then use here (7.12) and obtain

$$0 = \ddot{x}^\nu - \frac{1}{2}(v_\mu F^\nu_\lambda + v_\lambda F^\nu_\mu)\dot{x}^\mu\dot{x}^\lambda, \tag{7.15}$$

i.e.

$$\ddot{x}^\nu = +(v_\mu\dot{x}^\mu)F^\nu_\lambda\dot{x}^\lambda. \tag{7.16}$$

If one could replace v_μ with $\dot{x}_\mu$ (in which case $v_\mu\dot{x}^\mu$ would be the unity), Eq. (7.16) would represent the equation of motion of special relativity with Lorentz force, provided that one were to think of F as the electromagnetic 2−form up to an appropriate constant. In other words, the electromagnetic field would become associated with the torsion of spacetime [72], [64].

A torsion $F_{\nu\lambda}\dot{x}^\lambda$ actually is Finslerian. In the Finsler bundle, the 4-velocity is not a field on curves or redundant coordinates in the fibers. These coordinates now belong to the Finslerian base manifold, and they become velocity components on curves of that manifold that are natural liftings of spacetime curves. The proper way to write the torsion $F_{\nu\lambda}\dot{x}^\lambda$ in the Finsler bundle is $\Omega^0 = -F$. Since the group in the fibers of Finslerian spacetime is the rotation group in three dimensions, Ω^0 is an invariant; it does not acquire Ω^i terms in other frame fields.

Finslerian torsions have terms that look Finslerian and others that do not (but still are!). When the first ones are null, the equation of the autoparallels depends on Ω^0 but not on Ω^i. It takes the form of equation of motion with Lorentz force (the metric would still contribute with the gravitational force) [81].

9.8   Euclidean and Riemannian curvatures

The Euclidean and Riemannian curvatures coexist and are related through the contorsion, β^ν_μ, as follows:

$$d\omega - \omega \wedge \omega \equiv d\omega^\nu_\mu - \omega^\lambda_\mu \wedge \omega^\nu_\lambda = d(\alpha^\nu_\mu + \beta^\nu_\mu) - (\alpha^\lambda_\mu + \beta^\lambda_\mu) \wedge (\alpha^\nu_\lambda + \beta^\nu_\lambda). \quad (9.1)$$

This equation would take a far more complicated form if we expressed the contorsion in terms of the torsion. We reorganize (9.1) to read

$$d\omega - \omega \wedge \omega = d\alpha - \alpha \wedge \alpha - \beta \wedge \beta + (d\beta - \alpha \wedge \beta - \beta \wedge \alpha). \quad (9.2)$$

The notation introduced should be taken into account in order not to inadvertently set $\alpha \wedge \alpha$, $\beta \wedge \beta$ etc. equal to zero. These equations may be rewritten as

$$\Omega^\nu_\mu = \underline{\Omega}^\nu_\mu - (\beta \wedge \beta)^\nu_\mu + (d_\alpha\beta)^\nu_\mu, \quad (9.3)$$

with $(d_\alpha\beta)^\nu_\mu$ as components relative to $\mathbf{e}^\mu \wedge \mathbf{e}_\nu$ of the exterior(-covariant) derivative of β when computed as if α^ν_μ were the connection. Notice that $(\beta \wedge \beta)^\nu_\mu$ simply means $\beta^\lambda_\mu \wedge \beta^\nu_\lambda$.

Substitution of $\omega - \beta$ for α in the last parenthesis of (9.2) and (9.3) yields

$$d\omega - \omega \wedge \omega = d\alpha - \alpha \wedge \alpha + \beta \wedge \beta + (d\beta - \omega \wedge \beta - \beta \wedge \omega), \quad (9.4)$$

and

$$\Omega^\nu_\mu = \underline{\Omega}^\nu_\mu + (\beta \wedge \beta)^\nu_\mu + (d_\omega\beta)^\nu_\mu, \quad (9.5)$$

respectively. The $(d_\omega\beta)^\nu_\mu$ now are the components relative to $\mathbf{e}^\mu \wedge \mathbf{e}_\nu$ of the exterior(-covariant) derivative of β with respect to the actual Euclidean connection of the differentiable manifold.

In the tensor calculus, relations of this nature take very complicated forms. In order to continue dissuading readers from thinking that they would have an advantage in working with tensors in dealing with equations such as the preceding ones, we reproduce an equation of historical interest, sent by Cartan to Einstein [29]. Cartan wrote:

"Would you like to know how to express $R_{\alpha\beta}$ in terms of $\Lambda^{\gamma}_{\alpha\beta}$? One has

$$2R_{\alpha\beta} = \Lambda^{\beta}_{\underline{\alpha}\mu;\mu} + \Lambda^{\alpha}_{\underline{\beta}\mu;\mu} - \phi_{\alpha;\beta} - \phi_{\beta;\alpha} + \Lambda^{\rho}_{\alpha\underline{\mu}}\Lambda^{\beta}_{\underline{\mu}\rho} + \Lambda^{\rho}_{\beta\underline{\mu}}\Lambda^{\alpha}_{\underline{\mu}\rho} + S_{\alpha}S_{\beta}$$

$$-(\Lambda^{\beta}_{\underline{\alpha}\rho} + \Lambda^{\alpha}_{\underline{\beta}\rho})\phi_{\rho} + 2\Lambda^{\sigma}_{\alpha\rho}\Lambda^{\sigma}_{\underline{\beta}\rho} - g_{\alpha\beta}S_{\mu}S_{\underline{\mu}}." \tag{9.6}$$

This $R_{\alpha\beta}$ is our $\underline{R}_{\alpha\beta}$, i.e. for the LCC. The symbols used had been defined by Einstein in a paper on teleparallelism [39]. Underlined indices mean that they have been raised or lowered in the standard way through the metric (only the notation is not standard). Underlined subscripts must, therefore, be considered as raised, thus denoting contravariant components (Einstein's way of doing things in this regard lets one remember in what position a given index in a particular quantity was born). ϕ_{μ} is $\Lambda^{\alpha}_{\mu\alpha}$, and Λ is given as

$$\Lambda^{\alpha}_{\mu\nu} \equiv \Delta^{\alpha}_{\mu\nu} - \Delta^{\alpha}_{\nu\mu}, \tag{9.7}$$

with Δ introduced as

$$\delta A^{\mu} \equiv \Delta^{\mu}_{\alpha\beta}A^{\alpha}\delta x^{\beta}, \tag{9.8}$$

which is the replacement for

$$\delta A^{\mu} = -\underline{\Gamma}^{\mu}_{\alpha\beta}A^{\alpha}\delta x^{\beta} \tag{9.9}$$

involving the LCC. With the indices α, β, γ and δ being all different, S_{α} is $S^{\beta}_{\underline{\gamma\delta}}$, which in turn is defined as

$$S^{\alpha}_{\underline{\mu\nu}} \equiv \Lambda^{\alpha}_{\underline{\mu\nu}} + \Lambda^{\nu}_{\underline{\alpha\mu}} + \Lambda^{\mu}_{\underline{\nu\alpha}}. \tag{9.10}$$

We wish to warn inexperienced readers about misusing the annulment of the connection at a point. It implies $\beta = -\alpha$, but not the equality of $d\beta$ and $-d\alpha$. That would require the annulment of the connection also on a neighborhood of the point. Replacement of $\beta = -\alpha$ in $d\beta$ would be incorrect. In (9.2), it would force the Euclidean curvature to be zero at that point, which might be in contradiction with whatever assumptions might have been made about the Euclidean curvature in the first place.

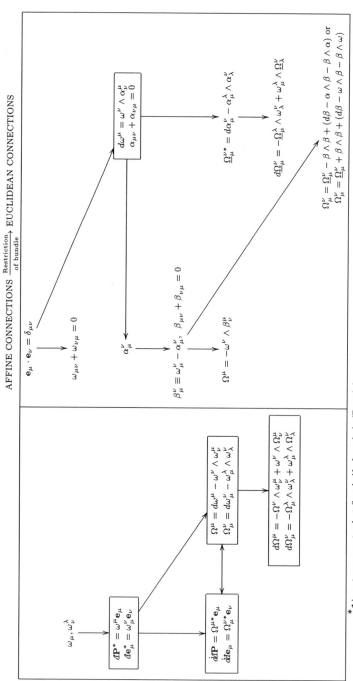

$*\tilde{d}$ is not an operator but first half of symbols $\tilde{d}\mathbf{P}$ and $\tilde{d}\mathbf{e}$.

$\dot{d}$: formal differentiation; $\dot{d}\dot{d}\neq 0$.

$\Omega^\mu \equiv R^\mu_{\nu\lambda}(\omega^\nu\wedge\omega^\lambda),\quad \Omega^\nu_\mu \equiv R^\nu_{\mu\lambda\pi}(\omega^\lambda\wedge\omega^\pi),\quad \underline{\Omega}^\nu_\mu \equiv \underline{R}^\nu_{\mu\lambda\pi}(\omega^\lambda\wedge\omega^\pi).$

Euclidean connections satisfy formulas on the right box in addition to the formulas on the left box.

Chapter 10

RIEMANNIAN PSEUDO-SPACES & RIEMANNIAN SPACES

10.1 Klein geometries in greater DETAIL

A Klein geometry is a geometry in the spirit of the Erlangen program in its modern interpretation. If a geometry is not of that type or an appropriate generalization thereof, the manifold in which it lives would be referred by Cartan as a pseudo-space. He actually referred to Riemannian manifolds in the first decades of their lives as pseudo-spaces. More on these pseudo-spaces to come in this and the next section.

A Klein geometry is nowadays defined as a pair (G, G_0) of a group G and a subgroup G_0 having a certain important technical property. There is not unanimity as to what this property should be (For details see [27], [67], [82]). We believe [27] to be the right one.

Cartan spoke of Klein geometries as the study of invariants under the transformations of a group. But it is clear from his writing, though implicitly, that he had in mind pairs of a group and a subgroup. He did not need to elaborate on the concept since, for the most part, he studied generalizations of specific Klein geometries. The groups G of those geometries have subgroups G_0 with the required property.

G_0 happens to be the subgroup constituted by all the elements of G that leave a point unchanged. The transformations in G_0 are present in the fibers of the generalized spaces. This is in contrast with transformations like translations, which are valid only in differential form in the generalizations (see the Preface). Later in his life Cartan wrote explicitly and at length about the role of group theory in modern differential geometry [21]. In his presentation, he mentioned G_0 repeatedly, but only exceptionally did he refer to the pair of group and subgroup.

In Euclidean geometry, G is the group of the so called displacements (translations and rotations, with the group of rotations as G_0). In affine geometry G is the group of the affine transformations (translations and linear transformations, with the linear transformations as G_0). In projective geometry, G is the projective group (translations and homographies, with the homographies as G_0). Again, the transformations in G_0 are the transformations from G that leave a point unchanged.

The G group is called the fundamental group of the corresponding geometry. One also speaks of the fundamental group of the space where the study of the properties of the figures takes place. With Cartan's own words [15]:

In each of these geometries and for expediency reasons, one attributes to the space where the figures under study are located the very properties of the corresponding group, or fundamental group; one thus speaks of "Euclidean space", "affine space", etc., instead of "space where one studies the properties of the figures that are left invariant by the Euclidean group, affine group, etc."

We proceed to explain what, in the modern view (read Cartan's), was wrong in Klein himself's view of his Erlangen program. In [21], Cartan stated that an analytic transformation from x^i to $\overline{x}^i$ transforms the differential form $g_{ij}dx^i dx^j$ into $\overline{g}_{ij}d\overline{x}^i d\overline{x}^j$. The formulas that show how to go from the x^i and the g_{ij} to the $\overline{x}^i$ and $\overline{g}_{ij}$ define an infinite Lie group depending on $n + [n(n+1)/2]$ variables. This is the group that Klein considered as underlying Riemann's geometry. We shall skip further details of this Kleinean view ("inappropriate" according to Cartan) of Riemannian geometry and of its evolution in the work of Veblen and of the school of Hamburg (per Cartan report [21]). After all, we are only interested in Cartan's view of how Riemannian geometry fits his generalization of what should have been Klein's view of geometry.

Cartan claimed that "the development of contemporary geometry" (again, actually meaning his own developments) also springs from Riemannian geometry, but in the stage of development achieved by this geometry after the discovery of the notion of parallelism [21] (Earlier in the paper, Cartan had given credit to Levi-Civita and Schouten for their almost simultaneous discovery of this notion).

Cartan further claimed that, in his work, the notion of group intervenes in a far more profound way than it have been the case. It continues to be the case with those who maintain that the group that defines Riemannian geometry is the infinite group of diffeomorphisms. He explained that this geometry involves notions of a Euclidean nature, such as distance, angle, area, volume, etc. He further stated that, although the notion of group would seem to be absent from this geometry, it is, however, present indirectly, since all these geometric notions have their origin in the group of Euclidean displacements. In fact, the frames constituted by vectors with origin at a point A of a Riemannian space (after 1917) are related by the group of rotations around A, and the theorems relative to these figures are the same as in Euclidean geometry. In other words, the rotations of frames (i.e. the elements of G_0) survive in finite form in Riemannian geometry. This is to be contrasted with the translations.

Still referring to generalizations such as Riemannian geometry (again, after 1917), Cartan stated that "the notion of parallelism permits equally to give a certain sense to the *infinitesimal* translations". The key word here is infinitesimal, which was not required in the case of rotations (we do not advocate the use of the term infinitesimal as freely as Cartan did almost a century ago, but there are cases, like the present one, where it helps to avoid clutter.) The group G also is the same group for the Klein geometries as for their respective generalizations, but only in differential form (Lie algebra) when the elements of G that are not in G_0 are concerned.

10.2   The false spaces of Riemann

This section deals with what Cartan called "false spaces of Riemann", term not to be confused with pseudo-Riemannian spaces, often used when the metric is not positive definite (Later in the chapter, the term Riemannian space will be another way to refer to differentiable manifolds endowed with a zero-torsion Euclidean connection). Cartan wrote [13]:

> "With his definition of parallelism (1), Levi-Civita was the first to succeed in making the *false* metric *spaces* of Riemann, if not true Euclidean spaces, which is impossible, at least *spaces with a Euclidean connection,* considered as collections of small pieces of Euclidean space . . ."

Emphasis is as in the original, where the terms false and spaces come together in *"faux espaces* metriques de Riemann." The reference (1) is to the famous Levi-Civita paper of 1917 [52].

At the end of the 19th century, there were two main developments in geometry. One was due to Riemann [61], [63] and the other one to Felix Klein [49]. Riemann's was based on the concept of distance. Klein's was based on the concept of group and, related to it, on the concept of geometric equality. We proceed to briefly cite from Cartan (ibid) on groups and geometric equality [13]:

> "The notion of group . . . became in geometry, thanks to F. Klein and S. Lie, a clasification principle which completed the removal of classical Euclidean geometry from the privileged position that it had occupied for such a long time; this geometry is in fact nothing but the study of those properties of figures that are conserved by a certain group of transformations (the displacements), and the axioms of geometric equality simply express the property of these transformations of constituting a group, as made evident by Poincare".

Cartan generalized the Klein geometries, i.e. the geometries of flat affine, Euclidean, conformal, projective, etc. spaces. Hence, this is not the same sense of generalization where a flat geometry generalizes another flat geometry (like affine geometry generalizes Euclidean geometry).

The false (i.e. the old) Riemannian geometry did not generalize the Kleinean Euclidean geometry. It generalized the metric part but not the affine part of the Euclidean structure because it did not have any affine content. It is only after 1917 that one can say that Riemannian spaces may be viewed "as if" made of small pieces of Euclidean space [15]. We said "as if" because, for example, very small pieces of a thin glass sphere may look flat, but the radius of the sphere remains "written" in every piece, regardless of size.

In his paper on the role of Lie group theory in the evolution of modern geometry, Cartan [21] emphasized the difference in viewpoint between Klein and Riemann as follows:

". . . whereas Riemannian geometry is a simple generalization of Euclidean geometry, Klein retains from the latter above all the notion of geometric equality, Riemann retains only the notion of distance. Pushed to their last consequences, the two view points are radically divergent: the notion of distance disappears from the more general Klein geometries and the notion of equal figures disappears from the more general Riemannian geometries; the notion of group ceases to be at the foundation of the Riemannian geometries . . ."

To be more specific, the quantities that we know as components of the differential form α^ν_μ were already used in the old Riemannian geometry, but they did not play then any affine role, and the concept of differential forms did not yet exist. In fact the general theory of affine and Euclidean (and conformal, and projective, etc.) connections was not formulated until the early 1920's by Cartan (See the references we have given of his work of those years). One must say, however, that affine considerations were implicit in Ricci and Levi-Civita's monograph of 1901 [60], and specifically in their derivation of the formula for the curvature in Riemannian spaces. They are, however, disguised as a search for tensorial quantites. This primitively-affine approach to the curvature was reproduced by Einstein in his work on general relativity [37].

In the same paper, Cartan [21] characterized the old Riemannian geometry as "the theory of the invariants of a quadratic differential form, $g_{i\nu}dx^i dx^\nu$, on n variables x^i with respect to the infinite group of analytic transformations performed on these variables". The obtaining of those invariants is the core of the problem of determining whether two geometric objects are equivalent, i.e. related by a coordinate transformation. Hence this is one of the so-called problems of equivalence, which Riemann solved [62]. As reported by Pauli in his book on the theory of relativity [57], Christoffel and Lipshitz also solved the same problem of equivalence, independently of each other and of Riemann.

In the next section we shall provide a more detailed summary than in section 7.5 of Cartan's derivation of the Riemannian curvature as an invariant of the metric, using again the method of equivalence. What makes it an invariant is that it is a differential form in a bundle. We are using the term bundle still less specifically than we have done so far in order to reflect the very informal way in which Cartan introduced it, as we are about to see.

10.3 Riemannian pseudo-geometry and method of equivalence

Riemann's paper of greatest geometric relevance after his seminal paper on the foundations of geometry is the one in which, for the first time in known mathematical history, a formula for the curvature of a manifold of dimension three or greater than three was derived [62], [63]. But it was not precisely a geometric paper, in spite of that result. It was a paper on heat transfer, written to respond to a problem posed by the French Academy, known as Parisian Academy at the time. The curvature appeared not as a geometric concept, but as a set of quantities whose being null satisfied the problem of equivalence that emerges in the theory of false spaces of Riemann. It did not look much as geometry. The body of propositions of Riemannian geometry was still in its infancy when Riemann died in 1866.

We now reproduce the main lines of Cartan's derivation [8] of the Riemannian curvature and related equations as an application of his technique to solve equivalence problems. Grossly speaking, Cartan's method of equivalence consists in differentiating a set of differential 1−forms in a bundle representative of a problem, then defining new quantities emerging from that differentiation and further differentiating and defining new concepts until one runs out of things to differentiate. For extensive treatment of the method of equivalence see [42].

The differential 1−forms representative of metrics are the most general ω^μ's that diagonalize it:

$$ds^2 \equiv g_{\mu\nu}(x)dx^\mu dx^\nu = \sum_\lambda \varepsilon_\lambda [\omega^\lambda(x,u)]^2 \tag{3.1}$$

where $\varepsilon_\lambda = \pm 1$. In 2-D Euclidean space, we have

$$\omega^1 = \cos\alpha dx - \sin\alpha dy, \quad \omega^2 = \sin\alpha dx + \cos\alpha dy. \tag{3.2}$$

There is only one "u coordinate" in this case, namely α. For the Lorentzian signature, the u coordinate is the coordinate of the corresponding Lorentz group.

The exterior differentials of the ω_μ's can be written in an infinite number of ways as

$$d\omega_\mu = \sum_\nu \varepsilon_\nu \omega_\nu \wedge \alpha_{\nu\mu}. \tag{3.3}$$

Equation (3.3) helps define differential forms $\alpha_{\nu\mu}$. It will be recognized as the first equation of structure when the torsion is zero, except that the torsion is not a concept needed to solve the problem of equivalence and the α's are not affine or Euclidean connections here.

The ds^2 does not depend either on coordinates u or their differentials. We shall designate by d_x and d_u the differentiation with respect to coordinates x and coordinates u respectively. ω_μ is a function of curves that does not depend on the du's. The x and u in superscripted parentheses refer to the terms respectively proportional to dx and du in differential 1−forms. Evidently:

$$\omega_\mu^{(u)} = 0, \quad d_u(ds^2) = 0. \tag{3.4}$$

The second of equations (3.4) can be expanded as

$$\sum_\mu \varepsilon_\mu \omega_\mu^{(x)} d_u \omega_\mu^{(x)} = 0, \tag{3.5}$$

where juxtaposition means tensor product. On the other hand, the first of equations (3.4) implies $d_x \omega_\mu^{(u)} = 0$ and, therefore,

$$d_u \omega_\mu^{(x)} = d_u \omega_\mu^{(x)} - d_x \omega_\mu^{(u)}. \tag{3.6}$$

Cartan further transforms $d_u \omega_\mu^{(x)}$, which he then substitutes in (3.5). This allows him to derive

$$\alpha_{\lambda\mu}^{(u)} + \alpha_{\mu\lambda}^{(u)} = 0 \tag{3.7}$$

as a necessary and sufficient condition for (3.3) to be satisfied. He thus proves that the skew-symmetric part of $\alpha_{\lambda\mu} + \alpha_{\mu\lambda}$ is linear in the dx^μ and, therefore, in the ω^μ.

Cartan then shows that the remaining freedom to choose the α's can be used to make $\alpha_{\lambda\mu} + \alpha_{\mu\lambda}$ equal to zero. Suppose we had a set $\alpha_{\lambda\mu}$'s such that its $\alpha_{\lambda\mu}^{(u)}$ part satisfied (3.7) and thus (3.3). Let $\underline{\alpha}_{\lambda\mu}$ be defined as

$$\underline{\alpha}_{\lambda\mu} = \alpha_{\lambda\mu} + \sum_{\nu=1}^{\nu=n} \epsilon_\nu \beta_{\lambda\mu\nu} \omega_\nu, \tag{3.8}$$

where β is arbitrary except for satisfying the condition

$$\beta_{\lambda\mu\nu} = -\beta_{\mu\lambda\nu}. \tag{3.9}$$

$\underline{\alpha}_{\lambda\mu}$ also satisfies (3.7) and (3.3). The sum $\underline{\alpha}_{\lambda\mu} + \underline{\alpha}_{\mu\lambda}$ then is

$$\underline{\alpha}_{\lambda\mu} + \underline{\alpha}_{\mu\lambda} = \alpha_{\lambda\mu} + \alpha_{\mu\lambda} + (\beta_{\lambda\mu\nu} + \beta_{\mu\lambda\nu})\omega^\nu, \tag{3.10}$$

Given the horizontality of $\alpha_{\lambda\mu} + \alpha_{\mu\lambda}$, one could use this fact to solve for the $\beta_{\lambda\mu\nu}$ that satisfies the system constituted by (3.9) and the equations

$$\alpha_{\lambda\mu} + \alpha_{\mu\lambda} + \beta_{\lambda\mu\nu} + \beta_{\mu\lambda\nu} = 0. \tag{3.11}$$

This system has a unique solution, which is obtained as per the solving of similar systems in the previous chapter. We thus have, dropping the from-now-on unnecessary underlining of $\underline{\alpha}_{\lambda\mu}$,

$$\alpha_{\lambda\mu} + \alpha_{\mu\lambda} = 0. \tag{3.12}$$

It is clear that just a number of the $n(n-1)/2$ are independent.

Cartan proceeds to differentiate (3.3) and to then substitute in the result the equations (3.3) themselves. He thus obtains

$$\omega^\mu \wedge (d\alpha_\mu^\nu - \alpha_\mu^\lambda \wedge \alpha_\lambda^\nu) = 0. \tag{3.13}$$

This leads him to define the differential form version of the Riemannian curvature as

$$d\alpha_\mu^\nu - \alpha_\mu^\lambda \wedge \alpha_\lambda^\nu. \qquad (3.14)$$

In principle, $\Omega_{\mu\nu}$ is a linear combination of terms $\omega^\rho \wedge \omega^\lambda$, $\omega^\mu \wedge \alpha_\nu^\sigma$ and $\alpha_\mu^\nu \wedge \alpha_\lambda^\pi$ since neither of the terms on the right hand side of (3.14) is horizontal. Cartan shows that the non-horizontal parts of $d\alpha_\mu^\nu$ and $-\alpha_\mu^\lambda \wedge \alpha_\lambda^\nu$ cancel each other out.

At this point, we abandon Cartan's treatment of this subject and concentrate on how his argument solves the equivalence problem of Riemann.

Suppose we are given a symmetric quadratic differential form. If it were a disguised form of the Cartesian metric (i.e. reducible to it by a coordinate transformation), we would compute its α_μ^ν and then $d\alpha_\mu^\nu - \alpha_\mu^\lambda \wedge \alpha_\lambda^\nu$. If this curvature differential form is not zero, the given metric is not the Cartesian metric, since it has to be zero in any coordinate system This is a consequence of being horizontal. The argument (for those familiar with the tensor calculus) is not different from the fact that if a tensor is zero with respect to some basis, it is zero with respect to any basis That is the necessary condition. It is also sufficient, but we shall not insist on this since the Riemannian curvature of the old Riemannian geometry is given by the same differential forms as the Euclidean curvature of the Levi-Civita connection. We have already dealt from the perspective of the theory of the moving frame with the sufficiency condition.

10.4 Riemannian spaces

A proper Riemannian space (meaning here "one which does not qualify as a Riemannian pseudo-space in the above sense of the word") is a differentiable manifold endowed with a zero-torsion Euclidean connection. If the torsion is zero, so is the contorsion. The Euclidean connection, ω_μ^ν, then coincides with its metric-dependent part, α_μ^ν, and is called the Levi-Civita connection (LCC). It then follows that the (Riemannian) curvature obtained in the process of solving the problem of equivalence of Riemannian metrics is the same as the curvature of the LCC.

All the equations valid for Euclidean connections apply to the LCC and, therefore, to Riemannian spaces. And all the equations that apply to affine connections with zero torsion also apply here. Thus the equations of structure are

$$0 = d\omega^\nu - \omega^\nu \wedge \alpha_\nu^\mu, \qquad (4.1)$$

$$\underline{\Omega}_\mu^\nu = d\alpha_\mu^\nu - \alpha_\mu^\lambda \wedge \alpha_\lambda^\nu, \qquad (4.2)$$

except that we have to add to them the distinctive mark of Euclidean connections, namely

$$\alpha_{\mu\nu} + \alpha_{\nu\mu} = 0. \qquad (4.3)$$

The argument, of course, is of a different nature from the one in the previous section, since the problem of equivalence has not been posed in the moving frame environment, where it is only implicit.

The affine extension of (4.3) is

$$dg_{\mu\nu} - \alpha^\lambda_\mu g_{\lambda\nu} - \alpha^\lambda_\nu g_{\lambda\mu} = 0. \qquad (4.4)$$

The coincidence of ω^ν_μ and α^ν_μ in this case means that $\Gamma_\mu{}^\nu{}_\lambda$ equals $\underline{\Gamma}_\mu{}^\nu{}_\lambda$. Hence, the autoparallels and the extremals coincide on Riemannian manifolds, and the stationary curves (extremals in particular) also satisfy the equation

$$0 = \frac{d\mathbf{u}}{ds}. \qquad (4.5)$$

As is the case for all Euclidean connections, we have

$$\underline{\Omega}_{\mu\nu} = -\underline{\Omega}_{\nu\mu}, \qquad (4.6)$$

and the first Bianchi identity becomes

$$\omega^\mu \wedge \underline{\Omega}^\nu_\mu = 0. \qquad (4.7)$$

From

$$\underline{\Omega}^\nu_\mu \equiv \underline{R}_\mu{}^\nu{}_{\lambda\pi}(\omega^\lambda \wedge \omega^\pi) = \frac{1}{2}\underline{R}_\mu{}^\nu{}_{\lambda\pi}\omega^\lambda \wedge \omega^\pi \qquad (4.8)$$

and (4.7), we readily get the annulment of a cyclic sum of components $\underline{R}_\mu{}^\nu{}_{\lambda\pi}$, namely

$$\underline{R}_\mu{}^\nu{}_{\lambda\pi} + \underline{R}_\lambda{}^\nu{}_{\pi\mu} + \underline{R}_\pi{}^\nu{}_{\mu\lambda} = 0, \qquad (4.9)$$

which is called the cyclic property, valid when (4.7) is satisfied.

To these algebraic properties we can of course add

$$\underline{R}_\mu{}^\nu{}_{\lambda\pi} = -\underline{R}_\mu{}^\nu{}_{\pi\lambda}, \qquad (4.10)$$

as for arbitrary affine and Euclidean connections. We also recall the property

$$\underline{R}_{\mu\nu\lambda\pi} = \underline{R}_{\lambda\pi\mu\nu}, \qquad (4.11)$$

first seen in section 2 of the previous chapter and valid for Euclidean connection whose torsion has zero exterior(-covariant) derivative.

The contractions $\underline{R}_\mu{}^\nu{}_{\lambda\nu}$ are designated as $\underline{R}_{\mu\lambda}$. They behave like the components of a tensor, which is the reason why one refers to $\underline{R}_{\mu\lambda}$ as Ricci tensor. It is symmetric:

$$\underline{R}_{\mu\lambda} \equiv \underline{R}_\mu{}^\nu{}_{\lambda\nu} = g^{\nu\pi}\underline{R}_{\mu\pi\lambda\nu} = g^{\nu\pi}\underline{R}_{\lambda\nu\mu\pi} = \underline{R}_\lambda{}^\pi{}_{\mu\pi} = \underline{R}_{\lambda\mu}. \qquad (4.12)$$

One defines Einstein's tensor as one whose components are

$$G_{\mu\lambda} = \underline{R}_{\mu\nu} - \frac{1}{2}g_{\mu\nu}\underline{R}, \qquad (4.13)$$

where $\underline{R}$ is $\underline{R}^\mu_\mu$. Once again, this is not the proper way to look either to the Einstein and Ricci objects, or to the so called energy-momentum tensor. They are naturally vector-valued differential 3−forms, to be further discussed in later sections.

In section 6 of chapter 8, we saw that the formula for the affine curvature differential 2−form gave rise to a concept of and formula for a curvature tensor. Then, in section 7, we saw the role of this tensor for determining the evolution of the tangent vector to a family of autoparallels as we move along and across the curves in the family. In section 11 of the same chapter, we specialized this result to the zero torsion case to obtain

$$\frac{\partial^2 \mathbf{n}}{\partial u^2} = t^\mu \underline{R}^\nu_{\mu\lambda\pi} t^\lambda n^\pi \mathbf{e}_\nu. \tag{4.14}$$

The only Euclidean connection with zero torsion is the Levi-Civita connection. Hence it is the only Euclidean connection to which Eq. (4.14) applies. We can use as parameter u the distance s on geodesics, which are also autoparallels in this case. We then have $t^\rho = \frac{dx^\rho}{ds} \equiv \dot{x}^\rho$, which allows us to write (4.14) as

$$\frac{\partial^2 \mathbf{n}}{\partial s^2} = \underline{R}^\nu_{\mu\rho\pi} \dot{x}^\mu \dot{x}^\rho n^\pi \mathbf{e}_\nu. \tag{4.15}$$

The parameter λ used in the definition of $\mathbf{n}$ could be any. We can choose it to be s, so that $\mathbf{n}$ will be a unit vector. This equation is then called the equation for geodesic deviation. It does not measure the separation of geodesics; the straight lines issued from a point in the Euclidean plane separate even though the curvature(s) is zero. Formula (4.15) rather speaks of the "acceleration" in the separation of the geodesics.

There is an infinite number of Euclidean connections on the same metric. Though they differ by their torsions, they all have the same set of geodesics (again, curves of stationary distance). Equation (4.15) thus applies to the geodesics of all the Euclidean connections, but not to their autoparallels unless, of course, we are dealing with the Levi-Civita connection.

10.5 Normal coordinates and annulment of connection at a point

Normal coordinates are typically introduced in Riemannian geometry to achieve the annulment of the connection at a point, Levi-Civita's in this case. It is clear that, for arbitrary Euclidean connections, no coordinate basis fields of tangent vectors makes it zero at a point. Indeed, corresponding to coordinate frame fields, we have $d\omega^\mu = ddx^\mu = 0$. If the connection were zero at a point, the right hand side of the first equation of structure would be zero. In other words, the torsion had to be zero at that point for annulment of the connection through normal coordinates. When the torsion is not zero, one still can achieve said annulment, as we saw in chapter 8, but not in terms of coordinate basis fields.

We proceed to obtain normal coordinates by means of a procedure that Cartan attributes to Riemann. Since it looks somewhat ad hoc, we first provide an example of the geometric ideas that underlie it.

Consider two perpendicular planes that contain the diameter through the point P of tangency of a plane π to a 2-sphere. Their intersection with π determines an orthonormal frame at P (up to orientation). Cartesian coordinates on π can now be assigned to points on the sphere in the neighborhood of P by way of a perpendicular projection. In this way the metric $dx^2 + dy^2$ on the plane is assigned to the neighborhood of P on the sphere. It is an approximation, the smaller the neighborhood the better.

Consider now a tangent orthonormal frame $(\mathbf{a}_i)$ at a point P of a differentiable manifold. Points M in an arbitrarily small neighborhood of P determine geodesics g through P. Let $\mathbf{u}$ be the unit tangent vector to g at P, directed towards M. Let c^i be the products $\mathbf{u} \cdot \mathbf{a}^i$ (If the metric is positive definite, these are the directional cosines). The normal coordinates of P are the quantities

$$x^\mu = c^\mu s, \tag{5.1}$$

where s is the length of the arc from P to M. Said better, the coordinate functions assign to each point M the quantities $c^\mu s$. The c^μ's are obtained, sure enough, by integrating the equations for the geodesics of the given metric, which thus enter the picture. It is clear that, on the geodesics tangent to the vectors $\mathbf{a}_i$, we have $x^\mu = s$.

Let us verify that the coordinates (5.1) are such that the Γ's become zero at P. We substitute (5.1) into the geodesic equation

$$\frac{d^2x^\mu}{ds^2} + \underline{\Gamma}'{}^\mu_{\nu\,\lambda} \frac{dx^\nu}{ds} \frac{dx^\lambda}{ds} = 0, \tag{5.2}$$

and obtain

$$\underline{\Gamma}'{}^\mu_{\nu\,\lambda} c^\nu c^\lambda = 0, \tag{5.3}$$

to be satisfied at P for all c's in a neighborhood. The arbitrariness of the c's implies that the $\underline{\Gamma}'{}^\mu_{\nu\,\lambda}$ are zero at P.

As an application of normal coordinates, we show property $\underline{R}_{\mu\nu\lambda\pi} = \underline{R}_{\lambda\pi\mu\nu}$ for the Riemannian curvature, originally obtained for arbitrary Euclidean connection in section 2 of the previous chapter. In terms of normal coordinates, we have, at the point where the connection becomes zero,

$$\Omega_{\mu\nu} = d\alpha_{\mu\nu} = (\underline{\Gamma}'_{\mu\nu\lambda,\pi} - \underline{\Gamma}'_{\mu\nu\pi,\lambda})(dx^\pi \wedge dx^\lambda) \equiv \underline{R}'_{\mu\nu\pi\lambda}(dx^\pi \wedge dx^\lambda), \tag{5.4}$$

with $\underline{R}'_{\mu\nu\lambda\pi} = -\underline{R}'_{\nu\mu\lambda\pi}$ and $\underline{R}'_{\mu\nu\lambda\pi} = -\underline{R}'_{\mu\nu\pi\lambda}$. Using equation (5.1) of the previous chapter, namely,

$$\underline{\Gamma}'^\pi_{\mu\lambda} = \frac{1}{2} g^{\nu\pi}(g_{\mu\nu,\lambda} + g_{\lambda\nu,\mu} - g_{\mu\lambda,\nu}), \tag{5.5}$$

we further get

$$\underline{R}'_{\mu\nu\lambda\pi} = \frac{1}{2}(g_{\nu\pi,\mu\lambda} + g_{\mu\lambda,\nu\pi} - g_{\mu\pi,\nu\lambda} - g_{\nu\lambda,\mu\pi}) \tag{5.6}$$

Symmetry of the the right hand side under exchange of pairs $(\mu\nu)$ and $(\lambda\pi)$ then implies (4.11). The horizontality of the curvature in turn implies that this result is valid in any frame field.

10.6 Emergence and conservation of Einstein's tensor

Contractions of Riemann's tensor lead to the discovery of a second rank tensor whose covariant derivative is zero. It is named the Einstein tensor. It was relevant at the time (but now only because of tradition) since that property of tensors was associated with conservation laws, which pertain to differential forms, not to tensors.

In affine-Klein and Euclidean-Klein (i.e. flat) spaces, conservation of vector or tensor-valued differential forms follows the annulment of the exterior(-covariant) derivatives. The non-scalar valuedness poses problems unless there are constant frame fields (teleparallelism), so that the integration can then be reduced to scalar-valued integration. This limitation applies to Riemannian geometry and, therefore, to general relativity if we assume that its connection is the LCC (It may actually happen that the connection is teleparallel and one mistakenly believes to be dealing with the LCC, as explained in section 8 of the previous chapter).

Let us report how the Einstein tensor emerges in the tensor calculus. One writes explicitly in terms of Christoffel symbols the covariant derivative with respect to the LCC connection of Riemann's curvature. One uses normal coordinates in order to make that equation more manageable. One writes down two more copies of it obtained by performing appropriate cyclic permutations of three of the five indices (See (6.1)). The right hand side of a suitable combination of the three equations so obtained is zero, and so must, therefore, be the same combination of the left hand sides. One thus obtains

$$\underline{R}^{\nu}_{\mu\,\lambda\pi;\rho} + \underline{R}^{\nu}_{\mu\,\pi\rho;\lambda} + \underline{R}^{\nu}_{\mu\,\rho\lambda;\pi} = 0. \tag{6.1}$$

A double contraction of indices ($\nu = \pi$, $\mu = \lambda$) yields

$$2\underline{R}^{\lambda}_{\rho;\lambda} - \underline{R}_{,\rho} = 0. \tag{6.2}$$

The Ricci tensor has a property which allows one to rewrite this as

$$0 = (\underline{R}^{\lambda\rho} - \frac{1}{2}g^{\lambda\rho}\underline{R})_{;\lambda} = G^{\lambda\rho}_{\;\;;\lambda}, \tag{6.3}$$

which is the announced result. Since all the displayed equations are tensorial, we may disregard our having used normal coordinates to obtain this result.

Differential forms allow one to simplify the above approach, while retaining its flavor. In the process, one can make clearer how these equations are connected with the conservation of the curvature, i.e. the second Bianchi identity for Riemann's curvature. Under the LCC, it can be written as

$$0 = d\underline{\mho} = \underline{R}^{\nu}_{\mu\,\lambda\pi;\rho}\omega^{\rho} \wedge \omega^{\lambda} \wedge \omega^{\pi}\mathbf{e}_{\mu} \wedge \mathbf{e}^{\nu}. \tag{6.4}$$

The remainder of the proof to get (6.3) remains unchanged. Because of the zero at the front of (6.4), we have ignored factors like $1/2!$ or $1/3!$ Notice also that one does not need to use normal coordinates.

Einstein's equations read, up to a constant factor,

$$G^{\mu\nu} = T^{\mu\nu}, \tag{6.5}$$

where $G^{\mu\nu}$ represents the components of the Einstein tensor and $T^{\mu\nu}$ is the energy-momentum tensor contributed by all interactions except the gravitational one. From (6.3) and (6.5), we get

$$T^{\mu\nu}_{\;;\nu} = 0. \tag{6.6}$$

Equation (6.6) states the conservation law of energy-momentum of all types except gravitational. But, again, it has very little significance in the absence of constant frame fields. Gravitational energy-momentum remains a mystery in Einstein's theory of gravity under the LCC.

10.7   EINSTEIN'S DIFFERENTIAL 3-FORM

The so called energy-momentum tensor is a bad concept. It should be replaced with the concept of energy-momentum vector-valued differential 3−form, or simply energy-momentum differential form. It is a 3−form (in spacetime!) because we integrate it on volumes. It is vector-valued because the result of the integration (when it makes sense) is a (four)vector.

The current of charge is a scalar-valued differential 3−form in *spacetime*, misleadingly represented as a (four)vector (see section 7 of chapter 4). The Hodge dual of the differential 3−form current of charge in the algebra of differential forms is a differential 1−form. We can associate with it a vector field, called the current (four)vector.

The energy-momentum tensor is to the energy-momentum (four)vector what the electromagnetic current is to the charge scalar. Those currents generalize the densities of (four)momentum and charge.

In the Einstein tensor, the first index refers to the components of the energy-momentum 4-vector. The other index constitutes a disguised form of stating the indices of differential 3−forms in 4-D by stating which index is missing. Equation (6.5) really says that a vector-valued differential 3−form constructed from Riemann's curvature is equal to energy-momentum 3−form.

Cartan [8] presents us with the components of "Einstein's vector-valued 3−form" **G** as

$$\mathbf{G} \equiv G^{\mu}\mathbf{e}_{\mu}, \tag{7.1}$$

where

$$G^0 \equiv \omega^1 \wedge \underline{\Omega}^{23} + \omega^2 \wedge \underline{\Omega}^{31} + \omega^3 \wedge \underline{\Omega}^{12}, \tag{7.2a}$$

$$G^1 \equiv \omega^0 \wedge \underline{\Omega}^{23} + \omega^2 \wedge \underline{\Omega}^{30} + \omega^3 \wedge \underline{\Omega}^{02}, \tag{7.2b}$$

$$G^2 \equiv \omega^0 \wedge \underline{\Omega}^{31} + \omega^3 \wedge \underline{\Omega}^{10} + \omega^1 \wedge \underline{\Omega}^{03}, \tag{7.2c}$$

$$G^3 \equiv \omega^0 \wedge \underline{\Omega}^{12} + \omega^1 \wedge \underline{\Omega}^{20} + \omega^2 \wedge \underline{\Omega}^{01}. \tag{7.2d}$$

Cartan's argument to get to those expressions is rather abstruse, involved as it is with invariants of a quadratic symmetric differential form. Notice that the indices on the left hand sides are the indices missing on the right hand side. Notice also the three cyclic permutations on the right hand side of each of these equations, as well as the cyclic permutations of the indices (1,2,3) in going from G^1 to G^2 and then to G^3.

We shall now see **G** from a perspective of differential forms that take values in the tangent *Clifford algebra* of spacetime. It requires some extra effort. Readers familiar with gamma matrices should have no difficulty to follow.

Recall that the subscript ν in $\underline{\Omega}_\nu{}^\mu$ and $\underline{R}_\nu{}^\mu{}_{\lambda\pi}$ corresponds to a linear function of vectors, which the metric allows us to replace with a vector field. We thus view Riemann's curvature as

$$\underline{U} = \underline{\Omega}^{\nu\lambda}(\mathbf{e}_\nu \wedge \mathbf{e}_\lambda) = \frac{1}{2}\underline{\Omega}^{\nu\lambda}\mathbf{e}_\nu \wedge \mathbf{e}_\lambda = \frac{1}{2}\underline{\Omega}_{\nu\lambda}\mathbf{e}^\nu \wedge \mathbf{e}^\lambda = \underline{\Omega}_{\nu\lambda}(\mathbf{e}^\nu \wedge \mathbf{e}^\lambda), \qquad (7.3)$$

where the exterior product matches the skew-symmetry of the two displayed indices in $\underline{\Omega}^{\nu\lambda}$. Einstein essentially built with the Riemannian curvature a vector-valued differential 3−form. For that, one has to get the 3−forms in Eqs. (7.2) from the 2−forms $\underline{\Omega}^{\nu\lambda}$. In other words, one has to get the vector-valued **G** from the bivector-valued curvature.

Consider the vector-valued differential 3−form $d\mathbf{P}(\wedge, .)\underline{U}$, where the symbol $\wedge$ in $(\wedge, \cdot)$ is for the exterior product in the algebra of differential forms, and the dot product is for the tangent algebra. Readers can verify with computations similar to those about to follow that $d\mathbf{P}(\wedge, .)\underline{U}$ does not yield the Einstein differential 3−form. However, let **Z** denote the unit $\mathbf{e}^0\mathbf{e}^i\mathbf{e}^j\mathbf{e}^k$ ($= \mathbf{e}^0\wedge\mathbf{e}^i\wedge\mathbf{e}^j\wedge\mathbf{e}^k$) of grade four in the tangent Clifford algebra of spacetime. **Z** is to (the tangent) Clifford algebra based on $(\mathbf{e}^0, \mathbf{e}^1, \mathbf{e}^2, \mathbf{e}^3)$ what w is to a Clifford algebra based on (dx, dy, dz) (See section 4 of chapter 6). Thus the differential form $\underline{U}\mathbf{Z}$ is the Hodge dual of $\underline{U}$ in this tangent algebra. We now have $(\mathbf{e}^0)^2 = 1$, $(\mathbf{e}^i)^2 = -1$.

We shall show that

$$\mathbf{G} = d\mathbf{P}(\wedge, \cdot)(\underline{U}\mathbf{Z}). \qquad (7.4)$$

But why should one care about an expression like this? We shall see at the end of section 7 how easily the conservation law of **G** is a consequence of (7.4). Another reason to care is as follows.

The differential form for an Einstein tensor in n dimensions would be a vector-valued $(n-1)$-differential form. But we still want energy-momentum to be a vector-valued differential 3−form for 3-volume integration in higher dimension (We have in mind a Kaluza-Klein space where propertime plays the role of fifth dimension). The expression $d\mathbf{P}(\wedge, \cdot)(\underline{U}\mathbf{Z})$ yields a vector-valued differential 3−form, independently of the value of n.

Finally, in order to deal with (7.4), one has to use Clifford algebra, thus taking us in the right direction if one wants to unify gravitation with quantum physics.

We return to the main argument. From (7.3), we readily get

$$\underline{U}\mathbf{Z} = \underline{\Omega}^{0i}\mathbf{e}_{0i}\mathbf{Z} + \underline{\Omega}^{ij}\mathbf{e}_{ij}\mathbf{Z} = -\underline{\Omega}^{0i}\mathbf{e}^{jk} + \underline{\Omega}^{ij}\mathbf{e}^{k0}, \qquad (7.5)$$

where $\mathbf{e}_{ij}$ is defined as $\mathbf{e}_i\mathbf{e}_j$ and thus as $\mathbf{e}_i \wedge \mathbf{e}_j$ in orthogonal basis fields. In (7.5), (i,j,k) is any cyclic permutation of (1,2,3). There is summation over the three pairs of the form (ij) and the three triples (ij,k). From (7.1), (7.4) and (7.5) follows that

$$G_\mu \mathbf{e}^\mu = \omega^\mu \mathbf{e}_\mu(\wedge, \cdot)(-\underline{\Omega}^{0i}\mathbf{e}^{jk} + \underline{\Omega}^{ij}\mathbf{e}^{k0}). \tag{7.6}$$

G^0 is then given by

$$G^0 = G_0 = \omega^l \mathbf{e}_l(\wedge, \cdot)\underline{\Omega}^{ij}\mathbf{e}^{k0} = \omega^1 \wedge \underline{\Omega}^{23} + \omega^2 \wedge \underline{\Omega}^{31} + \omega^3 \wedge \underline{\Omega}^{12}, \tag{7.7}$$

which is (7.2a).

The products $\mathbf{e}_0 \cdot \mathbf{e}^{0i}$, $\mathbf{e}_j \cdot \mathbf{e}^{ji}$ and $\mathbf{e}_k \cdot \mathbf{e}^{ki}$ yield $\mathbf{e}^i$. Hence, reading from (7.6), we form the combination

$$\omega^0 \mathbf{e}_0(\wedge, \cdot)\underline{\Omega}^{jk}\mathbf{e}^{i0} + \omega^j \mathbf{e}_j(\wedge, \cdot)\underline{\Omega}^{0k}\mathbf{e}^{ij} + \omega^k \mathbf{e}_k(\wedge, \cdot)\underline{\Omega}^{0j}\mathbf{e}^{ik}, \tag{7.8}$$

without summation over repeated indices or cyclic permutations. By using for (i,j,k) the three cyclic permutations of (1,2,3), we obtain the negatives of (7.2b)-(7.2d) multiplied by $\mathbf{e}^1$, $\mathbf{e}^2$, $\mathbf{e}^3$ respectively. These three unit vectors change sign when lowering indices. Equation (7.4) have thus been shown to yield (7.1)-(7.2).

At this point and in order to increase familiarity with the algebra involved here, we provide some simple concepts of the same. The identity

$$dxdy = \frac{1}{2}(dxdy + dydx) + \frac{1}{2}(dxdy - dydx) \tag{7.9}$$

applies to any type of product, for instance, to the tensor product (symbol not exhibited due precisely to the general validity of (7.9)). The Clifford product is the one where $\frac{1}{2}(dxdy + dydx)$ is identified with the dot product and $\frac{1}{2}(dxdy - dydx)$ with the exterior product. dx and dy are said to be of grade one, i.e. differential 1−forms. Their dot product lowers the grade and their exterior product raises the grade.

On general differentiable manifolds, we do not have Cartesian coordinates. We, however, have

$$\omega^\mu \omega^\nu = \omega^\mu \cdot \omega^\nu + \omega^\mu \wedge \omega^\nu, \tag{7.10}$$

with

$$\omega^\mu \cdot \omega^\nu \equiv \frac{1}{2}(\omega^\mu \omega^\nu + \omega^\nu \omega^\mu); \qquad \omega^\mu \wedge \omega^\nu = \frac{1}{2}(\omega^\mu \omega^\nu - \omega^\nu \omega^\mu). \tag{7.11}$$

If the ω^μ are those that orthonormalize the metric, they satisfy

$$\omega^\rho \cdot \omega_\mu = \delta^\rho_\mu. \tag{7.12}$$

We shall see in the next section that the components of the Einstein tensor are the components $G^{\alpha\mu}$ of the Hodge dual $\mathbf{G}z$ of $\mathbf{G}$ in the Clifford algebra of differential forms (called Kähler algebra) of spacetime, i.e.

$$\mathbf{G}z = G^{\alpha\mu}\mathbf{e}_\alpha\omega_\mu = G^\alpha{}_\mu\mathbf{e}_\alpha\omega^\mu, \tag{7.13}$$

where

$$z = \omega^0 w = \omega^0 \omega^1 \omega^2 \omega^3, \tag{7.14}$$

and where, again, we have adapted the expression for w to take into account that there are not Cartesian coordinate systems on general manifolds. Hence z is to a Clifford algebra based on $(\omega^0, \omega^1, \omega^2, \omega^3)$ what $\mathbf{Z}$ is to a Clifford algebra based on $(\mathbf{e}^0, \mathbf{e}^1, \mathbf{e}^2, \mathbf{e}^3)$. Regardless of whether the signature is (1,-1,-1,-1) or (-1,1,1,1), z^2 is -1.

At this point, we relate the $G^{\alpha\mu}$ to the components of the Einstein differential 3–form, components defined implicitly by

$$\mathbf{G} \equiv G^\alpha_{\mu\nu\pi}(\omega^\mu \wedge \omega^\nu \wedge \omega^\pi)\mathbf{e}_\alpha \equiv G^\alpha_{\mu\nu\pi}(\omega^{\mu\nu\pi})\mathbf{e}_\alpha. \tag{7.15}$$

From (7.13) and (7.15), we get

$$G^{\alpha\mu}\omega_\mu = G^\alpha_{\mu\nu\pi}(\omega^{\mu\nu\pi})z. \tag{7.16}$$

The product of $\omega^{\mu\nu\pi}$ with z picks the element of the basis of differential 1–forms with the only index ρ not present in $\mu\nu\pi$. We put $\mu\nu\pi$ in such an order that $(\mu\nu\pi\rho)$ be an even permutation of $(0, 1, 2, 3)$. From (7.12) and (7.16), we then have

$$G^{\alpha\rho} = \omega^\rho \cdot G^{\alpha\mu}\omega_\mu = G^\alpha_{\mu\nu\pi}\omega^\rho\left[(\omega^{\mu\nu\pi})z\right] = -G^\alpha_{\mu\nu\pi}\omega^{\mu\nu\pi}\omega^\rho z. \tag{7.17}$$

We next use that $\omega^{\mu\nu\pi}\omega^\rho$ is $\pm z$ and that z^2 is -1. The identification

$$G^{\alpha\rho} = G^\alpha_{\mu\nu\pi} \tag{7.18}$$

results.

10.8 Einstein's differential form: properties, components and Einstein's equations

The Einstein tensor is defined as $G^{\alpha\mu}\mathbf{e}_\alpha \otimes \mathbf{e}_\mu$ with the same components as the vector-valued differential 1–form $\mathbf{G}z$. In this section, we shall deal with the following subjects. First, we shall show the symmetry

$$G^{\alpha\rho} = G^{\rho\alpha} \tag{8.1}$$

without resort to the specific form of $G^{\alpha\mu}$ in terms of the components of Riemann's curvature. This reflects the fact that, in general, there should not be any need to compute components to derive the symmetry (8.1). Second, we shall compute those components by explicit calculation so that readers who could not follow all the preceding arguments do not harbor any doubts that the $G^{\alpha\rho}$ derived in the previous section are the same as in the tensor calculus. That also give us practice with Clifford algebra. We then deal with the conservation law from the perspective of differential forms and also show that, under the

LCC, the annulment of the covariant derivative of the Einstein tensor is equivalent to the annulment of $d\mathbf{G}$. Finally, we write the by now trivial expression for Einstein's equations in terms of differential forms.

The proof of (8.1) applies to similarly constructed second rank tensors from vector-valued differential 3−forms. Readers to whom this algebra may seem forbidding should just glance at the proof without strenuous attempt at understanding.

We write $G^{\alpha\rho}$ as

$$G^{\alpha\rho} = G^{\beta}_{\mu}\delta^{\alpha}_{\beta}\eta^{\rho\mu} = \omega^{\rho}\mathbf{e}^{\alpha}(\cdot,\cdot)G^{\beta}_{\mu}\mathbf{e}_{\beta}\omega^{\mu}, \tag{8.2}$$

and use (7.13) and (7.4) to obtain

$$G^{\alpha\rho} = \omega^{\rho}\mathbf{e}^{\alpha}(\cdot,\cdot)(\mathbf{G}z) = \eta_{\mu}\omega^{\rho}\mathbf{e}^{\alpha}(\cdot,\cdot)\{\left[\omega^{\mu}\mathbf{e}^{\mu}(\wedge,\cdot)\underline{(\mathbf{UZ})}\right]z\}, \tag{8.3}$$

where $\eta_{\mu} = (1,-1,-1,-1)$ or $(-1,1,1,1)$. Since $\underline{\mathbf{UZ}}$ equals $(1/2)\underline{R}_{\nu\lambda\pi\sigma}\mathbf{e}^{\nu\lambda}\omega^{\pi\sigma}\mathbf{Z}$, we further have

$$\left[\omega^{\mu}\mathbf{e}^{\mu}(\wedge,\cdot)\underline{\mathbf{UZ}}\right]z = \frac{1}{2}\underline{R}_{\nu\lambda\pi\sigma}\left[\mathbf{e}^{\mu}\cdot(\mathbf{e}^{\nu\lambda}\mathbf{Z})\right]\left[(\omega^{\mu}\wedge\omega^{\pi\sigma})z\right]. \tag{8.4}$$

A formula in Clifford algebra states that

$$(\omega^{\mu}\wedge\omega^{\pi\sigma})z = \omega^{\mu}\cdot(\omega^{\pi\sigma}z). \tag{8.5}$$

It would be out of place to go here into the proof of this formula. Suffice an informal verification. If μ equals either π or σ, both sides are evidently zero. If μ is different from both π and σ, $\omega^{\mu\pi\sigma}z$ is ω^{ρ}, where ω^{ρ} is the missing index. On the other hand $\omega^{\pi\sigma}z$ contains only the two indices other than π and σ, i.e. μ and ρ. The dot product with ω^{μ} yields ω^{ρ}.

Returning to the main argument, we have, from (8.3) and (8.4),

$$G^{\alpha\rho} = \frac{1}{2}\epsilon_{\mu}\underline{R}_{\nu\lambda\pi\sigma}\left\{\mathbf{e}^{\alpha}\cdot\left[\mathbf{e}^{\mu}\cdot(\mathbf{e}^{\nu\lambda}\mathbf{Z})\right]\right\}\left\{\omega^{\rho}\cdot\left[\omega^{\mu}\cdot(\omega^{\pi\sigma}z)\right]\right\}, \tag{8.6}$$

and, using the symmetry $\underline{R}_{\nu\lambda\pi\sigma} = \underline{R}_{\pi\sigma\nu\lambda}$, we get

$$G^{\alpha\rho} = \frac{1}{2}\epsilon_{\mu}\underline{R}_{\pi\sigma\nu\lambda}\left\{\mathbf{e}^{\alpha}\cdot\left[\mathbf{e}^{\mu}\cdot(\mathbf{e}^{\nu\lambda}\mathbf{Z})\right]\right\}\left\{\omega^{\rho}\cdot\left[\omega^{\mu}\cdot(\omega^{\pi\sigma}z)\right]\right\}. \tag{8.7}$$

We exchange the indices α and ρ in (8.8) and invert the order of the two parentheses to obtain

$$G^{\rho\alpha} = \frac{1}{2}\epsilon_{\mu}\underline{R}_{\nu\lambda\pi\sigma}\left\{\omega^{\alpha}\cdot\left[\omega^{\mu}\cdot(\omega^{\nu\lambda}\mathbf{Z})\right]\right\}\left\{\mathbf{e}^{\rho}\cdot\left[\mathbf{e}^{\mu}\cdot(\mathbf{e}^{\pi\sigma}z)\right]\right\}. \tag{8.8}$$

Since the contents of all these curly brackets are scalars (actually the numbers one or minus one), the right hand sides of (8.7) and (8.8) are the same and the symmetry (8.1) follows.

We now show what the $G^{\alpha\rho}$ we defined in the previous section are in terms of the components of Riemann's curvature. Multiplication of (7.1) by z and comparison with (7.13), yields $G^{\alpha}_{\mu}\omega^{\mu} = G^{\alpha}z$, and, therefore,

$$G^{\alpha}_{\rho} = \omega_{\rho} \cdot (G^{\alpha}z). \tag{8.9}$$

Consider, for instance, G^0_{0}. We start by expanding G^0, given by (7.2a):

$$G^0 = \omega^1 \wedge [\underline{R}^{23}_{0i}\omega^0 \wedge \omega^i + \underline{R}^{23}_{ij}(\omega^i \wedge \omega^j)] + \omega^2 \wedge [\underline{R}^{31}_{0i}\omega^0 \wedge \omega^i + \underline{R}^{31}_{ij}(\omega^i \wedge \omega^j)] +$$
$$+ \omega^3 \wedge [\underline{R}^{12}_{0i}\omega^0 \wedge \omega^i + \underline{R}^{12}_{ij}(\omega^i \wedge \omega^j)]. \tag{8.10}$$

If ρ in (8.9) is to be zero, G^0z must contain ω^0 as a factor for non-null multiplication with ω_0. Thus G^0 must not contain ω^0. That eliminates from consideration many terms in (8.10). On the other hand, the values of i and j are determined by the factor at the front of each square bracket in (8.10), since all three indices must be different for a trivector not to be zero, and all three of them are spatial indices. We choose the order of (i, j) so that we get cyclic permutations of (1,2,3). We thus get

$$G^0 = (\underline{R}^{23}_{23} + \underline{R}^{31}_{31} + \underline{R}^{12}_{12})\omega^1 \wedge \omega^2 \wedge \omega^3. \tag{8.11}$$

We can rewrite z as $\omega_1\omega_2\omega_3\omega^0$ and then as $-\omega_3\omega_2\omega_1\omega^0$ so that the equation $(\omega^1 \wedge \omega^2 \wedge \omega^3)z = -\omega^0$ becomes obvious. We thus get

$$G^0_0 = \omega_0 \cdot [-(\underline{R}^{23}_{23} + \underline{R}^{31}_{31} + \underline{R}^{12}_{12})\omega^0] = -(\underline{R}^{23}_{23} + \underline{R}^{31}_{31} + \underline{R}^{12}_{12}). \tag{8.12}$$

For G^{11}, we take into account (7.18) and thus consider the terms $\omega^0 \wedge \omega^2 \wedge \omega^3$ on the right hand side of (7.2b), i.e. in

$$\omega^0 \wedge [\underline{R}^{23}_{0i}\omega^0 \wedge \omega^i + \underline{R}^{23}_{ij}(\omega^i \wedge \omega^j)] + \omega^2 \wedge [\underline{R}^{30}_{0i}\omega^0 \wedge \omega^i + \underline{R}^{30}_{ij}(\omega^i \wedge \omega^j)] +$$
$$+ \omega^3 \wedge [\underline{R}^{02}_{0i}\omega^0 \wedge \omega^i + \underline{R}^{02}_{ij}(\omega^i \wedge \omega^j)] \tag{8.13}$$

to get

$$G^{11} = (\underline{R}^{23}_{23} + \underline{R}^{30}_{30} + \underline{R}^{20}_{20}) = -G^1_1. \tag{8.14}$$

We similarly get

$$G^{01} = -(\underline{R}^{02}_{12} + \underline{R}^{03}_{13}) = -G^0_1 \tag{8.15}$$

and

$$G^{12} = -(\underline{R}^{10}_{20} + \underline{R}^{13}_{23}) = G^1_2. \tag{8.16}$$

The other components of Einstein's tensor can be read almost directly by simply applying cyclic permutations to the last two. Notice that we have reached the Einstein tensor without going through the Ricci tensor

$$\underline{R}_{\nu\mu}\mathbf{e}^{\nu} \otimes \mathbf{e}^{\mu} \equiv \underline{R}_{\nu\mu\lambda}^{\lambda}\mathbf{e}^{\nu} \otimes \mathbf{e}^{\mu}. \tag{8.17}$$

The conservation law of the Einstein 3–form takes just the following one line computation when one uses Kähler's calculus. We indeed, have

$$d\mathbf{G} = d[d\mathbf{P}(\wedge, \cdot)(\underline{U}\mathbf{Z})] = d\mathbf{P}(\wedge, \cdot)[(d\underline{U})\mathbf{Z}] = 0, \tag{8.18}$$

since $\mathbf{Z}$ is a constant $0-$form, and, for the LCC, $d(d\mathbf{P})$ is zero. That allows us to ignore the effect of d on anything here but $\underline{\upsilon}$. We have also used $d\underline{\upsilon} = 0$.

In order to show that the annulments of $G^{\alpha\rho}_{;\rho}$ and $d\mathbf{G}$ are equivalent, we choose frame fields that, at any given point, annul the LCC. In that case, $G^{\alpha\rho}_{;\rho}$ becomes $G^{\alpha\rho}_{/\rho}$. On the other hand, we resort to (7.15) to obtain $d\mathbf{G}$. In the same frame fields, we have

$$d\mathbf{G} = dG^{\alpha}_{\mu\nu\pi}(\omega^{\mu\nu\pi})\mathbf{e}_\alpha = dG^{\alpha\rho}(\omega^{\mu\nu\pi})\mathbf{e}_\alpha = G^{\alpha\rho}_{/\rho}(\omega^{\rho\mu\nu\pi})\mathbf{e}_\alpha, \qquad (8.19)$$

since $d\mathbf{e}_\alpha$ is directly zero, and $d\omega^{\mu\nu\pi}$ is zero because each $d\omega^\beta$ satisfies $d\omega^\beta = \omega^\gamma \wedge \omega^\beta_\gamma = 0$. He have also resorted to (7.18). Hence $d\mathbf{G} = 0$ is equivalent to $G^{\alpha\rho}_{;\rho} = 0$ in constant frame fields (Had the connection not been Levi-Civita's, we could not have set $d\omega^\nu$ equal to zero, obstructed as we would have been by the torsion term in the first equation of structure). Since the result is tensorial, it is valid in any frame field. The proof is now complete.

Finally, in view of all the above, we write Einstein's equations with differential forms as

$$\mathbf{G} = \mathbf{\Pi}, \qquad (8.20)$$

where $\mathbf{\Pi}$ is the vector-valued differential $3-$form which is improperly treated as a tensor and known as the energy-momentum tensor. $\mathbf{\Pi}$ takes many specific forms depending on what parts of the physics it comes from.

10.9 The Einstein equations for Schwarzschild's metric with three arbitrary coefficients

In the literature, the Schwarzschild metric is first written with three arbitrary functions owing to spherical symmetry. They are then reduced to two by virtue of the option to choose the radial coordinate. This implies that the left hand side of, say, the third Einstein equation for the point mass (a fourth one is a repetition of the third one because of that symmetry) can be built in terms of the left hand sides of the first two. Such a relation would have to be matched by the same relation on the right hand sides in problems where these were not zero. We thus set to solve the point of mass problem with three arbitrary functions.

Why bring this up? We wish to start preparing the way for solving sophisticated, non-traditional problems involving Einstein's equations. Such would be the case, for instance, with a more sophisticated version of the Reissner-Nordstrom problem. To quote Einstein (p. 93 of [29]): "But no reasonable person believes that Maxwell's equations can hold rigorously. They are, in suitable cases, first approximations for weak fields." At this point, we do not have so much in mind a different set of Maxwell's equations, but a potentially different electromagnetic energy-momentum 3−form. Einstein's comment still applies, since a change in field equations would likely entail a change in electromagnetic energy-momentum.

We write the metric as

$$ds^2 = e^{2\nu(r)}dt^2 - e^{2\mu(r)}dr^2 - e^{-2\lambda(r)}r^2(d\theta^2 + \sin^2\theta d\phi^2). \qquad (9.1)$$

The independent non-vanishing α_μ^ν's pertaining to a basis of ω^μ's that orthonormalize (9.1) are readily obtained by inspection:

$$\begin{aligned}
\alpha_0^0 &= \nu' e^{\nu-\mu}dt, & \alpha_1^2 &= e^{-\mu}\Lambda d\theta, \\
\alpha_1^3 &= e^{\lambda-\mu}\Lambda \sin\theta d\phi, & \alpha_2^3 &= \cos\theta d\phi,
\end{aligned} \qquad (9.2)$$

where $\Lambda \equiv \lambda' r + 1$, and where primes are used to denote derivatives with respect to r. The components of the curvature 2−form follow:

$$\begin{aligned}
\Omega_0^1 &= e^{\nu-\mu}[\nu'' + \nu'(\nu' - \mu')]dr \wedge dt, \\
\Omega_0^2 &= -\nu' e^{\lambda+\nu-2\mu}\Lambda dt \wedge d\theta, \\
\Omega_0^3 &= -\nu' e^{1+\nu-2\mu}\Lambda \sin\theta dt \wedge d\phi, \\
\Omega_1^2 &= e^{\lambda-\mu}[(\lambda' - \mu')\Lambda + \lambda'' r + \lambda']dr \wedge d\theta, \\
\Omega_2^3 &= \sin\theta[-1 + e^{2\lambda-2\mu}\Lambda^2]d\theta \wedge d\phi, \\
\Omega_3^1 &= e^{\lambda-\mu}[(\lambda' - \mu')\Lambda + \lambda'' r + \lambda'] \sin\theta dr \wedge d\phi.
\end{aligned}$$

We proceed to compute the Einstein 3−form. As a consequence of spherical symmetry, G^3 does not yield anything that is not already contained in G^2. It will be omitted. The notation $d^{r\theta\phi}$ means $dr \wedge d\theta \wedge d\phi$, and so on. We thus obtain:

$$\omega^1 \wedge \Omega^{23} + \omega^2 \wedge \Omega^{31} + \omega^3 \wedge \Omega^{12} = d^{r\theta\phi}e^\mu(1 - e^{2\lambda-2\mu}\{\Lambda^2 + 2r[(\lambda' - \mu')\Lambda + \Lambda']\}), \qquad (9.3a)$$

$$\omega^0 \wedge \Omega^{23} + \omega^2 \wedge \Omega^{30} + \omega^3 \wedge \Omega^{02} = d^{t\theta\phi}\sin\theta e^\nu(1 - e^{2\lambda-2\mu}(\Lambda^2 + 2r\nu'\Lambda)], \qquad (9.3b)$$

and

$$\begin{aligned}
\omega^0 \wedge \Omega^{31} &+ \omega^3 \wedge \Omega^{10} + \omega^1 \wedge \Omega^{03} = \\
&= -d^{t\phi r}\sin\theta e^{\nu-\mu+\lambda}[\Lambda(\lambda' - \mu' + \nu') + (\nu'' + \nu'^2 - \nu'\mu')r + \Lambda'].
\end{aligned} \qquad (9.3c)$$

We equate the right hand sides of these equations to zero and simplify to obtain:

$$1 - e^{2\lambda-2\mu}\{\Lambda^2 + 2r[(\lambda' - \mu')\Lambda + \lambda'' r + \lambda']\} = 0, \qquad (9.4a)$$

$$1 - e^{2\lambda-2\mu}[\Lambda^2 + 2r\nu'\Lambda] = 0, \qquad (9.4b)$$

and

$$\Lambda(\lambda' - \mu' + \nu') + (\nu'' + \nu'^2 - \nu'\mu')r + \Lambda' = 0. \qquad (9.4c)$$

From the first two of these equations, we get:

$$\Lambda^{-1}\Lambda' = \mu' + \nu' - \lambda', \qquad (9.5)$$

which, after integrating, yields

$$-(\lambda - \mu - \nu) = \ln \Lambda + \text{ constant.} \tag{9.6}$$

Hence

$$\Lambda = e^{c+\mu+\nu-\lambda}, \tag{9.7}$$

where c is the integration constant.

We multiply (9.4b) by $e^{-2c-2\nu}$ in order to convert the coefficient of the bracket into Λ^{-1}. This allows us to write Λ as

$$\Lambda = \frac{2r\nu' e^{2\nu+2c}}{1 - e^{2\nu+2c}}. \tag{9.8}$$

From (9.7) and the definition of Λ, we get

$$\lambda' + \frac{1}{r} = \frac{2\nu' e^{2\nu+2c}}{1 - e^{2\nu+2c}}, \tag{9.9}$$

which integrated yields

$$e^{-\lambda} = (1 - e^{2\nu-2c})\frac{r}{r_0}, \tag{9.10}$$

where r_0 is another integration constant. Solving this for e^ν, we get

$$e^\nu = (1 - e^{-\lambda}\frac{r_0}{r})^{1/2} e^{-c} \tag{9.11}$$

and, from (9.7),

$$e^\mu = \Lambda e^{\lambda-\nu-c}. \tag{9.12}$$

The last two equations yield e^μ and e^ν as a function of λ, and become

$$e^\nu = \left(1 - \frac{r_0}{r}\right)^{1/2}, \quad e^\mu = e^{-\nu}, \tag{9.13}$$

for $\lambda = C = 0$. This is a classic result in general relativity.

It is worth noticing that equations (9.12) have been obtained from (9.4a)-(9.4b) without resort to (9.4c), which does not contribute anything new as we now show by direct proof of (9.4c) from (9.4a) and (9.4b).

We differentiate (9.4b) and multiply by $-\frac{1}{2}\Lambda^{-1}e^{2(\mu-\lambda)}$ ($\Lambda = 0$ is non-physical). Using (9.5), which was implied by (9.4a)-(9.4b), we replace $\Lambda^{-1}\Lambda'$ in the very last term of the equation so obtained. We thus get

$$(\lambda' - \mu')\Lambda + (\lambda' - \mu')2r\nu' + \Lambda' + \nu' + r\nu'' + r\nu'(\mu' + \nu' - \lambda') = 0. \tag{9.14}$$

We shall now show that (9.14) coincides with (9.4c). Equation (9.14) has six first level terms on the left hand side. The third term is the last one on the left hand side of the equation (9.4c). The sum of the fourth, fifth and sixth terms of (9.14) can be rewritten as

$$r(\nu'' + \nu'^2 - \mu'\nu') + \nu' - r\nu'\lambda' + 2r\nu'\mu'. \tag{9.15}$$

The term with the parenthesis in (9.15) is another one of those present in (9.4c). If we add the other three terms in (9.15) with the second term in (9.14), we get $\nu'(1 + r\lambda')$. This equals $\nu'\Lambda$, which added to the first term in (9.14) yields the first term in (9.4c). Hence equation (9.4c) is the differentiated (9.4b) multiplied by $-\frac{1}{2}\Lambda^{-1}e^{2(\mu-\lambda)}$. This implies a corresponding relation between the right hand sides of Einstein's equations, a strong constraint.

In Einstein's theory of gravitation, energy-momentum tensors are brought to the field equations from other parts of the physics (Incidentally, not all energy-momentum tensors are symmetric; Einstein's tensor is). In a hypothetical unification of the interactions, the Einstein tensor might be just a term at par with the other ones in the same equation, the separation into left and right hand sides being artificial. Spherical solutions at very short distance might not even exist. The ubiquitous spin does, after all, break the spherical symmetry. The final chapter on the theory of gravitation —which certainly is in the right track with Einstein's theory— may have not yet been written.

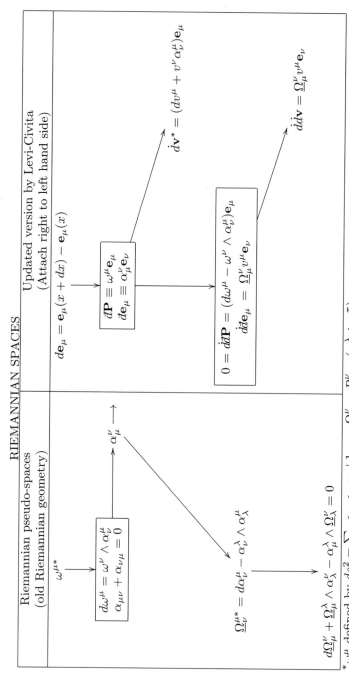

Part V

THE FUTURE?

Chapter 11

EXTENSIONS OF CARTAN

11.1 INTRODUCTION

Differential forms constitute the most powerful language for differential geometry. É. Cartan used them to develop his theory of connections in the early 1920's [11], [12], [14]. But those objects have a life of their own —even larger than differential geometry itself— since they can be endowed with structure additional to what is required in that geometry. The purpose of this and the next two chapters is to show that some largely overlooked mathematics of the 19th and 20th centuries help one understand how to further extend the field of differential geometry.

Riemann and Felix Klein are the two father figures in the evolution of geometry in the 19th century (see Section 8.1). But the big picture is incomplete without a third father figure, namely Hermann Grassmann ([43] is a publication in English of his main work; see appendix B for a biographical sketch). His ideas were so advanced for his time that their relationship to geometry still is largely missed. His relevance is such that we have found it pertinent introducing his biography in Appendix B.

The relevance for differential geometry of Grassmann work, which is predominantly algebraic, lies in that there is only one point of separation between algebra and geometry. Although a zero is absent in affine space (studied by geometry), the latter can be identified with a vector space (studied by algebra) if we choose an arbitrary point to play the role of zero. But even the "zero points" are not arbitrary in differential geometry, since they are the points of tangency of tangent spaces to differentiable manifolds. One cannot, therefore, give too much importance to the separation of algebra and geometry.

In a scenario where differential geometry were a main field of research but only rudiments of algebra were known, the latter discipline would have emerged as a branch of differential geometry. Since there are branches of algebra far

215

removed from differential geometry, we may view the latter as the sector of the algebra-geometry edifice that concerns itself with differentiations and integrations, i.e. the operations that allow one to generalize the elementary or Klein geometries. The logical and historical order, however, is that extensions of algebra lead to extensions of the concept of differentiation, which in turn lead to extensions of standard differential geometry.

One such extension leads to Finsler bundles and, related to them, to a canonical Kaluza-Klein (KK) theory that converts the theory of moving frames into something richer. In Finsler bundles, the pair of groups (G, G_0) is replaced by a triple of them (G, G_0, G_{00}). Thus, for instance, the special relativistic spacetime structure would be seen as represented by the Poincaré, Lorentz and $SO(3)$ groups.

In KK theory, points play a role additional to the one they play in the theory of moving frames, which use the points of the manifold to anchor themselves to it. This is a passive role. In the Kaluza-Klein extension of which we shall speak in this chapter, points also represent point particles that move around independently of any frames. In this way, one does not need to view the trajectory of a particle as a succession of origins of frames, as is the case in the theory of the moving frame. So, G plays here the role of subgroup.

Another algebraic extension with differential geometric implications would be the replacement of a derivative based on exterior algebra with one based on Clifford algebra. In parallel to our "introducing" the concepts of affine curvature, torsion etc. through exterior differentiation of vector fields (See section 5.7), we could conceive in principle a more comprehensive geometry based on the replacement of exterior differentiation with a Clifford-algebra-based differentiation. Also, linear algebra and Lie algebra would be immersed in Clifford algebra. This would be a still unexplored Kähler generalization of the theory of moving frames.

Finally, there is another aspect to the KK extension of geometry: one should view symmetries in a new light. The 5-D structure now is the main one. The groups of symmetry cannot be the same in subspaces that appear similar but are not actually so in a profound way. One should look at spacetime symmetry as a manifestation in spacetime subspace of some symmetry pertaining to the 5-D KK space. The manifestation of that symmetry may take another form in space-propertime subspace. We deal with that in the last section of this chapter.

11.2   Cartan-Finsler-CLIFTON

In this book, we have seen the affine and Euclidean brands of differential geometry as theory of moving frames. Their Klein geometries are ostensibly related to corresponding finite Lie groups and algebras. The frame bundles are of dimension $n + n^2$ and $n + [n(n - 1)/2]$, respectively in the affine and Euclidean cases. In contrast, the crucial dimension in the tensor approach to differential geometry is n and its integer powers, both in the affine and Euclidean cases. This is related to the fact that the tensor calculus implicitly deals with what we know to be sections of frame bundles, but where the last ones are otherwise absent.

The tensor calculus precedes Cartan's theory of connections by two long decades. But worse is the fact that the absence of frame bundle considerations also occurs in most modern approaches to the classical sector of differential geometry. Thus, the key element in Finsler geometry is seen to be, incorrectly in our opinion, Finslerian metrics, not Finsler frame bundles. But geometry is what geometers do. I respect that view, but it may be of little interest in areas of geometry that should concern physicists.

I shall try to show here that there is a better way. I must first report, however, the large amount of publications on sophisticated physical issues by H. Brandt using a type of Finsler geometry that does not use frame bundles. But it at least relates the tangent bundle to sophisticated physics. The title "Finslerian Quantum Field Theory" of one of his papers [2] speaks by itself and contains the almost complete list of his Finsler related papers.

Cartan's seminal paper on the Cartan-Finsler connection [19] uses the moving frame method, but the relation to a frame bundle and to the Cartan-Klein program is not clear. A more explicit approach to the type of Finsler geometry we advocate is an important paper by Chern of 1948 on this subject [25], but much less so in his work of the last two decades of his life. This may have been due to the evolution of his interest towards global issues, which are connection-independent. The present author's approach to Finsler geometry falls within the Cartan-Klein program. It subordinates the metric to the connection, and is again based on the moving frame method [73], [74], [75], [76]. See also Deicke [30], [31]. The concept of Finsler connection can be introduced in affine geometry [73], i.e. independently of the concept of metric. As a by-product, we may have Finsler connections on Riemannian and pseudo-Riemannian metrics.

The breakthrough on affine-Finsler connections is due to Clifton, with prodding by the present author. I brought to his attention, at a time when he had acquired a deep understanding of the Cartan-Finsler connection [19], that his Finsler practice overlooked the philosophy of the Cartan-Klein program. In Cartan's approach, the character of a geometry is granted to it by the connection (remember geometric equality), not by the metric (remember Riemannian pseudo-spaces). And there is not a major reason in physics to consider at this point metrics other than the Lorentz metric, even if there were a physically suitable preferred frame and the connection of spacetime were Finslerian. I challenged Clifton to first introduce Finslerian affine frame bundles, which would then be restricted to obtain bundles for Finsler-Euclidean connections. He delivered [73] (readers can find in that reference why Clifton should have been the main author of [73]-[75] but was not).

The mathematical significance (or at least the utility) of Finsler frame bundle geometry lies in that it brings up the question of what is the elementary geometry underlying the Cartan-Clifton branch of Finsler geometry. Inspection of the formal definition of Clifton of affine-Finsler connections [73] leads one to conclude that, as already said, there is a triple of groups, and not just (G, G_0).

The physical significance of the restriction of the affine-Finsler bundle to pseudo-Riemannian metrics of Lorentzian signature is that the geometry of the spacetime of special relativity emerges as the Klein geometry underlying connections on those bundles. Said better, there is a way to look at such a spacetime other than as being pseudo-Riemannian. This in turn yields a new way of looking at 4-velocities, which is crucial for developing KK theory canonically generated by Lorentz-Finsler geometry. This in turn brings up the subject of canonical KK geometry for high energy physics (See next section).

Yang-Mills theory is based on auxiliary bundles, not directly related to the tangent bundle. Unfortunately, modern approaches to the classical sector of differential geometry for physicists care little about frame bundles. This author cared and found that the Lorentz force (abstraction made of proportionality constants) is unavoidable in the Finsler frame bundle. One does not even have to look for the connection whose autoparallels are given by the equation of motion with Lorentz force; it is unavoidable in the Finsler frame bundle with metric of Lorentz signature. In addition to the gravitational force contributed by the metric, Lorentz-Finsler connections contribute with an acceleration (thus force) that takes the form of —and only of— the Lorentz force when only the least explicitly Finslerian terms in the equations of structure are considered [79], [81].

11.3 Cartan-KALUZA-KLEIN

Connections on Finsler frame bundles represent geometry as a theory of moving frames with velocities located on the base space rather than on the fibers of the standard frame bundles. This feature helps to solve a problem which may be said to have been intimated by Cartan, though he may have overlooked returning to it. We proceed to explain.

In his paper on Einstein's equations [8], Cartan obtained the equations of structure of 3-D Euclidean space as consequence of the following. He had a particle and a frame to which he referred the coordinates of the particle. He left the point fixed and combined a differential translation and a differential rotation of the frame. He differentiated the equation that states the differentials of the coordinates of a fixed point in terms of that translation and rotation. Further manipulation allowed him to obtain the equations of structure. Moving the point was not considered. If one did, an equivalence of active and passive translations might be invoked in Euclidean space, but not in general. Be as it may, the motion of the particle in connection with that derivation brings to the fore that the particle might have a role to play in the equations of structure, which is not presently the case.

The motion of particles can be integrated into the core of geometry through a KK space where the propertime of the particle constitutes the fifth dimension. The former 4-velocity is then the unit vector associated with it. Not surprisingly, Cartan himself introduced in 1923 a formulation of several branches of the physics in terms of a five dimensional space, the dimensionality being disguised by his use of Plückerian coordinates [11]. His bivectors come in two classes. Six of them he calls bivectors and belong to the spacetime subspace. The other four he calls sliding vectors, which intimates the association of the fifth dimension with the particle. See [77].

This "canonical" KK space has very significant features, which seems to indicate that there is still much to learn about the flat spacetime of special relativity. Let e_0 denote the unit vector in the laboratory's time direction, and let u denote the unit vector in the propertime direction. Consider the subspaces spanned by (e_0, e_1, e_2, e_3) and (e_1, e_2, e_3, u). In special relativity, the spacetime bases are orthonormal but the bases (e_1, e_2, e_3, u) of space-propertime subspaces are not orthogonal since the dot products $e_i \cdot u$ must be $u_i \gamma$, where γ is $(1 - u^2)^{-1/2}$. Needless to say that, in this KK perspective, the dynamics of a quantum system will concern the subspace spanned by (e_1, e_2, e_3, u), rather than the one spanned by (e_0, e_1, e_2, e_3). The lack of orthonormality does not bode well for special relativity in this KK context (One can always orthonormalize, but the replacement for u will no longer result in a tangent to the trajectories of particles). But there is a surprise.

Consider the preferred frame alternative to special relativity with absolute relation of simultaneity that still complies with the key experimental results of Michelson-Morley, Kennedy-Thorndike and Ives-Stilwell experiments [84]. The timelike unit vector is no longer perpendicular to the (e'_1, e'_2, e'_3) subspace of this alternative, but the vector u is, like e_0 is orthogonal to (e_1, e_2, e_3) in the relativistic case.

In this KK scenario, comparison of the space-propertime subspace for SR and for the just mentioned alternative appears to indicate that the good working of the relativistic quantum mechanics that is based on the orthonormal bases (represented in the relations among the gamma matrices) appears to be an argument in favor of preferred frame physics rather than special relativity.

Since special relativity works so well, any preferred frame proposal will appear unattractive in principle given the concomitant non-orthogonality of e'_0 to (e'_1, e'_2, e'_3) in the preferred frame scenario. That this is not a problem has been well understood by those who support the thesis of conventionality of synchronizations, which states that special relativity and the preferred frame alternative being considered are indistinguishable from each other. Though conventionalists may have carried their claim too far, there is nevertheless much truth in it. In the preferred frame scenario, one cannot hold a synchronization consistent with the spacetime transformations with absolute simultaneity [84]. Clocks in our real world are not at rest or in slow motion with respect to a hypothetical preferred frame. They are automatically synchronized as per Einstein's synchronization procedure. The failure to hold an absolute synchronization does not invalidate the possibility of a physics based on an absolute relation of simultaneity, but masks most or all of its effects. For more details and references see [84].

There are several other powerful reasons advocating the aforementioned KK space. Let us point here at a physical one. The last sentence of the previous section speaks of the possible association of electromagnetic field with torsion through autoparallels in Finsler frame bundles. Let v^μ be the components of the 4-velocity of a particle. As stated at the end of section 9.7, a torsion of the form $\Omega^\mu = -v^\mu F$ is an incorrect way of writing what should be considered as an expression pertaining to the Finsler frame bundle, where v^μ is not a field on a curve but redundant coordinates on the Finslerian base manifold. The v^μ become the components of the 4-velocity on natural liftings of spacetime curves. More specifically, the 4-velocity is e_0 in the Finsler bundle, an invariant under the SO(3) group of the fibers of that bundle.

We translate this finding to the KK space. F becomes the component of the torsion in the direction of $\mathbf{u}$ in the KK space. But, as in standard KK theory, $\mathbf{u}$ is not just one more vector; the fifth direction is different. KK space is unlike spacetime, where all transformations (translations and isometries) not involving null directions are on an equal footing up to issues of signature. Transformations in a KK space involve the fifth direction differently from the others. In our KK theory, boosts of particles are no longer the isometries known as hyperbolic rotations in spacetime but in time-propertime subspace of the KK space (there is no problem with the three independent components of the velocity; the connection takes care of that, as it does in Finsler geometry where $d\mathbf{u}$ is de_0). Cartan's theory of moving frames is once again a magnificent tool to explore extensions of classical differential geometry of interest for physics.

11.4   Cartan-Clifford-KÄHLER

This book must have made clear by now that differential geometry is about exterior differentiation. More precisely, it is calculus of vector-valued differential forms where the underlying algebra of the differential forms proper —i.e. disregarding the accompanying valuedness algebra — is exterior algebra. On the other hand, the Kähler calculus is based on Clifford algebra of differential forms.

Exterior differentiation is contained in Kähler differentiation, but not conversely; there is less information in the exterior derivative than in the Kähler derivative. And yet integrability of a system is based only on an integrability condition about exterior differentiation even though the complementary information provided by interior differentiation is absent. It looks as if, in the integration process, we get for free the information (contained in the interior derivative) lost by exterior differentiating. An incipient study of this issue was undertaken elsewhere [80]. That paper intimated that two algebras, whatever their nature, are deeply intertwined, as we have also seen in this book. At that time, we did not provide specifics of how that takes place. The details are becoming increasingly clear.

The exterior product of two consecutive sides of a parallelogram gives some information about the latter. But, on a curved surface, parallelograms in the ordinary sense of the word do not exist. We may integrate a scalar-valued differential 2−form $f(x^1, dx^2)dx^1 \wedge dx^2$ on a region of a curved surface, but this information is not equivalent to the information contained in the exterior product of two vectors. The latter speaks in particular of the spatial orientation in Euclidean space of the figure formed by the two vectors. In the case of the curved surface, the aforementioned integration gives only a number, but not an orientation in space. For that, one would need to integrate a bivector valued differential 2—form.

Consider a system of differential form equations for a quantum mechanical structure that has symmetries. If the differential forms are endowed with a Clifford rather than a exterior algebra structure, one may expand members of the algebra of differential forms in terms of members of the ideals that those symmetries define in the algebra. The implications of such expansions [48], [83] may be relevant for the growth of differential geometry, as we now explain.

For angular momentum, Kähler builds a left ideal with the idempotent $\frac{1}{2}(1 + idx \wedge dy)$ (be aware that $\frac{1}{2}(1 + dx \wedge dy)$ is not an idempotent). But we do not need the unit imaginary if we take the idempotent to be $\frac{1}{2}(1 + dx \wedge dy\mathbf{a}_1\wedge\mathbf{a}_2)$, also written as $\frac{1}{2}(1 + dxdy\mathbf{a}_1\mathbf{a}_2)$. The concomitant phase shifts that give the dependence on the angular coordinate are now to be viewed as $\exp[(1/2)\phi\mathbf{a}_1\mathbf{a}_2]$, where, again, $\mathbf{a}_1\mathbf{a}_2$ plays the role of the unit imaginary. Similarly, the unit imaginary in the energy operator and in the exponent of phase shifts associated with time translations would be replaced by the unit vector in the propertime (time) direction in the space-propertime (respectively spacetime) subspaces of the aforementioned KK space.

Enter Cartan. He was not quite as explicit with regards to Clifford algebras, in part due to the fact that the appears not to have discussed them except in his long paper of 1908 on Complex Numbers [5] (Incidentally, this is a very good paper to learn of the great variety of different products introduced by Grassmann). But he made an extensive use of exterior valued differential forms, which involves also a second exterior algebra. If he had extended the two exterior algebras in those forms to Clifford algebras, he would have expanded differential geometry beyond the exterior calculus, bringing idempotents and ideals into the realm of differential geometry. A rapprochement of the respective mathematics of general relativity and of quantum mechanics would have ensued.

To say the least, we have indicated in this section how one can expand Cartan's theory of moving frames into new realms along the lines suggested by the work of Kähler through his exploitation of the Clifford algebra of differential forms.

11.5   Cartan-Kähler-Einstein-YANG-MILLS

The titles of the previous sections in this chapter had only names of mathematicians. We shall now involve the names of Einstein and Yang-Mills for further development of differential geometry directly related to the tangent bundle. Could it not be that the auxiliary bundles of present day physics —though not any auxiliary bundle that one could think of— are directly related to the tangent bundle? And what would that differential geometry be like?

We have intimated in this book that the Finsler frame bundle allows for the matching of a piece of the torsion with the electromagnetic field. The torsion and, more interestingly, the contorsion have many more degrees of freedom than needed to accommodate that field, but apparently not in a way that would involve the $U(1) \times SU(2) \times SU(3)$ group. In other words, it is not at all clear where the weak and strong interactions are. Certainly they are not in spacetime symmetry if treated in the trite way in which the paradigm makes use of it.

We need Riemann's curvature for gravitation. The clue as to how to satisfy this need in a way compatible with a connection other than LC's lies in the study of conservation laws that we undertook in section 8.10. It leads one to the conclusion that there is not true conservation on generalized manifolds without TP. Unless we have trivial TP (like in Klein geometries), the torsion and thus the contorsion, β, are not null. Since β is the difference between ω and the LC's differential form α, it is valued in the same Lie algebra as they are. β is going to play the role of the Yang-Mills connection. It certainly is not a connection, which is better than if it were one, being as it is part of *the* connection. The other part, α, represents the gravitational interaction and both parts are joined in a very natural and strong way.

What is the connection that can accommodate all that? And what has all this to do with Einstein? In the late 1920's, he proposed TP for the connection of spacetime. Cartan advised him that he incorporate into his system of equations the first Bianchi identity and the curvature. Had Einstein had a better understanding of differential geometry, he would have realized how well did Cartan's suggestions fit his thesis of logical homogeneity of physics and differential geometry [38].

Einstein abandoned his TP project in frustration. Cartan did not go on his own and, in any case, the state of calculus and of the geometric art was not ripe for the task. Indeed, at the time of his correspondence with Einstein on TP, Cartan had not yet done his seminal work on Finsler geometry [19], which was crucial for Clifton's acquiring the frame bundle vision of which we spoke in this chapter. In addition, another element was needed in the opinion of this author: the Kähler calculus and concomitant vision of quantum physics (See chapter 13). Cartan nevertheless discussed the general characteristics of Einstein's attempt at unification with teleparallelism [18].

The Kähler calculus is needed for the KK framework, which in turn is needed for Yang-Mills in this geometric perspective. KK space allows one for the same basic symmetry to manifest itself as Lorentz symmetry in the spacetime subspace and as $U(1) \times SU(2)$ symmetry in in the space-propertime subspace. In this subspace, $SU(2)$ would replace $SO(3)$ because we would be dealing with the action of rotations on spinors as members of ideals of space-propertime Clifford algebra.

Regarding $U(1)$, we already saw its emergence in the Finsler bundle as $\Omega^0 = -F$. This becomes $\Omega^4 = -F$ in the fifth dimension of the KK space since the $\mathbf{e}_0$ of the Finsler bundle becomes the unit vector $\mathbf{u}$ tangent to the propertime lines. KK space provides for a representation of electromagnetism freer from spacetime frames than its representation in Finsler geometry is. But there is more to the role of the Kähler calculus.

The Kähler view of quantum mechanics that he saw emerge from his calculus moves the focus on the symmetries of basic quantum mechanical equations to the symmetries that can be put together in solutions of those equations. In other words, the idempotents that define the ideals that embody those symmetries come to the fore. From the ansatz for the solutions with time translation and rotational symmetry of the Dirac equation with electromagnetic coupling emerges the ansatz for the solutions of equations with $U(1) \times SU(2)$ symmetry. Notice in the electromagnetic case that those solutions do not directly the $U(1)$ symmetry, but indirectly through its entanglement with proper energy, and almost as directly with spin. In other words, we are seeing in the ansatz for solutions with symmetry the relation of $U(1) \times SU(2)$ to the Poincaré group.

Regarding $SU(3)$, Schmeikal has shown that the *tangent Clifford algebra* of spacetime may accommodate quarks and the generators of $SU(3)$ through the use of mutually annulling primitive idempotents [66]. Bring in Kähler's treatment of quantum mechanics with differential forms —and, again, what his methods say about the solution of exterior equations for physical systems endowed with time translation and rotational symmetry. That treatment provides in a natural way one such family of mutually annulling primitive idempotents, but in a *Clifford algebra of differential forms*. That result requires only scalar-valued differential forms. More general valuedness will obviously generate many more families of idempotents to play similar roles.

In TP, we do not need to make $U(1) \times SU(2) \times SU(3)$ consistent with (pseudo)-Euclidean curvature —usually but erroneously called affine curvature— since all tangent spaces to the base manifold can be identified. The ideals at different points of spacetime or, more pointedly, space-propertime, can be similarly identified.

The auxiliary bundles of Yang-Mills theory may not be needed after all, since tangent bundle geometry in the sense of our KK space may achieve the same goals, but more naturally. One should be able to see how the two parts of the connection play their respective roles when one further develops the many lines of research that the theory of moving frames lays in front of us. And be aware of the fact that the fifth dimension is not something totally new. It is a new way of looking at propertime. It was there as long as there clocks, say hydrogen atoms, and almost as long as there was time itself.

Chapter 12

UNDERSTAND THE PAST TO IMAGINE THE FUTURE

12.1 Introduction

The first four decades of the nineteenth century witnessed the work of such mathematical luminaries as Gauss in calculus and geometry, and Cauchy with his complex calculus and his theory of algebraic keys. Retrospectively, I consider their very important works the last ones of a past era. For this author, a new era starts in 1844 with the publication of Hermann Grassmann's Opus Magnum. Titled *Die Ausdehnungslehre*, it is also known as the Calculus of Extension. But it is mainly about algebra. All this thus speaks of the intimate relation between algebra, calculus and geometry. Grassmann's publications are very difficult to understand, but have fortunately been interpreted. In the opinion of the present author, the Grassmann era is far from being over. In order to understand why, we shall explain the present state of the areas of algebra, calculus and geometry that were, directly or indirectly, impacted by his work. We present a brief biography of this extraordinary thinker in one of the appendices.

12.2 History of some geometry-related algebra

In his paper "The tragedy of Grassmann", distinguished French mathematician Dieudonné put Grassmann's mathematical work [43] in perspective [32]. He pointed out that in 1844 (ten years before Riemann's famous lecture at Göttingen on the foundations of geometry!), Grassmann was developing mathematics for spaces of higher dimension and constructing objects of different grades [43]. Specifically, he was dealing with multivectors and all sorts of geometric products at a time when the theory of vector spaces comprised only addition of

225

vectors and their multiplication by scalars.

A year earlier, William Rowan Hamilton had discovered the concept of quaternions. He produced an algebra with an interesting multiplication table with three independent units of square minus one, and one unit of square plus one, namely the real unit number. But only in 1853 took place his publication titled *Lectures on Quaternions*, which was to him what Die Ausdehnungslehre was to Grassmann.

In 1873, William Kingdom Clifford started to publish his algebraic work nowadays known as Clifford algebra. For his scattered contributions to this subject see [26]. He gracefully acknowledged his great debt to Grassmann. Clifford's work allows a deep comparison, though indirect, of the impacts of the algebraic works of Grassmann and Hamilton. The algebra of quaternions is isomorphic to the Clifford algebra Cl^2. But it is just one algebra. On the other hand and to say the least, Grassmann produced a system of algebras (the exterior algebras) applicable to any number of dimensions, and much more. See his biography. But he was less specific than Hamilton when a quadratic form is involved. Followers of Grassmann and Hamilton became bitter adversaries during the nineteenth century.

The work of Grassmann thus is incomparable far more relevant than Hamilton's and more broad than Clifford's. Although quadratic forms are of the essence in the geometric work of Riemann, there is very little relation of his work to that of Clifford and Hamilton, which was algebraic rather than differential-geometric.

It is important to be aware of the fact that the exterior and interior products can be viewed as shorthand for combinations of Clifford products, as we already saw in chapter 3. Hence, it is totally unnecessary to complicate matters by thinking in terms of structures where Clifford algebra would be a substructure, and exterior algebra with a dot product added to it would be another substructure.

It seems to this author that the main example of further development of Grassmann's ideas not having to do with quadratic forms has to do with projective algebra/geometry. In the projective realm, Peano further developed those ideas in 1888 [58]. See the interesting opinion of Gian-Carlo Rota and collaborators in 1995 on the impact of Grassmann, in the context of their own work in the line of thinking of Grassmann and Peano [1].

Following custom, we shall use the term Clifford algebra when it is built upon a tangent vector space or a tangent module. And we shall use the term Kähler algebra when it is built upon a module of differential 1-forms [46], [48]. But we shall use the term Clifford product in both cases.

The stated developments in algebra are not the only ones in the epoch that we are considering. But, in our view, they are the most relevant ones in connection with the convergence of calculus and geometry.

12.3   History of modern calculus and differential forms

There is some calculus in the mist of Grassmann's algebraic work (See a couple of articles in electrodynamics and mechanics in the compilation of English translation of much of his work [43]). But going beyond this brief statement would only distract us from his incomparably more important algebraic work. Something similar but to a lesser degree could be said of the work of Clifford that touches calculus.

Other than the complex variable calculus of Cauchy, which goes back to the beginning of the nineteenth century, the only calculus of that century that is in heavy use up to this day is the vector calculus. It was conceived almost simultaneously by the physicist Gibbs and the engineer Heaviside in the 1880's, using ideas from both Grassmann and Hamilton. We refer readers to the book "A History of Vector Analysis" by Michael J. Crowe [28], where they will find more empathy for Gibbs and Heaviside than they will get from the present author. Alas, there is in the vector calculus more of Hamilton than of Grassmann since Hamilton only cared about four units $(1, \mathbf{i}, \mathbf{j}, \mathbf{k})$.

Then came the exterior calculus of differential forms, proposed by É. Cartan in 1899 [4] in the context of solving systems of differential equations. It is underlaid by exterior algebra, built by exterior product of copies of the module of differential 1−forms for differentiable manifolds of arbitrary dimension. This new calculus was in fact much more than just a new calculus in the ordinary sense of the word, as that paper contained material on differential forms that is only seen in treatises on exterior differential systems (any system of differential equations, ordinary or partial, can be written as an exterior system). This is not strange since Cartan was standing on the shoulders of such luminaries on the subject as Lie, Frobenius and Pfaff.

This work by Cartan on differential forms constituted a paper within a paper. It is likely that it would have obtained a much faster traction in the literature if it had been published separately from the rest of the paper, thus emphasizing the appearance of a new calculus.

Cartan was virtually the only one to use this tool for a few decades. One decisive application of his calculus is constituted by his 1908 paper on Lie groups [6]. His 1922 book *"Lectures on Integral Invariants"* [9] is a monument to the extraordinary power of exterior scalar-valued differential forms, specially in the field of classical mechanics.

As we shall intimate, one can have a more comprehensive algebra of scalar-valued differential forms (Kähler's), and, correspondingly, a still better calculus. Out of respect for unfortunate tradition, we shall use the term differential forms when they are considered as elements of their exterior algebra, rather than as members of more comprehensive ones. When the algebra being used is Clifford's (see section 3), we shall preferentially refer to them as clifforms.

In the meantime, the tensor calculus emerged from a collection of scattered results spanning four decades [60]. It filled a gap, since the vector calculus does not apply to Euclidean spaces of arbitrary dimension, and to Riemannian spaces of any dimension except dimension one. The best version of the traditional theory of curves and surfaces in 3-D Euclidean space used the vector calculus. Similarly, the tensor calculus became the calculus in Riemannian geometry.

In 1934, Erich Kähler produced what is nowadays known as the Cartan-Kähler theory of exterior systems [45]. Fast forward to 1960-62. In those years, Kähler published his work on the calculus that bears his name. He called it the interior calculus. The name exterior-interior calculus would be more appropriate. Kähler developed it only for manifolds endowed with a Riemannian metric and Levi-Civita connection, but the generalization to other connections is trivial [78].

Kähler used his calculus in 1961 to obtain the fine structure of the hydrogen atom with scalar-valued clifforms [47]. More importantly, he used these to show how the concept of spinor emerges in solving equations of differential forms, even if the equations are not restricted to the ideals to which the spinors belong [47]. In 1962 [48], he presented the general theory more comprehensively than in 1960 [46], and used it to get the fine structure more expeditiously than in 1961 [47]. A very important and yet overlooked result that he obtained is that antiparticles (at least in his handling of the electromagnetic interaction) emerge with the same sign of energy as particles [48].

At its most general, he considered tensor-valued differential forms, for which he did not provide applications. But even the restriction of his calculus to scalar-valued differential forms produces amazing results where spin emerges not as an internal property of particles attached to orbital angular momentum, but at the same level with it.

Amazing is the naturalness of his treatment of relativistic quantum mechanics through his Kähler-Dirac equation or, said better, Kähler equation, which supersedes Dirac's. With that equation and electromagnetic coupling, antiparticles emerge with the same sign of energy as particles.

The author of this book has shown that computations in relativistic quantum mechanics with Kähler's calculus using simply scalar-valued clifforms are much easier and less contrived than with Dirac's calculus [83]. The expansion of that equation using the mass of the electron as the dominant energy term gives rise to the classical form of the electromagnetic Hamiltonian as in the Pauli equation. In short succession, the same process gives rise to the Hamiltonian at the level of the fine structure, without resort to Foldy-Wouthuysen transformations. In retrospect, these transformations are not physically significant.

The Kähler calculus of scalar-valued differential forms is to quantum mechanics what the calculus of vector-valued differential forms is to differential geometry, and thus to general relativity. An obvious next step is to find applications of the Kähler's calculus involving valuedness other than scalar.

12.4   History of standard differential GEOMETRY

We are concerned with developments in differential geometry that took place after the first publication, in 1844, of Grassmann's seminal work on algebra. Thus our historical report on this subject starts with Bernard Riemann's planting in 1854 the seeds of what would later become the "program" of Riemannian geometry. We have used the term program to put Riemannian geometry in the same bag with geometries where the main concept is the concept of distance.

Nothing much happened for a decade, when Riemann tackled the problem of whether two quadratic differential forms are related by a coordinate transformation. It is a problem of a type called problem of equivalence and, roughly speaking, there is a method to solve such a problem called method of equivalence. It was not a very well defined method at the time, which is why we may speak of the different versions of it that, according to Pauli [57], were used by Riemann, Christoffel and Lipshitz to independently obtain what is now called Riemann's curvature. Its annulment constitutes a necessary and sufficient condition to give a positive answer to the specific problem given at the beginning of this paragraph.

Fast forward to the year 1872, when the mathematician Felix Klein launched his famous Erlangen program. We speak about it at greater length in our earlier book "Differential Forms for Cartan-Klein Geometry", subtitled "Erlangen Program with Moving Frames". Before buying it, consider that the contents of that book and the present one is largely common. But we have added the appendices, totally renewed the preface and chapters 1, 12 and 13, and added sections on Clifford algebra in chapter 4, as well as a few scattered sections in other chapters.

The Erlangen program was based on the concept of group. It was Henri Poincaré who, according to Élie Cartan, interpreted Klein's program as being equivalently about geometric equality [15]. In the modern interpretation —to the extent that it has adopted Cartan's views on geometry— Klein's program is to the concept of geometric equality what Riemann's program is to the concept of distance.

Klein spaces/geometries are not nowadays what Klein himself considered as spaces/geometries. Not every Klein space of that time is a Klein space today. For example, the sphere is not the receptacle of a geometry in the modern sense of Klein space. The modern Klein spaces are, so to speak, flat spaces. Cartan spaces are their modern generalizations in the direction that Cartan provided.

Cartan spoke of a Klein geometry as the study of invariants under the transformations of a group. Late in his life, Cartan wrote explicitly about the role of group theory in modern differential geometry [21]. He had in mind pairs of a group G and a subgroup G_0, subject to some condition about which there is not unanimity. But he did not need to elaborate on some further condition to be satisfied since, for the most part, he studied specific types of generalized geometries, which satisfied it, whichever we think is the relevant one.

Cartan mentioned G_0 repeatedly, but only exceptionally did he refer explicitly to the pair of group and subgroup, although this structure is otherwise very clear, though only implicit, in his writings.

The G groups are called the fundamental groups of the geometries. One also speaks of the fundamental group of the space where the study of the properties of the figures takes place. Let us repeat Cartan's words [15]:

> In each of these geometries and for expediency reasons, one attributes to the space where the figures under study are located the very properties of the corresponding group, or fundamental group; one thus speaks of "Euclidean space", "affine space", etc., instead of "space where one studies the properties of the figures that are left invariant by the Euclidean group, affine group, etc."

G_0 happens to be the subgroup constituted by all the elements of G that leave a point unchanged. All transformations in G_0 are present in the fibers of the generalized spaces. This is in contrast with transformations like translations, which are valid only in differential form in the generalizations, and only in the base manifold.

In Euclidean geometry, G is the group of the so called displacements (translations and rotations, with the group of rotations as G_0); in affine geometry G is the group of the affine transformations (translations and linear transformations, with the linear transformations as G_0); in projective geometry, G is the projective group (translations and homographies, with the homographies as G_0).

The emergence of the tensor calculus, of which we spoke in the previous section, may be seen as the next step in the development of differential geometry, except for the advance in the formulation of the classical theory of curves and surfaces that had taken place not much earlier with the emergence of the vector calculus. In contrast with the method of equivalence, the tensor calculus already had a distinctive differential-geometric flavor. Due to this, many physicists of past decades would view the differential geometry of Riemannian spaces as virtually synonymous with the tensor calculus used by those authors. Such would not be, however, the case, for instance, with the algebraic, more formal presentation of the same calculus by Lichnerowicz [53].

Significant is the fact that the derivation of the expression for the Riemannian curvature (derivation later adopted by Einstein in his work on general relativity) has more flavor of exterior calculus —disguised as antisymmetrization to remove certain terms— than of method of equivalence.

Levi-Civita was disciple of Ricci, but it is the former who gets greater exposure in Cartan's writings because of the connection that bears his name. It was proposed in 1917, and constitutes the point of contact of the Klein and Riemann programs in the following sense. Until the Levi-Civita connection, there was no concept of geometric equality in Riemannian geometry, i.e. of equality of vectors at different points. To be precise, there is no geometric equality in the Levi-Civita connection either, but we may at least speak of geometric equality of tangent vectors in the neighborhoods of lines. True geometric equality over finite regions of a manifold only exist if the manifold is endowed with teleparallelism.

In 1922, É. Cartan published a series of Notes to the French Academie of Sciences announcing what was to become over several years a series of papers on his theory of moving frames. This is a theory which applies in principle to all Klein geometries. We have dealt in this book with the affine and Euclidean cases, both elementary and generalized. Cartan unified the Riemann and Klein views of geometry through a generalization of Klein's Erlangen program of 1872. Dieudonné speaks of this generalization as follows (reproduced from the introduction of a book by Gardner [42]):

". . . it is fitting to mention the most unexpected extension of Klein's ideas in differential geometry. He had envisaged groups of isometries in Riemannian spaces as a possible field of study of his program, but in general a Riemannian space does not admit any isometries except for the identity transformation. By an extremely original generalization, É. Cartan was able to show that here as well the idea of "operation" still plays a fundamental role; but it is necessary to replace the group with a more complex object, called the "principal fiber space"; one can roughly represent it as a family of isomorphic groups, parameterized by the different points under consideration; . . ."

Regrettably, Cartan's results are presented nowadays in ways totally different from his original one. But these modern ways do not exhibit the essence of Cartan's program, how his work generalizes Klein's, and how, in the process, Riemann's program is absorbed into it.

As we have made abundantly clear, Cartan approached his generalization of Klein's geometries as a problem of integrability, whose language is the language of differential forms. When Cartan dealt with what others consider to be tensor-valuedness, he was actually dealing with exterior products of tangent vector bases or of fields thereof. Something that looks like a skew-symmetric (tangent) tensor, need not be a tensor. It may be a multivector. The key as to whether one is dealing with one or the other lies with the products and differentiations that we perform with and on them. Suffice to say that the tensor product of skew-symmetric tensors does not yield in general another skew-symmetric one.

Cartan informally introduced, in addition to the exterior algebra of differential forms, an exterior algebra of valuedness. He extended to it the action of his d operator of differentiation, which is the reason why he (and Kähler) use the term exterior differentiation when others use the term exterior covariant differentiation. Readers who would like an approach to this extension more formal than in this book should consult Flanders [41].

Next in line would be Cartan's work on Finsler geometry. We must backpedal a little bit to 1918. It was Paul Finsler who, under the guidance of his doctoral supervisor Constantin Carathéodory, studied the geometry of curves and surfaces where a concept of metric more general than Riemann's was defined (This concept had been briefly broached by Riemann himself). But this was a generalization of the old Riemannian geometry, so to speak, i.e. one without connections. Thus the wide use that has been made of the name of Finsler gives him implicitly far more credit than he deserved. In fact the first Euclidean connection (i.e. metric compatible affine connection) on a Finsler metric is due to Cartan in 1934 [19]. It is known as the Cartan-Finsler connection. It is to Finsler metrics what the Levi-Civita connection is to Riemannian metrics.

It is troublesome that modern Finsler geometers work in what amounts to sections of a bundle but generally fail to define what the bundle is. If they did, they would have to consider what are affine-Finsler frame bundles, and what the restriction of such bundles that one could call metric-Finsler bundles is. Not even Cartan did explicitly deal with this issue.

Consider the standard spacetime of special relativity. We can push it to its Finsler bundle. In other words, we can reorganize the frames over its sphere bundle or bundle of directions. A set of coordinates in the base space could be (t, x^i, u^j), the u^j's being velocity coordinates. The group in the fibers is the group of rotations in 3-D. We shall name it G_{00}. We thus have the triple (G, G_0, G_{00}). G_0 (Lorentz group) now plays a diminished yet still relevant role. Its significance can be clearly ascertained in dealing with alternatives to special relativity. It is the bridge to the frames of relevance in those alternatives.

In one way or another, there are three groups in the geometry of special relativity when viewed from a Finsler bundle perspective: the Poincaré group, its standard Lorentz subgroup, and the latter's largest subgroup of rotations, $O(3)$. But this has not been formalized and, in any case, transcends the scope of the present book.

Before we continue with Finsler geometry, consider a mid-century development of significance for differential geometry. It is the already mentioned Kähler calculus. It is of special importance for quantum physics, but it does not impede that, in the process, the basic magnitudes of differential geometry emerge even without resort to issues of geometric nature. He disposed of issues like covariant differentiation by ansatz, the only connection he considered being the LCC. He did not deal with bundles, nor with where do connections take their values. But, to say the least, differential geometry and quantum mechanics now share in his differential forms a common mathematical language. Hence, he contributed to the development of geometry, though the implications of his work are not yet seen.

A relatively little known but very significant differential topologist by the name of Yeaton H. Clifton enters the picture now (The famous global differential geometer S.-S. Chern told this author of his great respect for Clifton as a mathematician). In 1989, I made progress with the issue of geometrizing the equation of motion of special relativity with Lorentz force in a tangent bundle context. The connection had to be viewed as Finslerian to make sense, though I had obtained it without explicitly resorting to Finsler geometry. I communicated this result to Clifton. It greatly impressed him because he had tried himself several classical-geometric options without finding anything that satisfied him. I thus earned his respect and we started a collaboration where he provided genial response to my questions.

My main question was: if Euclidean connections (i.e. what are usually called metric compatible affine connections) are restrictions of affine connections by virtue of restriction of the bundles on which affine connections live, of what affine connections are the usual Finsler connections (i.e. based on more general concepts of distance) restrictions? After recognizing that he had not asked himself that question, he immediately provided a series of great results. A main one was a theory of affine-Finsler bundles, i.e. not requiring a metric structure. In particular, such connections could be restricted by the Lorentz metric structure, which is (pseudo-)Riemannian. In such a structure, one can accommodate a magnitude such as the electromagnetic field, whose components have two indices, within the torsion, which has three.

One really appreciates the relevance of Clifton's work when one studies the modern presentations of differential geometry, specially the theory of connections, and then reads the papers by Cartan where that theory is presented for the first time. One would think that one is dealing with two different theories. The modern presentations appear to have been motivated by the fact that Cartan's work is often considered not to be rigorous. Clifton made it rigorous through a few definitions and a still smaller number of theorems. The problem of specializing that to the simple pre-Finslerian affine and Euclidean connection to make them rigorous was then a simple exercise that I took care of. We have dealt with this "special case" on its own in section 13 of chapter 8, and section 2 of chapter 12.

To summarize the significance of Clifton, let us say that the understanding that he had of Cartan's work permitted him to formulate very elegantly a theory of affine Finsler bundles, i.e. Finsler geometry without a concept of distance. It is just a simple extension of the theory of affine and Euclidean connections. As a by-product, we now also have a rigorous formulation for the theory of affine and Euclidean connections with the moving fame method, which was thought to be impossible, at least if one does not change its original flavor.

Finally, let me ask a question for honest differential geometers. Can they formulate a theory of connections on Finsler bundles with the modern methods?

12.5   Emerging unification of calculus and geometry

In the tensor calculus, only the lowest ranks of tensor-valuedness emerge. This seems to indicate that general tensor algebra, which is infinite dimensional, is only indirectly relevant, as mother of finite dimensional quotient algebras: exterior, Clifford, ... There are other reasons to ignore tensor valuedness, which will be presented in the next chapter. They are the algebras of the Cartan and Kähler calculi. The latter author did not, however, produce applications even for vector valuedness. His scalar-valued applications are, however, of great importance for the development of physics, as we shall see in the next section.

Geometry, on the other hand, is based on the connection equations,

$$d\mathbf{P} = \omega^\mu \mathbf{e}_\mu, \qquad d\mathbf{e}_\mu = \omega_\mu^\nu \mathbf{e}_\nu \tag{5.1}$$

(See section 5.7). The algebra here is the Lie algebra, which is where the connection takes its values. More relevant for present purposes is that we are dealing here with vector valuedness and exterior (covariant) differentiation. Cartan's differential geometry might be extended if one found how to get interior differentiation into it in a natural way. It was not necessary in our development. Some new concept would be necessary to make its introduction natural. For the moment, it would appear that an approach between calculus and geometry must come from the side of the former.

The key equation of the Kähler calculus is [46]

$$\partial u = a \vee u. \tag{5.2}$$

When a is the appropriate scalar-valued differential form, it replaces the Dirac equation with electromagnetic coupling. He did not speak in 1960 of a corresponding equation for tensor-valued input $\mathbf{a}$. The equation

$$\partial \mathbf{u} = \mathbf{a} \vee \mathbf{u}, \tag{5.3}$$

where the Clifford algebra refers to the differential forms themselves (Kähler algebra), remains valid regardless of valuedness. When $\mathbf{a}$ is not a scalar function, $\mathbf{u}$ cannot be of homogeneous valuedness (say tensor valuedness of definite rank) since it increases the rank of the right hand side, which ∂, on the left hand side, does not. The valuedness of the right and left hand sides of (5.3) would not be the same except for inhomogeneous valuedness extending all the way to infinity.

Kähler dealt with tensor-valuedness $\mathbf{a}$ in his 1962 paper by introducing a contrived Kähler equation that reads

$$(\partial u)_{\mu...\nu}^{\lambda...\pi} = a_{\mu...\nu r...s}^{\lambda...\pi p...g} u_{p...g}^{r...s}. \tag{5.4}$$

We do not need to learn the concepts that go into the making of Eq. (5.4). Suffice to notice that the rank of the valuedness-tensor input $a_{\mu...\nu r...s}^{\lambda...\pi p...g}$ is double the rank of $u_{p...g}^{r...s}$. Equation (5.2) is not so restricted. Both of them thus have their own un-inviting features.

Kähler stated in 1962 that (5.4) becomes (5.2) in the specific case of scalar-valuedness. That is a true but misleading statement since he could have postulated (5.3), thus avoiding (5.4). He could then have said that both equations (5.3) and (5.4) coincide for scalar-valued input. Rather than concern ourselves with the possible reason(s) behind Kähler's 1962 choice (or lack thereof), it is more important to consider what post-scalar valuedness should a physicist be interested in.

A notable alternative is Clifford valuedness, as we have made abundantly clear in [88], where it emerges naturally, as shown in a preprint that we submitted to the general physics section of the arXiv [89] (It was redirected to the theory sector of high energy physics (where we do not have a sponsor). The Clifford valuedness of differential forms there emerges naturally. We now have Clifford valued differential forms as the actors for the potential unification of calculus and geometry. This potentiality now has to become realized and thus bring us closely in principle to the unification of quantum mechanics and general relativity.

12.6   Imagining the future

The histories of mathematics and physics are full of accidents that have made them have their present shapes, full of avoidable deficiencies. Replacing them is the road map to the future. The replacements exist, some of them overlooked for more than a century.

By accidents in mathematics, we mean that the historical order of mathematical discovery did not correspond to what, retrospectively, could be considered as the logical order. Examples are Cauchy's calculus, largely or totally unnecessary, as we have recently shown. Or the adoption of vector algebra, which is peculiar because it is specific to three dimensions. Or the view of the geometry of Lorentzian spaces as pseudo-Riemannian rather than Finslerian (because of giving more importance to the metric than to the structure of the bundle). Or the disconnect of differential geometry from the language of physicists due to the attempt to make Cartan rigorous through the use of forceps.

The accidents in physics are largely a consequence of the accidents in mathematics, when the *not so good* was chosen because the *better* did not yet exist. An example is the adoption of the LC connection at a time when it was the only connection known. Another example is the Dirac calculus, adopted when the Kähler calculus was not yet known.

Those accidents need not have constituted a problem except for the fact that inertia and the lack of interest in the foundations of physics and/or mathematics impeded the adoption of the right option when it eventually became available.

Assume that differential geometers had focussed their interest more on further developing the moving frame method than on making it rigorous in contrived ways. They might have realized that modern differential geometry is —regardless of who we dress it— a theory of just moving frames. It should be a theory where particles and fields should play a more central role. After all, frames are for referring particles and fields to them.

I have recently shown that, when this is done, a 5-D Kaluza-Klein space without compactification of the fifth dimension emerges [89]. The fifth dimension, τ, is one which, on trajectories of particles, becomes their proper time. The (x^i, τ) subspace is the natural arena for quantum physics. $U(1) \times SU(2)$ is then shown to appear by simply using Kähler's calculus in relativistic quantum mechanics and replacing the unit imaginary with appropriate real magnitudes of square minus one [89]. In work in preparation, $U(1) \times SU(2) \times SU(3)$ will be shown to be just a continuation of the process by which $U(1)$ becomes $U(1) \times SU(2)$. We give the underlying idea in the last section of the next chapter.

Chapter 13

A BOOK OF FAREWELLS

13.1 Introduction

In the previous two chapters, we have made the point that there is a large amount of unfinished work in the lines of geometry and calculus created by Cartan and Kähler. In this chapter, we make the presentation of our next book, where new lines of mathematical development are outlined when not developed. How can physics make progress with its most difficult issues while ignoring such related mathematics?

The case we try to make in it is contained in the title "Differential forms: farewell to vector, tensor, Cauchy and Dirac calculi". In order to be able to bid farewell to them, we must have a replacement. We show that the Kähler's calculus of differential forms not only replaces but actually supersedes them.

13.2 Farewell to vector algebra and calculus

Practitioners of Clifford algebra know that it replaces vector algebra with advantage. In this section, we summarize our attempt to make that point in our next book. Since the subject of that book is not Clifford algebra per se, its treatment there of this algebra will be only as comprehensive as needed for replacing vector algebra, and for other objectives of this book, like replacing the vector and Cauchy calculi.

Clifford algebra is not an algebra created by mathematicians to make physicists' life miserable. It has to be considered as the true Euclidean algebra, valid in arbitrary dimension. Standard vector algebra is valid only in dimension 3. There is not a similar algebra in other dimensions, though there is a vector product in seven dimensions. But it is a different type of algebra since, perpendicular to the plane of two vectors, there is a whole five dimensional space.

Whenever at least one of two factors in a Clifford multiplication is of grade one, the dot and, after surgery, vector products come together. Why surgery? The vector product $\mathbf{a} \times \mathbf{b}$ is the composition of the exterior product, $\mathbf{a} \wedge \mathbf{b}$, with which we have already become familiar in terms of differential forms, and Hodge duality (See section 6.4). The surgery consists in not taking the dual of $\mathbf{a} \wedge \mathbf{b}$.

Vector calculus is horrible for several reasons. One of them is that its curl is based on the vector product. So, we do not have a curl in other dimensions. Another reason is that it uses tangent vectors where it should use differential forms. One more is that it often uses more structure than needed to solve a problem, say a metric structure. Still another one is that one can do so little with it that it has to be complemented with all the other calculi that we also think of replacing with differential forms. The Kähler calculus —based on Clifford algebra of differential forms— replaces tangent vectors and tangent-valued operators with differential forms whose coefficients are respectively functions and operators.

The replacement of much of the vector calculus with the calculus of differential forms is nothing new. It can be found in different books on the exterior calculus, say for obtaining curl and divergence in curvilinear coordinates. In the Kähler calculus, they are given by the exterior $(d\alpha)$ and interior $(\delta\alpha)$ derivatives of a differential $1-$form, not of a vector field. The two operations are just part of one, $\partial \; (= d + \delta)$ (See equation (5.1) of the previous chapter). It is easy to check that, if α is a differential $1-$form in Euclidean 3-space, $d\alpha$ is a differential $2-$form with the same components as the curl of the vector field that has the same components as α. And $\delta\alpha$ is the scalar known as the divergence of that vector field. They are much easier to work with, actually without resort to Hodge duality. In addition, ∂ acts on differential forms of any grade and of any valuedness. Kähler differentiation also applies to vector fields, but not to yield the curl and the divergence of a vector field in the vector calculus.

Kähler defines the interior "derivative" in terms of the LCC on the differentiable manifold. He then shows that his definition coincides what in the modern literature is defined as the co-derivative, which is obtained in terms of the metric. Kähler's definition, however, may be used with any Euclidean connection.

Neither ∂ nor δ satisfy the (present form of the) Leibniz rule, meaning that they are not derivatives in the modern sense of the word. However, following Kähler, we shall still use the term derivative. It is only a matter of "relaxing the concept of Leibniz rule", as it had been relaxed in the past with the emergence of the exterior derivative.

The foregoing considerations make a case for farewell to vector calculus, after first getting a glimpse of what it is. Let it be clearly understood that we do not advocate not to teach the vector calculus, but that one do so only to the point where the transition to the exterior calculus is easiest. In any case, due to the rather generalized deterioration in academic standards, one does not get much more than a glimpse of vector calculus in most courses on the subject. In addition, that knowledge can be recycled.

13.3   Farewell to calculus of complex VARIABLE

The alternative to the Cauchy calculus that we propose is a direct application of the Clifford algebra of differential forms in the plane. We give the defining relations both in Cartesian and polar coordinates:

$$dx^2 = dy^2 = d\rho^2 = 1, \qquad d\phi^2 = \frac{1}{\rho^2}. \tag{4.1}$$

Hence,

$$dxdy = -dydx, \qquad (dxdy)^2 = -1. \tag{4.2}$$

As far as algebra is concerned, the complex looking inhomogeneous differential form

$$z \equiv x + ydxdy, \tag{4.3}$$

plays in the real plane the role that the complex variable $x + iy$ plays in the complex plane. It emerges from the relation between $d\phi$ and dy:

$$d\phi = \frac{xdy - ydx}{x^2 + y^2} = \frac{x - ydxdy}{x^2 + y^2}dy = \frac{1}{x + ydxdy}dy = \frac{1}{z}dy, \tag{4.4}$$

where we have used that

$$(x + ydxdy)(x - ydxdy) = x^2 + y^2, \tag{4.5}$$

by virtue of the second equation 4.2.

We proceed to provide an outline of the argument for the theorem of residues. Given the integral $\int_{-\infty}^{\infty} F(x)dx$, consider the differential $1-$form

$$\alpha \equiv F(x + ydxdy)dx \tag{4.6}$$

on a closed curve formed with the x axis of the real plane and a semicircle Γ at infinity. If $\int_\Gamma \alpha \longrightarrow 0$ as the radius of the semicircle becomes greater and greater, we may then write

$$\int_{-\infty}^{\infty} F(x)dx = \oint \alpha. \tag{4.7}$$

Given the pull-back of α to polar coordinates (as in the physics literature, we ignore the pull-back symbol),

$$\alpha = h(\rho, \phi)d\rho + j(\rho, \phi)d\phi, \tag{4.8}$$

j is

$$j = \rho^2(\alpha \cdot d\phi) = \rho^2 \left[F(x + ydxdy)dx \right] \cdot (\frac{1}{z}dy), \tag{4.9}$$

in terms of Cartesian coordinates.

Let us use the symbol "$*$" to indicate reversion of the order of dx and dy, which changes the sign of $dxdy$, as when replacing i with $-i$ in complex conjugation. With F standing for $F(z)$, the right hand side of (4.9) further becomes

$$j = \frac{\rho^2}{2}\left[F\frac{1}{z^*}dx\,dy + F^*\frac{1}{z}(dx\,dy)^*\right] = (Fzdx\,dy)_0 = (-Fz)_2, \qquad (4.10)$$

where the subscripts zero and two pick up the scalar part and the coefficient of the 2−form part respectively. The last one is equivalent to picking the imaginary part of $F(z)z$.

Let us take the next step in a simplified way. If the contour integral has a pole of first order, one gets the integration circle go to zero and, under appropriate conditions that we shall ignore here, one pulls $(-Fz)_2$ out of the integral, which then becomes just $\oint d\phi$ and, therefore, 2π. The well known theorem of residues as it applies to first order poles results.

If the pole is of higher order, one proceeds as in Cauchy's theory. As if that were not enough, the additional theoretical developments necessary for special purposes in the Cauchy calculus now become much simpler. It should not be surprising since, again, differential forms constitute the language of integration. In particular, the equivalence of the concepts of analytic (power series expansion), holomorphic (existence of derivatives) and the Cauchy-Riemann conditions become much more transparent since one does not need to create a concept of differentiation with respect to the z of the Cauchy calculus or to the new z. Differentiation with respect to real variables suffices.

Retrospectively, differentiations and integrations with respect to a complex variable constitute a less than ideal treatment of problems whose solutions should be dealt with the language of differential forms. Real integrals presently performed with Cauchy's calculus should just be viewed with the right perspective, namely as an extension of a corollary of Stokes theorem of the real calculus in regions of the real plane that are not simply connected.

Two papers by this author deal with this subject rather comprehensibly [86], [87]. The second of these extends the theory just described, so as to deal also with integrals where the result is of the type $u + vdxdy$. They are accessible to everybody free of charge. Our next book should pick up the contents of those two papers and provide numerous additional worked examples, so that one can teach in a far shorter and more transparent way in the real plane what nowadays goes by the name of calculus of complex variable.

13.4   Farewell to Dirac's CALCULUS

Dirac's calculus and theory was a fantastic achievement of the arguably second greatest physicist of the twentieth century; his work may have had greater impact than even Einstein's. One could hardly have done better at the time. But that was then. When compared with Kähler's calculus, born a few decades later, Dirac's leaves a lot to be desired. That will be shown in a chapter in our next book, where the main purpose will be to show that the advantage of Kähler's calculus lies in its use of differential forms where one now uses gamma matrices.

Of special interest is the treatment of the Lie derivative of those forms from the original and overlooked perspective of the treatments of Lie differentiation of the subject by Cartan [9], Kähler [46], [48] and Slebodzinski [69].

The extensive use made of Dirac's calculus has taught physicists to navigate its waters, but some difficult spots remain. We specially mean spurious negative energy. Through his replacement of the Dirac equation with his own equation, Kähler showed the emergence of antiparticles without resort to negative energy solutions [48]. He also showed that there is nothing mysterious about spin. As he put it [48]:

The spin of the electron will be interpreted as the necessity to represent the state of an electron by a state differential rather than by a state function.

It is not the intention of this author to reproduce in his next book the Kähler calculus in all its glory, but enough of it in order to show by comparison how unnecessarily cumbersome, and misleading at certain points, the Dirac theory can be.

We proceed to give some examples additional to the ones just mentioned. This author has shown that the obtaining of the "post Pauli-Dirac" approximation of the Kähler equation with electromagnetic coupling is little more than a one page development, half of it taken by the statement of results. It is done without resort to Foldy-Wouthuysen transformations and without the spurious interpretation of small components as representing negative energy contributions [83]. Neither Foldy-Wouthuysen, nor anybody else for that matter, appears to have addressed the fact that the same development of the equation should apply to antiparticles, except for the sign of the charge.

At the same time as one shows that the Foldy-Wouthuysen work is misleading, one interprets the $-mc^2$ term as the result of adding a spurious term $-2mc^2$ to the true rest energy mc^2 [83]. One cannot ignore the problem of negative energies in the Dirac theory on the grounds that it does not arise in the very successful quantum field theory, since the latter provides corrections, but not the basic values that it corrects.

Related to the foregoing quotation, Kähler could have been more specific and stated in the same breath that the action of the Lie derivative on differential forms is at the root of the emergence of spin. If he had raised more forcefully and in the English language this issue, he might have launched a reassessment of the use of the Dirac theory in high energy physics. In particular, he would have promoted Einstein's view of particles as concentrations of the field

The way in which one tinkers with the Dirac equation (perhaps unavoidably because of limited sophistication of the Dirac calculus) for obtaining free particle solutions is at the root of the spurious emergence of (particle, antiparticle) pairs with opposite sign of the energy. Kähler should have emphasized that the solutions to his differential equation are in principle members of his algebra, and not only members of those algebras' ideals. The spinor solutions are only specific solutions, not general ones. A key issue and source of interesting questions is what symmetries can be put together in the same ideal, given the idiosyncrasies of products of idempotents corresponding to different symmetries. A little bit of this can be seen in section 6.

To summarize, Dirac's theory *may be conceptually wrong* in spite of the tremendous success of the high energy theory built upon it. Solutions of the Kähler equation pertaining to the algebra of clifforms may be the primordial concept; the solutions in ideals, thus representations of particles as spinors, are derived concepts. One by-product is that also quarks are a derived concept by a mechanism similar to the one by which particles emerge from the primordial field. These statements are supported in papers being prepared. It is the intention of this author to put into the book that we are announcing here the mathematics needed for understanding these claims.

13.5   Farewell to tensor calculus

For completeness purposes, a chapter of the book been announced will provide a summary of the main part of the present book. Its purpose is to explain why one should bid farewell to the tensor calculus and to approaches to differential geometry that draw significantly from it. It will emphasize the points about to be mentioned in this section and should put together some of the perspectives that non-expert readers should have acquired from reading the present book.

Tensor calculus *lacks true geometric pedigree*. Indeed, tensor fields relate to sections of the frame bundle, not to the frame bundle itself. But the frame bundle, not the sections, represent the groups that define a Klein geometry.

Tensor calculus *emphasizes as geometric what is not*, a point made by Cartan when he argued against viewing Riemannian geometry as the geometry of an infinite Lie group [21]. Yang-Mills theory involves connections valued in finite Lie algebras of auxiliary bundles. Connections in non-Yang-Mills differential geometry also take values in finite Lie algebras. This should be emphasized in order to avoid being mislead into viewing by the infinite group of diffeomorphisms as underlying the geometry.

Tensor calculus *belies the nature* of the objects that it tries to represent. Tensors belong to a graded algebra, but only the lowest ranks of tensors emerge in geometry. Ranks higher than five do not appear. Tensor algebra usurps the roles of its quotient algebras.

Tensor calculus *subverts the logical order of concepts*. Its most common version makes very little emphasis on tensor algebra and focusses on how things transform. But the transformation properties are a consequence of the underlying tensor algebra, which most authors for physicists barely define.

As a consequence, *results happen in a haphazard way*. Such is the case for instance with the argument to "discover" curvature in the approach of Ricci and Levi-Civita to Riemannian geometry, approach that Einstein adopted. Curvature there emerges not as a concept, but as a set of symbols that transform in a particular way.

One more respect in which tensor calculus fails to make justice to the foundations of geometry is its casual treatment of issues of integrability and equations of structure, specially of the first equation of structure. It thus produces the wrong statement that the torsion is the skew-symmetric part of the connection, which is correct only in coordinate bases.

Tensor calculus is *too coarse*, as it lacks discriminating power. It fails to distinguish the different nature of objects whose components transform in the same way. As a result, one speaks of different differentiations (covariant, exterior, ...) of a mathematical object like, for example, w^i.

Tensor calculus is *too rigid*. It makes Finsler geometry too cumbersome, due in part to the fact that the dimension of the space of tangent vectors and of the module of differential 1−forms are different in that geometry.

Tensor calculus is *inimical to Klein geometries, bundles and Lie algebras, not to mention vector equality*. Tensor calculus is one whose time is long past.

13.6 Farewell to auxiliary BUNDLES?

This author claims that auxiliary bundles are unnecessary if one simply uses the right mathematics. The key lies in recognizing the implications of viewing quantum mechanics from the perspective of Kähler's rather than Dirac's calculus. A new vision of quantum mechanics emerges. First, let us make the case for the Kähler calculus and concomitant quantum theory.

1. This calculus is the natural extension of the Cartan calculus, as it takes into account not only exterior differentiation (curls if you will) but also interior differentiation (divergences) with differential forms.

2. It is more comprehensive than the Dirac calculus since the spinors (eight real components) are replaced with objects of 32 real components. The Kähler equation —replacement for the Dirac equation— is a most natural one (read unavoidable) in the Kähler calculus.

3. It only uses scalar-valuedness to take care of what the Dirac calculus does with vector-valuedness, even though other valuednesses are allowed and should be used.

4. In it, spin is at par with orbital angular momentum and need not be forced into being by invoking it as an internal property of particles. Spin, being about rotations, cannot be identified with torsion, which is about the breaking of integrability of the translation part of the connection. Kähler shows that spin is simply about knowing how to perform partial derivatives of differential forms when two systems of coordinates are involved.

5. In it, particles and antiparticles emerge with the same sign of energy. This is a direct consequence of splitting into two pieces the differential forms that satisfy the Kähler equation.

6. Foldy-Wouthuysen transformations are unnecessary to obtain the classical electromagnetic Hamiltonian beyond the Pauli terms. In spite of what is claimed by its practitioners, those transformations do not separate particles

from antiparticles but odd and even parts of the primordial field in which an electron or a positron contain the dominant amount of energy in that field.

7. The difference of $2mc^2$ between the actual energy mc^2 of a positron and the energy $-mc^2$ that the Dirac theory assigns to it in the Dirac theory is easily explained by the same process as in #3.

Once we have made the case for it, let us see its implications for the foundations of quantum physics.

A. The field is the primordial concept. In a Stern-Gerlach experiment, we are certainly changing the orientation of spin, but this is a derived concept. What changes is the field itself of which the electrons form part. It entails the change in the concentrations of the energy of the field that we identify as the electrons in the beam.

B. Particles emerge from the primordial field as its components (in special cases) when decomposed into members of ideals. In a process in which a pair is created, or when particles come out together in a deep inelastic experiment, these are entangled ab initio by virtue of the fact that they emerge from the decomposition of some state of the primordial field.

C. Hence, an internal property from a particle perspective is standard propertime property from a field perspective, and is, therefore, a spacetime property.

D. $U(1)$ and Poincaré symmetry come together in the splitting of the algebra of primordial fields into ideals. Hence, charge emerges in a spacetime context. In other words, charge emerges from the inner dynamics of bundles related to the tangent bundle, not related to auxiliary bundles. See item (a) below.

E. The symmetries that define particles are not symmetries of the equations of the primordial field, or even of the primordial field itself. They are properties of spinor solutions of those equations. In other words, they are symmetries of members of certain ideals. Hence, we should look at the symmetries of the idempotents that define those ideals. The symmetries that we can embody simultaneously into idempotents matter, not the symmetries of the equations that those idempotents satisfy, which are always larger.

F. The proper way to do rotations of spinors is through tangent Clifford algebra. The valuednesses of the differential forms in the theory of the moving frame —valuednesses of which we spoke in section 5 of chapter 12— is now extended into a Clifford valuedness. Connections have to take care of that, which makes the arena of a corresponding theory of connections the same as the arena for Kähler's theory beyond scalar-valuedness, and thus beyond the electromagnetic interaction.

G. When the unit imaginary is replaced by appropriate real geometric quantities, the other so called internal symmetries start to emerge in a spacetime related context. It is "spacetime related context" and not simply "spacetime context", since the theory of the moving frame on manifolds is not all that there is to the theory of connections. Particles and physical fields are not represented in depth in the theory of moving frames. A space-propertime subspace of a 5-D Kaluza-Klein space without compactification comes to the fore for quantum mechanics [89].

H. This Kaluza-Klein space exists only in an appropriate "preferred frame context", even though the Lorentz transformations retain their value for significant specific purposes. The Lorentz metric, now expressed as a Clifford algebra equation, plays the role of natural lifting condition for curves in time-space-propertime.

I. Preferred frames together with a metric canonically determine teleparallel connections, not Levi-Civita connections. Both types of connections are equally consistent with the same metric, as only the relations of affine type change. They favor teleparallelism since it entails equality of vectors at a distance, which Levi-Civita connections do not. In addition, teleparallelism brings new degrees of freedom, embodied in the torsion.

Let us now show how Clifford algebra and the Kähler's theory based on it take us beyond the electromagnetic interaction when we let the mathematics speak.

As a first step and by considerations related to solutions with time translation and rotational symmetries, the factor $e^{im\phi/\hbar - iE_0 t/\hbar}$ in solutions with proper values of energy and angular momentum for electrons and positrons now has to be replaced with

$$e^{im\phi/\hbar - iE_0 t/\hbar}\frac{1}{2}(1 \pm idt)\frac{1}{2}(1 * idxdy), \tag{6.1}$$

where the star means $\pm$ and $\mp$. So, a starting point to find such proper solutions is to write these in the form

$$u = e^{im\phi/\hbar - iE_0 t/\hbar}p \vee \tau^{\pm} \vee \epsilon^*, \tag{6.2}$$

where

$$\tau^{\pm} \equiv \frac{1}{2}(1 \pm idxdy), \qquad\qquad \epsilon^{\pm} \equiv \frac{1}{2}(1 \mp idt), \tag{6.3}$$

and where p is a differential form that depends only on $(\rho, z, d\rho, dz)$ [47]. Idempotents (6.3) represent the two properties of particles that are determined by representations of the Poincaré group, mass and spin, here accessed on different grounds. Key also is the fact, not obvious if one has not stated the proof of #5, that charge comes intrinsically united with energy, since the two signs in $\epsilon^{\pm}$ correspond to the two signs of charge [48]. Hence, in the electromagnetic solutions which are proper functions, $U(1)$ is present in the factor $\epsilon^{\pm}$, but there is the additional factor $\tau^{\pm}$ in order to make particles.

We shall cut the story short and propose that $U(1) \times SU(2)$ is represented by the 12 idempotents

$$\varepsilon^{\pm}\tau^*_{yz}, \qquad \varepsilon^{\pm}\tau^*_{zx}, \qquad \varepsilon^{\pm}\tau^*_{xz}, \tag{6.4}$$

where

$$\tau^{\pm}_{ij} \equiv \frac{1}{2}(1 \pm \mathbf{a}_i\mathbf{a}_j dx^i dx^j), \qquad \varepsilon^{\pm} \equiv \frac{1}{2}(1 \mp \mathbf{w}d\tau), \tag{6.5}$$

in the same way as $\tau^{\pm} \vee \epsilon^*$ represents the electromagnetic interaction. i and j are two consecutive numbers ordered as in $1, 2, 3, 1$, and $\mathbf{w}$ is the unit vector dual to proper time.

One cannot, however, represent particles as in (6.2) using (6.5) at the same time. The proper venue for such a possibility requires further steps that involve ab initio quarks. To obtain then it suffices to multiply (6.4) by another idempotent that takes three values. But let us stop there. We have said enough for making the point that there is a future in bidding farewell to all calculus except a further enriched calculus of differential forms. With these, we approach the realization of Einstein's thesis of logical homogeneity of theoretical physics and tangent bundle differential geometry at the same time as calculus also becomes tangent bundle geometry, and this geometry becomes calculus. If this is not the germ of a farewell to auxiliary bundles, what is?

Appendix A

GEOMETRY OF CURVES AND SURFACES

A.1 Introduction

Curves and surfaces are differentiable manifolds of dimensions one and two respectively, except if they self-intersect. In this case we would cut them into pieces at the intersections, in order to have manifolds. Doing so changes the global properties of the set in question, not the local ones, except, again, at the intersections. The Gauss-Bonnet theorem is an exception, meaning that it is (or can be made into) a result of global differential geometry. We are interested in local properties. Connections are local concepts.

A more important difference is that we shall deal with 1-D and 2-D differentiable manifolds embedded in 3-D Euclidean space, as opposed to those manifolds on themselves. Think of the concept "embedded" as if it were "immersed" in the *ordinary sense of the term* (Technically, embedding is an injective immersion). But forget about the technical meanings and look at the issue in the following simple terms.

In the theory of curves and surfaces, we do not think of them as we do when dealing with differentiable manifolds of dimensions one and two; a curve would then have zero curvature and torsion since differential two forms are zero in dimension one. And yet we speak of curvature and torsion of curves that are not zero in general. So, we are dealing with alternative concepts of torsion and curvature, since vectors that are not tangent, specially and specifically normal vectors, are also considered. The rule to relate vectors at different points of a curve or a surface —thus the differentiation of vector fields— is the connection of 3-D Euclidean space (This connection, being trivial in terms of constant frame fields, does not make its presence be felt).

On surfaces, the metric of 3-D Euclidean space can be expressed in terms of just two coordinates, but at the price of making the quadratic form more complicated than the Cartesian one. Computations on surfaces are not, however, that trivial. In general, the nature of the issues considered involves frame fields adapted to the surface, not constant frame fields. As for the standard concept of (Levi-Civita) connection on a surface, it is here a matter of neglecting the normal component in the differentiation of tangent vectors to the surface.

As has been the case through the text, we shall continue to use the moving frame method. One has not yet taken full advantage of it. We do better than appears to be the case in the literature by the use of surface-adapted 3-D frame fields that are canonically determined by the diagonalization of the second fundamental form of surfaces.

We shall see that the Frenet frame field is canonically determined by curves and that the "geodesic frame field" is canonically determined by the pair of curve and surface. There is the coordinate frame field on surfaces, canonically determined by the pair of surface and coordinate system on it. There are also the canonical frame fields of 3-D Euclidean space, i.e. the constant orthonormal frame fields denoted as $(\mathbf{i}, \mathbf{j}, \mathbf{k})$. Missing is the canonical frame field of an embedded surface.

For comparison purposes, there will be some paragraphs which are not proper part of the theory of curves and surfaces with the theory of moving frames. They illustrate the cumbersomeness of the by now almost abandoned traditional approach. If we removed those comparisons, our treatment would be more than two pages shorter.

A.2 Surfaces in 3-D Euclidean space

A.2.1 Representations of surfaces; metrics

One way of giving a surface is in the form

$$\mathbf{x} = x^i(u, v)\mathbf{a}_i, \tag{2.1a}$$

where $\mathbf{a}_i$ is a fixed basis, $\mathbf{i}, \mathbf{j}, \mathbf{k}$. Equivalently, we have

$$x = x(u, v), \quad y = x(u, v), \quad z = z(u, v), \tag{2.1b}$$

for a given domain of the parameters. These actually are coordinates on the surface. We may also give it as

$$f(x, y, z) = 0. \tag{2.2}$$

The connection between (2.1) and (2.2) can be established as follows. We solve for u and v in two of the equations (2.1b) and substitute in the third of those equations, then taking all terms to the left hand side. On the other hand, we go back from (2.2) to (2.1) by writing, for instance,

$$x = u, \quad y = v. \tag{2.3}$$

Substitution of (2.3) in (2.2) allows us to view z as an implicit function of x and y. Of course, one may then use other parametrizations through an invertible system of equations

$$u' = u'(u, v), \qquad v' = v'(u, v). \tag{2.4}$$

The metric of 3-D Euclidean space in terms of Cartesian systems is

$$ds^2 = \sum_i (dx^i)^2, \qquad i = 1, 2, 3. \tag{2.5}$$

where

$$dx^i = x^i{}_{,\alpha} \, du^\alpha, \qquad \alpha = 1, 2 \tag{2.6}$$

and where $x^i = (x, y, z)$ and $u^1 = u$, $u^2 = v$. A subscript after a comma means partial differentiation. The variable with respect to which the differentiation takes place is apparent in any such expressions. Once these operations have been performed, any trace of x, y, z, dx, dy and dz has disappeared and we are left with

$$ds^2 = g_{\alpha\beta}(u^\gamma) du^\alpha du^\beta, \tag{2.7}$$

which is also written as

$$ds^2 = E du^2 + 2F du dv + G dv^2. \tag{2.8}$$

It is the called the first fundamental form. Clearly

$$E = g_{11}, \quad F = g_{12}, \quad G = g_{22}. \tag{2.9}$$

The positive definiteness of the metric implies

$$EG - F^2 > 0, \tag{2.10}$$

as we are about to show.

(2.8) is the way in which one expresses distance on curves, which is actually computed as

$$\sqrt{E\left(\frac{du}{d\lambda}\right)^2 + 2F\left(\frac{du}{d\lambda}\right)\left(\frac{dv}{d\lambda}\right) + G\left(\frac{dv}{d\lambda}\right)^2} \, d\lambda, \tag{2.11}$$

where λ is a parameter on the curve. In general, we can take the parameter to be v itself. (2.11) then becomes

$$\int \sqrt{E\mu^2 + 2F\mu + G} \, dv, \tag{2.12}$$

where μ is du/dv. The equation

$$y = Ex^2 + 2Fx + G \tag{2.13}$$

is a parabola. If the metric is positive definite, this is an upright parabola with vertex above the x axis. It does not cut the line $y = 0$. Thus the roots of $Ex^2 + 2Fx + G = 0$ must be imaginary, which implies

$$EG - F^2 > 0. \tag{2.14}$$

A.2.2   Normal to a surface, orthonormal frames, area

Vectors normal to a surface are not tangent vectors. Hence, they cannot be given as a linear combination of tangent vectors, but rather in terms of, for example, the basis $\mathbf{a}_i = \mathbf{i}, \mathbf{j}, \mathbf{k}$. Given the surface $f(x, y, z) = 0$, its tangent plane at (x_0, y_0, z_0) is obtained by substituting $x^i - x_0^i$ for dx^i in $f_{,i}\, dx^i = 0$, i.e.

$$f_{,i} \cdot (x^i - x_0^i) = 0, \tag{2.15}$$

where the dot is used for product, thus preempting taking $x^i - x_0^i$ as argument of $f_{,i}$. Let $g(x, y, z)$ be defined as $-f(x, y, z)$. The equation $g(x, y, z) = 0$ represents the same surface as $f(x, y, z) = 0$. But the unit vectors

$$\frac{\sum f_{,i}\, \mathbf{a}_i}{\sqrt{\sum (f_{,i})^2}}, \qquad \frac{\sum g_{,i}\, \mathbf{a}_i}{\sqrt{\sum (g_{,i})^2}}$$

are opposite of each other. Whether we take one or the other as the unit normal to the plane is not significant at this point.

Differentiating (2.1), we get

$$d\mathbf{x} = dx^i(u, v)\mathbf{a}_i = \mathbf{x}_{,u}du + \mathbf{x}_{,v}dv. \tag{2.16}$$

$\mathbf{x}_{,u}$ and $\mathbf{x}_{,v}$ are defined by these relations. They constitute a tangent coordinate vector basis field. It is not, therefore, orthonormal in general. The equation of the plane at the point of coordinates (u_0, v_0) is given by

$$\mathbf{x} = \mathbf{x}_0 + \mathbf{x}_{,u} \cdot (u - u_0) + \mathbf{x}_{,v} \cdot (v - v_0), \tag{2.17}$$

where $\mathbf{x}_0$, $\mathbf{x}_{,u}$ and $\mathbf{x}_{,v}$ are evaluated at (u_0, v_0). The dots at mid height have been put for the same reason as in (2.15). In moving frame notation, they are the equivalent of

$$d\mathbf{P} = \omega^i \mathbf{e}_i, \qquad \mathbf{P} = \mathbf{P}_0 + X^i \mathbf{e}_i. \tag{2.18}$$

The following is a necessary step to launch the moving frame method once the equation of a surface has been given. Let $(\hat{\omega}^1, \hat{\omega}^2)$ be any pair of differential 1-forms that satisfies

$$ds^2 = g_{\alpha\beta}(u, v)du^\alpha du^\beta = (\hat{\omega}^1)^2 + (\hat{\omega}^2)^2. \tag{2.19}$$

A first (retrospectively very important) consequence of the normalized form of the metric is that $\hat{\omega}^3$ is zero on the surface. Hence, the pull-back to the surface of the torsion of Euclidean space (torsion which, of course, is zero) reads

$$0 = d\hat{\omega}^3 = \hat{\omega}^\alpha \wedge \hat{\omega}_\alpha^3 = \Gamma_{\alpha\beta}^3 \hat{\omega}^\alpha \wedge \hat{\omega}^\beta, \tag{2.20}$$

where, for the last step, we have used that the pull-back of $\hat{\omega}_\alpha^3$ is a linear function of only $\hat{\omega}^1$ and $\hat{\omega}^2$. We immediately get $\Gamma_{12}^3 = \Gamma_{21}^3$, which we prefer to write as

$$\Gamma_{32}^1 = \Gamma_{31}^2 \tag{2.21}$$

(on the surface).

We now proceed to diagonalize the expression

$$g_{\alpha\beta}du^\alpha du^\beta \equiv g_{11}du^2 + 2g_{12}dudv + g_{22}dv^2, \tag{2.22}$$

by first obtaining

$$g_{\alpha\beta}du^\alpha du^\beta - \frac{1}{g_{11}}(g_{11}du + g_{12}dv)^2 = dv^2 \frac{g_{22}g_{11} - g_{12}^2}{g_{11}}, \tag{2.23}$$

as per Eq. (4.51) of chapter 3. We thus have

$$g_{\alpha\beta}du^\alpha du^\beta = \left(\frac{g_{11}du + g_{12}dv}{\sqrt{g_{11}}}\right)^2 + dv^2 \frac{g_{22}g_{11} - g_{12}^2}{g_{11}}, \tag{2.24}$$

and

$$\hat{\omega}^1 = \frac{g_{11}du + g_{12}dv}{\sqrt{g_{11}}}, \qquad \hat{\omega}^2 = dv \frac{(g_{22}g_{11} - g_{12}^2)^{1/2}}{\sqrt{g_{11}}} \tag{2.25}$$

and, further,

$$\hat{\omega}^1 \wedge \hat{\omega}^2 = (g_{22}g_{11} - g_{12}^2)^{1/2} du \wedge dv = \sqrt{EG - F^2}\, du \wedge dv. \tag{2.26}$$

Clearly, there is an infinite number of pairs $(\hat{\omega}^1, \hat{\omega}^2)$ that orthogonalize the metric of the surface. (2.25) is just one of them. We shall use this freedom to determine *frame fields adapted to surfaces*.

A.2.3 The equations of Gauss and Weingarten

The connection equations of Euclidean space are simply $d\mathbf{a}_i = 0$ in terms of a Cartesian frame field. The equations of Gauss and Weingarten are the connection equations of 3-D Euclidean space when the frame field is constituted by $\mathbf{x}_{,u}$, $\mathbf{x}_{,v}$ and the unit vector $\mathbf{N}$ perpendicular to the plane that they determine.

At a time when the concept of connection equations did not exist, Gauss wrote the differentiation of the tangent frame field $(\mathbf{x}_{,u}, \mathbf{x}_{,v})$ as

$$\mathbf{x}_{,u,u} = \Gamma_{11}^1 \mathbf{x}_{,u} + \Gamma_{11}^2 \mathbf{x}_{,v} + e\mathbf{N}, \tag{2.27a}$$

$$\mathbf{x}_{,u,v} = \Gamma_{12}^1 \mathbf{x}_{,u} + \Gamma_{12}^2 \mathbf{x}_{,v} + f\mathbf{N}, \tag{2.27b}$$

$$\mathbf{x}_{,v,v} = \Gamma_{22}^1 \mathbf{x}_{,u} + \Gamma_{22}^2 \mathbf{x}_{,v} + g\mathbf{N}, \tag{2.27c}$$

which reflects the state of the art of the time. $d(\mathbf{x}_{,u})$ is a differential 2-form on the surface, valued in 3-D Euclidean vector space. It can be written as the following bilinear combination

$$d(\mathbf{x}_{,u}) = \Gamma_{1\alpha}^1 du^\alpha \mathbf{x}_{,u} + \Gamma_{1\alpha}^2 du^\alpha \mathbf{x}_{,v} + \Gamma_{1\alpha}^3 du^\alpha \mathbf{N} \tag{2.28}$$

(and similarly for $d\mathbf{x}_{,v}$). Here we can read that

$$\mathbf{x}_{,u,u} = \Gamma_{11}^1 \mathbf{x}_{,u} + \Gamma_{11}^2 \mathbf{x}_{,v} + \Gamma_{11}^3 \mathbf{N}, \tag{2.29}$$

(and similarly for $\mathbf{x}_{,u,v}$ and $\mathbf{x}_{,v,v}$).

The two Weingarten equations are

$$\mathbf{N}_{,u} = \frac{fF - eG}{EG - F^2}\mathbf{x}_{,u} + \frac{eF - fE}{EG - F^2}\mathbf{x}_{,v}, \tag{2.30a}$$

$$\mathbf{N}_{,v} = \frac{gF - fG}{EG - F^2}\mathbf{x}_{,u} + \frac{fF - gE}{EG - F^2}\mathbf{x}_{,v}. \tag{2.30b}$$

In terms of orthonormal frames, we would replace

(a) $\mathbf{x}_{,u}$ and $\mathbf{x}_{,v}$ with $\hat{\mathbf{e}}_\alpha$,

(b) (e, f, g) with Γ's as stated above. We shall later see how they were introduced originally.

(c) (E, G, F) with $(1, 1, 0)$.

Notice the great simplification that then results. Further simplification follows in section 4 when we adopt the canonical frame field of surfaces.

A.3   Curves in 3-D Euclidean space

A.3.1   Frenet's frame field and formulas

Curves in E^3 are 1-dimensional manifolds, if we eliminate the points where the curve intersects itself. They can be given as the intersection of two surfaces

$$f(x, y, z) = 0, \qquad g(x, y, z) = 0, \tag{3.1}$$

or as

$$x = x(u), \quad y = y(u), \quad z = z(u), \tag{3.2}$$

which is the developed form of

$$\mathbf{x} = x^i(u)\mathbf{a}_i. \tag{3.3}$$

The relation between (3.1) and (3.2) is obtained by solving for two of the Cartesian coordinates in (3.1), and then changing parameter if one so wishes. In the opposite direction, we eliminate u in (3.2)

The distance on the curve between two points on the same is obtained by integration of

$$ds = \sqrt{\left(\frac{dx}{du}\right)^2 + \left(\frac{dy}{du}\right)^2 + \left(\frac{dz}{du}\right)^2}\, du. \tag{3.4}$$

Frenet's frame is the 3-D frame field adapted to a curve that, if we require it to be orthonormal, is canonically determined by the curve.

The first element of the frame is the unit tangent vector $\mathbf{t}$,

$$\mathbf{t} \equiv \frac{d\mathbf{x}}{ds} = \frac{d\mathbf{x}}{du}\frac{1}{ds/du}, \tag{3.5}$$

with ds/du obtained from (3.4). $\mathbf{t}$ is obviously a unit vector

$$\mathbf{t} \cdot \mathbf{t} = 1, \qquad (3.6)$$

since $d\mathbf{x} \cdot d\mathbf{x}$ is ds^2. The tangent line to a curve at its point $\mathbf{x}_0$ is

$$\mathbf{x} = \mathbf{x}_0 + \mathbf{t}s, \qquad (3.7)$$

s being assigned the value zero at $\mathbf{x}_0$.

From (3.6), we define a curvature vector $\boldsymbol{\kappa} \equiv \frac{d\mathbf{t}}{ds}$. We then have:

$$\mathbf{t} \cdot \boldsymbol{\kappa} = \frac{d\mathbf{t}}{ds} \cdot \boldsymbol{\kappa} = 0 \qquad (3.8)$$

which means that the equations

$$\boldsymbol{\kappa} = \kappa \mathbf{n} = \frac{1}{\rho} \mathbf{n} \qquad (3.9)$$

define $\mathbf{n}$, κ and ρ if we demand that $\mathbf{n}$ be a unit vector. This vector is the second element of a right handed orthonormal frame. One refers to κ and ρ respectively as curvature and radius of curvature. Because of the sought orthonormality of the frame, the coefficient of $\mathbf{n}'$ relative to $\mathbf{b}$ will be the negative of the coefficient of $\mathbf{b}'$ relative to $\mathbf{n}$. Thus:

$$\mathbf{n}' = -\kappa \mathbf{t} + \tau \mathbf{b}, \qquad \mathbf{b}' = -\tau \mathbf{n}. \qquad (3.10)$$

τ is called torsion of the curve.

Equations (3.9) and (3.10) are called the Frenet equations. $\mathbf{b}$ is perpendicular to the plane of plane curves. It is then of constant direction and actually a constant vector, since it is of unit size. τ is, therefore, zero. Plane curves are torsionless. On the other hand, a helix has torsion since the plane of $\mathbf{t}$ and $\mathbf{n}$ keeps changing. This plane is called the osculating plane.

A.3.2 Geodesic frame fields and formulas

We are going to study surfaces through curves on them. It is then useful to consider 3-D right handed orthonormal frames $(\hat{\mathbf{u}}_i = \mathbf{t}, \mathbf{u}, \mathbf{N})$. $\mathbf{u}$ is a vector in the tangent plane and $\mathbf{N}$ is usually taken on the opposite side (of the tangent plane) to $\boldsymbol{\kappa}$ (see example in next paragraph). We shall refer to this frame field as the geodesic frame field of the curve on the given surface.

One defines the *normal and geodesic curvatures* κ_n and κ_g of the curve at a point by means of the decomposition

$$\mathbf{t}' \equiv \frac{d\mathbf{t}}{ds} = \boldsymbol{\kappa} \equiv \boldsymbol{\kappa}_n + \boldsymbol{\kappa}_g = -\kappa_n \mathbf{N} + \kappa_g \mathbf{u}. \qquad (3.11)$$

We follow Struik and use the subscript "$_n$" even though the projection is on $\mathbf{N}$, not on $\mathbf{n}$ (see (3.9)). We have introduced the minus sign in front of $\kappa_n \mathbf{N}$ so that

κ_n be positive under the most usual convention when choosing the direction of the normal to the sphere. Thus the normal $\mathbf{n}$ to a parallel is inwards, whereas the normal at the same point to the sphere itself is usually taken to be outwards. This example is also useful in connection with how the radii of curvature of the parallel and of the sphere are related. Recall that the magnitudes of the curvatures and their associated radii are the inverses of each other.

Orthonormality determines the $\mathbf{t}$ component of $\mathbf{N}'$. We thus have,

$$\mathbf{N}' \equiv \frac{d\mathbf{N}}{ds} = \kappa_n \mathbf{t} + \tau_g \mathbf{u}, \tag{3.12}$$

which defines the coefficient τ_g, called geodesic torsion of the curve. Using again orthonormality, we obtain

$$\mathbf{u}' \equiv \frac{d\mathbf{u}}{ds} = -\kappa_g \mathbf{t} - \tau_g \mathbf{N}. \tag{3.13}$$

A.4 Curves on surfaces in 3-D Euclidean space

A.4.1 Canonical frame field of a surface

We clearly have

$$\kappa_n = -\boldsymbol{\kappa} \cdot \mathbf{N} = -\frac{d\mathbf{t}}{ds} \cdot \mathbf{N} = \mathbf{t} \cdot \frac{d\mathbf{N}}{ds} = \frac{d\mathbf{x}}{ds} \cdot \frac{d\mathbf{N}}{ds}. \tag{4.1}$$

$d\mathbf{x} \cdot d\mathbf{N}$ is a quadratic symmetric differential form on du and dv:

$$d\mathbf{x} \cdot d\mathbf{N} = e\,du^2 + 2f\,du\,dv + g\,dv^2. \tag{4.2}$$

It is called the *second fundamental form*. We do not have any use for this form, as there is a much better way to study normal curvature through the introduction of the concept of canonical frame field of a surface.

Let $\hat{\mathbf{u}}_\alpha$ denote any orthonormal, counterclockwise 2-D frame tangent to the surface and let $\hat{\mathbf{u}}_3$ be $\mathbf{N}$. Since only zero forms and 1-forms live on a curve, the last ones are multiples of each other. Formal division by a differential 1-form is justified and we get

$$\kappa_n = \frac{\hat{\omega}'^\alpha \hat{\mathbf{u}}_\alpha}{ds} \cdot \frac{\hat{\omega}_3'^i \hat{\mathbf{u}}_i}{ds} = \frac{\hat{\Gamma}_{31}'^1 (\hat{\omega}'^1)^2 + (\hat{\Gamma}_{32}'^1 + \hat{\Gamma}_{31}'^2)\hat{\omega}'^1 \hat{\omega}'^2 + \hat{\Gamma}_{32}'^2 (\hat{\omega}'^2)^2}{ds^2}. \tag{4.3}$$

Equation (4.3) applies to a one-dimensional manifold of frames related at each point by rotations around the normal to the surface. We reserve the use of unprimed quantities for the orthonormal frame $(\hat{\mathbf{e}}_1, \hat{\mathbf{e}}_2, \hat{\mathbf{e}}_3)$ that satisfies

$$\hat{\Gamma}_{32}^1 = -\hat{\Gamma}_{31}^2, \tag{4.4}$$

and, because of (4.4), also

$$\hat{\Gamma}_{32}^1 = \hat{\Gamma}_{31}^2 = 0. \tag{4.5}$$

The frame so defined, $(\hat{\mathbf{e}}_i')$ is unique up to reversion of axes, since (4.4) is an equation applied to a one-parameter set. This will become clear in the following, at the same time as we indicate how to obtain it. We shall refer to it as the *canonical frame field of the surface.* The normal curvature then takes the very simplified form

$$\kappa_n = \frac{\hat{\Gamma}_{31}^1 (\hat{\omega}^1)^2 + \hat{\Gamma}_{32}^2 (\hat{\omega}^2)^2}{ds^2}. \tag{4.6}$$

A.4.2   Principal and total curvatures; umbilics

Inspection of (4.6) shows that the extreme values of κ_n are

$$\kappa_1 = \hat{\Gamma}_{31}^1, \qquad \kappa_2 = \hat{\Gamma}_{32}^2 \tag{4.7}$$

when computed in the $\hat{\mathbf{e}}_i$ frame field. The equations

$$\hat{\omega}^\alpha = 0, \tag{4.8}$$

give each one of the *directions of principal curvature* (ds^2 reduces to $(\hat{\omega}^2)^2$ and $(\hat{\omega}^1)^2$ in those directions). For any other curve, the direction is determined by the ratio $\hat{\omega}^1/\hat{\omega}^2$ for that curve.

Lines of curvature is the term used to refer to those curves which go in the direction of principal curvature at all their points. They are obtained by integrating the differential equations (4.8), one for each curve.

The product $\kappa_1\kappa_2$ is called the *Gauss or total curvature* K of the surface.

Points where $\hat{\Gamma}_{31}^1$ equals $\hat{\Gamma}_{32}^2$ do not determine a value for κ_n, which thus is the same in all directions. Those points are called *umbilics*. All maximum circles through a point on a sphere have the same normal curvature. All points of a sphere are umbilical. On a ellipse, there are four umbilics if all three axes are different. The number of umbilics is a function of the surface and is usually finite.

Readers who would feel more comfortable with a more conventional approach to extreme values of the normal curvature need only formally divide (4.3) by $(\hat{\omega}^2)^2$. It then takes the form

$$\kappa_n = \frac{a\lambda^2 + b\lambda + e}{\lambda^2 + 1}, \qquad \lambda \equiv \frac{\hat{\omega}^1}{\hat{\omega}^2}. \tag{4.9}$$

Equating the derivative to zero, we get $2a\lambda = 2e\lambda$. If a is different from e, the ratio λ equals zero, which means that we get the curve $\hat{\omega}^1 = 0$. The parametrization in (4.9) excludes the curve $\hat{\omega}^2 = 0$, which happens to also yield one of the two extreme values of κ_n. Of course, it is obtained by defining λ as $\hat{\omega}^2/\hat{\omega}^1$. If a equals e, the equation $2a\lambda = 2e\lambda$ is satisfied for any λ, which means that we are at an umbilical point.

A.4.3   Euler's, Meusnier's and Rodrigues'es theorems

We may write (4.6) as

$$\kappa_n = \kappa_1 \frac{(\hat{\omega}^1)^2}{ds^2} + \kappa_2 \frac{(\hat{\omega}^2)^2}{ds^2} = \kappa_1 \cos^2 \theta + \kappa_2 \sin^2 \theta. \tag{4.10}$$

This equation is known as *Euler's theorem*.

Since the input that each curve provides in Eq. (4.10) is just its direction, a corollary known as *Meusnier's theorem* immediately follows. It reads: all curves through a point tangent to the same direction have the same normal curvature.

In the canonical frame field, the third connection equation reads

$$d\mathbf{N} = \Gamma_{31}^1 \mathbf{e}_1 \hat{\omega}^1 + \Gamma_{32}^2 \mathbf{e}_2 \hat{\omega}^2. \tag{4.11}$$

Consider $\hat{\omega}^2 = 0$, i.e., Eq. (4.11) then becomes

$$d\mathbf{N} = \Gamma_{31}^1 \mathbf{e}_1 \hat{\omega}^1 = \kappa_1 \mathbf{t} \; ds = \kappa_1 d\mathbf{x}, \tag{4.12}$$

and similarly for the $\hat{\omega}^2$ line of curvature with κ_2. Hence we have in both cases

$$d\mathbf{N} - \kappa d\mathbf{x} = \mathbf{0}. \tag{4.13}$$

This equation is known as *Rodrigues* equation. Comparison of $d\mathbf{N} = \kappa_i \mathbf{t} \; ds$ with (3.12) implies *Rodrigues'es theorem*: the geodesic torsion is zero for the lines of curvature and only they.

A.4.4   Levi-Civita connection induced from 3-D Euclidean space

We shall use the symbols D and d for the exterior (covariant) derivative under the connections of the Euclidean space and of a surface embedded in it, respectively. Needless to say, DD is zero but dd is not. D induces on the surface the Levi-Civita connection, as we shall now see.

For any orthonormal frame field of E^3, we have $\hat{\omega}_3^3 = 0$ and

$$D\hat{\mathbf{e}}_\alpha = \hat{\omega}_\alpha^j \hat{\mathbf{e}}_j = \hat{\omega}_\alpha^\beta \hat{\mathbf{e}}_\beta + \hat{\omega}_\alpha^3 \hat{\mathbf{e}}_3, \qquad D\hat{\mathbf{e}}_3 = \hat{\omega}_3^\alpha \hat{\mathbf{e}}_\alpha. \tag{4.14}$$

Because of orthonormality and, thus, skew-symmetry, the only independent $\hat{\omega}_\alpha^\beta$ different from zero is $\hat{\omega}_1^2$.

Let us verify that $\hat{\omega}_\alpha^\beta$ so introduced is indeed the Levi-Civita connection of the surface. It is a Euclidean connection because, in addition to skew-symmetry, its torsion is zero. Indeed, in any frame field we have

$$0 = d\hat{\omega}^\alpha - \hat{\omega}^i \wedge \hat{\omega}_i^\alpha = d\hat{\omega}^\alpha - \hat{\omega}^\beta \wedge \hat{\omega}_\beta^\alpha - \hat{\omega}^3 \wedge \hat{\omega}_3^\alpha \tag{4.15}$$

for the first two torsion components in Euclidean space. Since the last term of (4.15) is zero (because the $\hat{\omega}^3$ of the surface is zero), we obtain the desired result.

A.4.5   Theorema egregium and Codazzi equations

We now compute $DD\hat{e}_\beta$ modulo $\hat{e}_3$. We easily get

$$0 = DD\hat{e}_\beta = (d\hat{\omega}_\beta^i - \hat{\omega}_\beta^j \wedge \hat{\omega}_j^i)\hat{e}_i = (d\hat{\omega}_\beta^\alpha - \hat{\omega}_\beta^\gamma \wedge \hat{\omega}_\gamma^\alpha - \hat{\omega}_\beta^3 \wedge \hat{\omega}_3^\alpha)\hat{e}_\alpha, \quad mod \; \hat{e}_3. \quad (4.16)$$

The null contents of the parenthesis, specialized for $\beta = 1$, allows us to write the Levi-Civita curvature of the surface as

$$\Omega_1^2 = d\hat{\omega}_\beta^\alpha - \hat{\omega}_\beta^\gamma \wedge \hat{\omega}_\gamma^\alpha = \hat{\omega}_1^3 \wedge \hat{\omega}_3^2 = -\hat{\omega}_3^1 \wedge \hat{\omega}_3^2 =$$

$$= -\hat{\Gamma}_{3\alpha}^1 \hat{\Gamma}_{3\beta}^2 \hat{\omega}^\alpha \wedge \hat{\omega}^\beta = -(\hat{\Gamma}_{31}^1 \hat{\Gamma}_{32}^2 + \hat{\Gamma}_{32}^1 \hat{\Gamma}_{31}^2)\hat{\omega}^1 \wedge \hat{\omega}^2. \quad (4.17)$$

In the canonical frame field of the surface, we have $\hat{\Gamma}_{32}^1 = \hat{\Gamma}_{31}^2 = 0$. Hence

$$\Omega_1^2 = -\kappa_1 \kappa_2 \hat{\omega}^1 \wedge \hat{\omega}^2 = -K \; \hat{\omega}^1 \wedge \hat{\omega}^2. \quad (4.18)$$

Since the Levi-Civita connection of the surface only depends on its metric, this equation contains the statement that the total curvature is a bending invariant. In other words, it does not change under deformations of the surface that do not stretch it. This statement is known as Gauss'es egregium theorema (excellent theorem). It also tells us that the Gauss curvature coincides with the only independent component of Ω_1^2 in terms of $\hat{\omega}^2 \wedge \hat{\omega}^1$ (the sign is a mater of what unit differential 2-form one chooses). Thus Ω_1^2 is independent of rotations in the tangent plane.

The Codazzi equations

$$\frac{\partial e}{\partial v} - \frac{\partial f}{\partial u} = e\Gamma_{12}^1 + f(\Gamma_{12}^2 - \Gamma_{11}^1) - g\Gamma_{11}^2, \quad (4.19a)$$

$$\frac{\partial f}{\partial v} - \frac{\partial e}{\partial u} = e\Gamma_{22}^1 + f(\Gamma_{22}^2 - \Gamma_{12}^1) - g\Gamma_{12}^2 \quad (4.19b)$$

are a cumbersome way of writing the contents of the remaining curvature equations. Since e, f and g are Γ's for coordinate frame fields (recall Eqs. (4.2) and that $d\mathbf{N}$ is $d\mathbf{e}_3$), one recognizes in (4.19) the pattern of curvature equations. It is not worth deriving them since we understand what they are. Interested readers should just make f equal to zero and make appropriate translations of symbols e and g.

A.4.6   The Gauss-Bonnet formula

We develop (4.18) further by integrating on any simply connected surface:

$$-\iint K \; \hat{\omega}^1 \wedge \hat{\omega}^2 = \iint \Omega_1^2 = \iint d\hat{\omega}_1^2 = \oint \hat{\omega}_1^2. \quad (4.20)$$

Remember that Ω_1^2 is the curvature of the surface, not one of the components curvature of 3-D Euclidean space.

Recall the 3-D geodesic frame field $(\hat{\mathbf{u}}_i = \mathbf{t}, \mathbf{u}, \mathbf{N})$. By itself, $(\mathbf{t}, \mathbf{u})$ is a 2-D orthonormal frame defined by the pair of a surface and a curve on it. It is not dual to the canonical basis of soldering forms $(\hat{\omega}^1, \hat{\omega}^2)$. Let $\hat{\omega}_1^2$ be the LC connection on the surface related to the same frame field as $(\hat{\omega}^1, \hat{\omega}^2)$.

At each point of a curve, the differential of $\mathbf{u}$ can be viewed as composed of the horizontal part given by $\hat{\omega}_1^2$, and a vertical part, $d\phi$, where ϕ is the angle that relates the canonical frame field of the surface to the geodesic frame field of a particular curve. Similarly for $\hat{\omega}_2^1$ ($= \hat{\omega}_1^2$), which is more closely related to $d\mathbf{u}$. Inspection of (3.13) allows us to write

$$\kappa_g ds = \hat{\omega}_1^2 + d\phi, \tag{4.21}$$

and

$$-\iint K \, \hat{\omega}^1 \wedge \hat{\omega}^2 = \oint \kappa_g ds - \oint d\phi. \tag{4.22}$$

So, finally,

$$\oint \kappa_g ds = -\iint K \, \hat{\omega}^1 \wedge \hat{\omega}^2 + 2\pi, \tag{4.23}$$

which is the Gauss-Bonnet formula.

In the case of the plane, K being zero, $\oint \kappa_g ds$ is 2π on any closed curve, provided it is just a one loop curve. It might be $2n\pi$ for an n loop closed curve. Or it might be zero if after going around one loop and just before closing, it returns on almost the same path before closing where it started.

Result (4.23) is local in principle, in the sense that it is not a global result. But a global result can be obtained if one "cuts" a surface in such a way that it remains connected and all of it is inside a closed curve, and then applies the theorem. On the right hand side of (4.23), we may have, instead of 2π, the value 0 (sphere) or 2π (torus), etc. It depends on what is the curve needed to enclose the whole surface. For detailed applications of this theorem, we refer the uninitiated in classical differential geometry to the beautiful book by Struik [70].

This global result can in turn be converted into a topological result by acceptable deformation of the surface which does not change the left hand side of (4.23). This does not change the $2n\pi$ on the right either. Hence $-\iint K \, \hat{\omega}^1 \wedge \hat{\omega}^2$ is a topological invariant, which certainly is an amazing result, "easily" obtained with differential forms.

For another application consider polygons. Their sides may be curvilinear; in fact they will if the surface is not plane. For (4.23) to apply, we would have to smooth the vertices so that there is no discontinuity in the tangent to the curve, i.e. so that it has derivatives and thus a concept of curvature. Such a smoothing would be like the exit ramp of a high speed highway, but tightly placed to the actual vertex. The integral $\int d\phi$ on any such a smoothing represents the angle of the change of direction at the vertex, counting counterclockwise. Thus, for the smoothed polygon, we have

$$\oint_{smth\ plgon} \kappa_g ds = \oint_{polygon} \kappa_g ds + \sum_i \phi_i, \qquad i = 1, ..., n \tag{4.24}$$

where n is the number of sides of the polygon. We would thus replace the left hand side of (4.23) with the right hand side of (4.24), and then move $\sum_i \phi_i$ to the right hand side of the equation. In this way, we obtain

$$\oint_{polygon} \kappa_g ds + \iint K \, \hat{\omega}^1 \wedge \hat{\omega}^2 = 2\pi - \sum_i \phi_i, \qquad (4.25)$$

which is the form in which, because of connection with ancestral geometry and topology, the theorem is stated.

Let us speak of one such simple connection. Those changes of direction are exterior angles, formed by the prolongation of a side with the next. They are related to the interior angles, θ_i, by $\phi_i = \pi - \theta_i$. Consider a standard triangle (rectilinear sides) in the plane. The left hand side of (4.25) because κ_g does not change in the straight parts and K is zero in the plane. The right hand side yields

$$0 = 2\pi - \sum_{i=1}^{3} (\pi - \theta_i). \qquad (4.26)$$

We thus get the well known result that the sum of the three interior angles is π.

A.4.7 Computation of the "extrinsic connection" of a surface

Given a surface in 3-D Euclidean space, Cartan's method for obtaining by inspection the 3-D connection adapted to it (or by any other method for that matter) is not applicable. The reason is that the surface by itself does not determine a frame field in Euclidean space. For example, a plane is a coordinate surface (x^3 constant) for an infinite number of coordinate systems, thus for an infinity of coordinate frame fields, among them the Cartesian, cylindrical and spherical ones.

Given a surface, the basis vectors $\mathbf{x}_{,u}$ and $\mathbf{x}_{,v}$ are well defined in terms of $(\mathbf{i}, \mathbf{j}, \mathbf{k})$. In normalizing the metric, we obtain a new frame field by means of

$$du \, \mathbf{x}_{,u} + dv \, \mathbf{x}_{,v} = \hat{\omega}'^{\alpha} \hat{\mathbf{e}}'_{\alpha} \qquad (A.4.2)$$

where the $\hat{\omega}'^{\alpha}$ are the ones given as $\hat{\omega}^{\alpha}$ in (2.19). This allows us to obtain $\hat{\mathbf{e}}'_{\alpha}$ and $\hat{\mathbf{e}}'_3$ (the latter by vector product), and their differentials, all in terms of $\mathbf{x}_{,u}$, $\mathbf{x}_{,v}$, $\mathbf{x}_{,u} \times \mathbf{x}_{,v}$ du and dv. The $\hat{\Gamma}'$ can be read only after du and dv are replaced by their expressions in terms of the ω'^{α}. We do not need to read them but just do that replacement for the next step. We shall thus have $d\hat{\mathbf{e}}'_3 = \omega'^{\alpha}_3 \hat{\mathbf{e}}'_{\alpha}$, the $\hat{\omega}'^3_3$ being zero for orthonormal frame fields.

We finally come to the canonical frame field, $\hat{\mathbf{e}}_i$. Clearly $\hat{\mathbf{e}}_3 = \hat{\mathbf{e}}'_3$. Hence $d\hat{\mathbf{e}}_3 = d\hat{\mathbf{e}}'_3 = \hat{\omega}'^{\alpha}_3 \hat{\mathbf{e}}'_{\alpha}$. We replace the $\hat{\mathbf{e}}'_{\alpha}$ in terms of the $\hat{\mathbf{e}}_{\beta}$. The coefficients of $d\hat{\mathbf{e}}_3$ in terms of the $\hat{\mathbf{e}}_{\beta}$ are the sought $\hat{\omega}^{\alpha}_3$. But, in order to read the $\hat{\Gamma}^{\alpha}_{3\beta}$, we have to express the $\hat{\omega}'^{\alpha}$ present in $\hat{\omega}'^{\beta}_3$ with their expressions in terms of the $\hat{\omega}^{\alpha}$.

Recall that the principal curvatures are $\hat{\Gamma}^1_{31}$ and $\hat{\Gamma}^2_{32}$. There is no great advantage in this computation over a more traditional one. But the logic and structural simplicity allows one to more readily recollect the process to derive anything of relevance, and the principal curvatures in particular.

Appendix B

"BIOGRAPHIES" ("*PUBLI*" *GRAPHIES*)

B.1 Elie Joseph Cartan (1869–1951)

B.1.1 Introduction

Since this report is not a biography proper, we refer readers to one by J. J. O'Connor and E. F. Robertson [56] as complement. In this appendix, we are concerned only with providing a perspective of Cartan's work.

This author believes that in matters of a mathematic-scientific nature, Cartan is the greatest human mind to ever walk this earth, way above anybody from recent centuries (Poincaré, Einstein, Gauss, Euler, Newton, etc.). We specified "recent" in order to make the comparison meaningful. We shall make the case for such assertions in part by combining authoritative opinions on different areas; nobody is in a position to emit them on all the areas in which Cartan involved himself. And in part by showing that Cartan sprinted during all his productive life, which extended until a few years before he died of, reportedly, a long illness. We shall have proved all of this by the end of this report of his work. At the very least, I shall have unearthed for you much of the work by Cartan about which you probably were unaware.

The version owned by this author of his Complete Works lists 184 publications, at least ten of which are bona fide books. We know of a few other publications. For instance, a paper with his son Henri is not counted among them.

As we shall see, some of Cartan's greatest contributions were buried as introductions to technical papers on subjects of lesser fame. For comparison purposes, recall that Ricci and Levi-Civita wrote a paper on the tensor calculus in 1901 [60]. In 1899 Cartan published his creation of the far superior exterior calculus (recognized as such only in recent decades) as an introduction to his paper "About some differential expressions and the Pfaff problem" [4]. A mathematician would have got his name in the history of mathematics just for creating this calculus, It is now the basis of differential geometry and of the theories of Lie groups and exterior differential system. Through its use, Cartan also showed that the study of a space at the infinitesimal level can give information about its behavior in the large. In other words, he created a tool not only for differential geometry, differential equations and Lie groups, but also for global geometry and topology (see Gauss-Bonnet in previous appendix). Let us proceed systematically.

Our starting point will be the classification that has been made of most of his papers into three large areas in his Complete Works [24]. They are algebra, differential equations and differential geometry. Other papers reproduced there can be classified within the following areas, where Cartan claimed to also have published: complex numbers, topology, integral invariants and mechanics, and relativity. See his "Notice sur les travaux scientifiques", which he wrote in 1931, to be found as the first paper in [24].

B.1.2   Algebra

In algebra, he classified Lie groups and Lie algebras, both real and complex, simple and semisimple. He gave their representations. He buried his discovery of spinors —like he did so often with other important discoveries— in a 1913 paper on projective groups, which included the study of the linear representations of simple groups [7]. It was only after Dirac showed their relevance in physics that spinors received explicit attention in Cartan's work [23].

Cartan was a major contributor to the field of Clifford algebras, not only because of spinors but also because of the so called multiplicity of 8 and triality. But there is much more, having to do with method and language, which facilitate the making of further discoveries. Let us hear from the late philosopher and mathematician Rota and collaborators Barnabei and Brini [1]. Writing in 1985 about Grassmann, who was another supergenius, they said this:

> "There is strong evidence to indicate that most main algebraists of the time, such as Gordan, Capelli, Hermite, Cayley, Silvester and even Hilbert ... did not realize the sweeping extent of Grassmann's discovery and its relevance in invariant theory.
> The epigons of invariant theory in this century, such mathematicians as Turnbull, Aitken, Alfred Young, Littlewood and even Hermann Weyl, perpetuated the same sin of omission, and one finds in these authors' work scattered rediscoveries and partial glimpses of ideas that could have made to bloom, had the authors used even only the notation of exterior algebra."

(Yes, David Hilbert and Hermann Weyl missed exterior algebra). Rota et al. then proceed to give credit to: Clifford, Schröder, Whitehead, Cartan and Peano. They then credit Cartan specifically with:

"He realized the usefulness of the notion of exterior algebra in his theory of integral invariants, which was later to turn into the potent theory of differential forms."

We shall speak of his work on integral invariants further below. But, in view of what has just been said, does he not qualify as the greatest algebraist of his time, like Grassmann was before him?

B.1.3 Exterior differential systems

Any system of ordinary or partial differential equations can be written as an exterior differential system. Around the turn of the century, I asked another participant in a meeting on differential geometry of the South East of the USA the following question. How is the field of partial differential equations structured? He responded something like this "The work in this field is comprised of two parts. One part is the Cartan-Kähler theory of differential systems [45]. The other part is made of everything else that has been written on the subject."

In the eighties, some of the top experts in that field (Bryant, Chern, Gardner, Goldschmidt and Griffiths [3]) joined forces to work through and appreciate this branch of Cartan's work, the result of which was their work "Exterior differential systems". One of them, the late professor Gardner, wrote also a monograph on Cartan's method of equivalence, method which belongs to this area of mathematics. In the introduction, he had this to say [42]:

"Elie Cartan's approach was motivated by his work on infinite pseudogroups....In 1908, in has to be one of the most remarkable papers in mathematics, 'The subgroups of the continuous groups of transformations', Elie Cartan formulated and desribed a procedure ... He did this in just 25 pages out of a 137 paper..."

Those 25 pages are a paper within a paper.

B.1.4 Genius even if we ignore his working on algebra, exterior systems proper and differential geometry

We now add to them other examples of genial work from his book on integral invariants [9]. Just that book, if it were explained and used, would by itself make Cartan one of the greatest mathematicians of all time and one of the foremost mathematical physicist of his time.

The pace is dizzying. Chapter 1 in that book starts by obtaining the equations of motion from the variational principle. By page 4 he has already obtained the Hilbert integral and by page 7 he has derived the equations of motion from the Hilbert integral. By the end of the chapter, page 16, he has derived Jacobi's theorem and has started discussion of the canonical form of the equations of motion in the presence of perturbation forces.

For another great example of the caliber of this book, let us jump to its chapter 9, page 81. In four pages, he states the concept of Lie operator acting on functions (which he calls infinitesimal transformation), he extends it to the ring of differential forms, explains its impact for obtaining first integrals, gives two ways in which that operator generates new differential invariants from a given one, relates those two and produces in the process formulas to generate those invariants.

In the same book, Cartan provides a wealth of applications to fluids, optics, rigid body dynamics, many body problems, standard issues in the foundations of classical mechanics (like generalizations of the theorem of Poisson-Jacobi and of the Poisson-Jacobi brackets).

Fluids are treated like part of classical mechanics, i.e. following the particles in their motion rather than in terms of what happens at any given point. Results emerge in quick succession.

In the same book and as an application of Lie operators to the n-body problem, Cartan deals with standard potential energy where the exponent of the radius in the denominator is left undetermined. In four pages, which are barely enough to write down his results, he obtains 11 differential invariants (whose integrals are integral invariants, one of them resulting in the Jacobi integral). From them he obtains a wealth of first integrals and even obtains the theorem that states the invariance of the product of the square of the angular momentum and the total energy, both in the movement around the center of mass. In one page he obtains from first principles the invariance of the magnitude of the velocity of the points in a moving rigid body.

After one graduates from the study of the book on Mechanics by Landau and Lifshitz [51], you may be said to be prepared for a new level of comprehension of the subject with that book by Cartan.

B.1.5   Differential geometry

Gardner continued as follows the contents of the citation we have made of him in subsection 1.3 [42].

> "The impact of these 25 pages on Klein's program was eloquently described by J. Dieudonne' when speaking of Cartan: 'Finally, it is fitting to mention the most unexpected extension of Klein's ideas in differential geometry...'

Consider the statement by Hermann Weyl in 1938, which we reproduce from the same Gardner reference:

> "'Cartan is undoubtedly the greatest living master in differential geometry.'"

Why? Cartan reinterpreted, unified and generalized the programs of Klein and Riemann in geometry. Using his discoveries on group and algebras and on the integrability of differential systems, he created the theories of affine, metric, Euclidean, projective, conformal connections, etc. He indeed showed that virtually all other efforts in geometry fit into his scheme. In the process, he gave the differential form of Euclid's postulates in spaces of any number of dimensions.

Cartan also authored the most important paper in Finsler geometry, which gave rise to the connection now known as the Cartan-Finsler connection. He created the concept of fiber spaces. In fact he worked with such spaces (equivalently bundles) since very early in his professional life, and it took half a century for the concept to be understood. Credit for bringing such an understanding is rightly given to Ehresmann as far as mathematicians are concerned [36]. But physicists know very well what they call the set of inertial frames, as it has a well known and understood bundle structure. They only need to learn the technical names.

The method of the moving frame carries his name because he is credited with having created it. This is incorrect. Several French mathematicians had already developed and used it, as Cartan himself reported. But his extreme virtuosity on this subject allowed him to create frames appropriate for arbitrary Lie groups, and then study them with that method.

Many physicists who often claim acquaintance with his work have not even understood his view of how Riemannian geometry is to be seen, namely as pertaining to finite Lie groups rather than the infinite Lie group of coordinate transformations. This is why they fail to see how gravitation (which they incorrectly associate with the infinite group of diffeomorphisms) could be unified with the other interactions without resort to far fetched ideas. Another example of much formalism and little substance in much of the modern literature: even technical books fail to give nice examples of spaces with torsion, in spite of the fact that there are very simple examples (one of them given by Cartan himself in 1924, namely what we have called the Columbus connection).

B.1.6   Cartan the physicist

Cartan published much on physical issues. We have already spoken of his work in classical mechanics. He also worked in other areas of (classical) physics, work that has been totally ignored. It may have been due to its sophistication (it was written in terms of differential forms), to being in French and to having to do with foundational problems that his work on physics has been totally ignored. If it had been studied and understood, he would have been considered a great physicist.

Take electrodynamics. Just his discussion of electrodynamics in a few pages of his 1924 paper (second in the series) on the theory of affine connections supersedes similar work in a paper of 1934 by Van Dantzig presented by Dirac to and published in Proceedings of the Cambridge Philosophical Society [71]. In those few pages Cartan discusses what is the right way of writing Maxwell's equations, whether these equations depend on connection, and whether electrodynamics itself (there are the energy equations in addition to Maxwell's) depends on connection. He even gives the form of the electromagnetic energy-momentum tensor in a kind of Kaluza-Klein space that he introduces in an informal way. In fact, in 1923-1924 he wrote all the fundamental physics of the time (except the emerging quantum mechanics) in terms of differential forms, with implications that he did not mention as to the emptiness of general covariance as a general principle, since it is just a heuristic help.

Consider the following. Most relativists still do not understand the conservation law of vector-valued forms, although he gave it in his 1922 on Einstein's equations. This is of particular importance in cosmology. With the Levi-Civita connection, it does not make sense to integrate those forms because that connection does not allow for path-independence identification of the tangent spaces at different point. If one still obtains good results it is because a metric together with a preferred frame determines a teleparallel connection. Cosmologists, without knowing it, are using the teleparallel connection determined by the metric they use and the frame of reference comoving with matter at the largest scales. But then space has torsion, which people with a very poor knowledge of geometry and of Lie derivatives say it is not present in spacetime. If they just knew where to look.

Cartan's potentially greatest contribution to physics may still be realized if and when the Cartan-Einstein unification were to become one day (part of) the unified theory that everybody wishes for. Indeed, Cartan already explained to Einstein (but the latter did not understand) the difference between formulating a geometric unified theory as a physicist and as a demiurge, specifically in teleparallelism. To be a demiurge means that one demands of the field equations that they prove that the universe is what one postulates it to be, and not just that the universe be consistent with the postulate. In the case of teleparallelism (TP), it means that the field equations must imply TP, and not simply ignore the affine curvature because it is zero, which is what Einstein was doing. Cartan did inform on this to Einstein (and actually showed to him a couple of equations in this regard!, but the latter did not listen). The irony is that if the equations of structure of some space are made field equations of the physics, that is the purest form of implementation of Einstein's thesis of logical homogeneity of geometry and theoretical physics.

B.1.7   Cartan as critic and mathematical technician

Cartan did in passing and more efficiently work for which other authors had received credit, like Slebodzinski on Lie differentiation [68] and Weitzenbock [90] and [91] on teleparallelism. Whitney is famous for his work on embedding [92]. Although the subject covered is far from being the same, Cartan's comparable work on embedding is just one more of his papers and not even mentioned in the literature.

He had the vision of mathematics that allowed him to incontestably say what was wrong with other people's programs regardless of their stature, as we retrospectively know. Thus, for example, he had the perspective and authority to use the term "the false spaces of Riemann" (he proceeded immediately to speak of their redemption by Levi-Civita). In less dramatic terms, he expressed pointed criticisms and/or limitations of the work of other mathematicians, comments which history has proved to be correct. Such is the case with his comments on the work of Weitzenbock, Ricci and a Finsler geometer whose name we prefer not to mention. See also numerous examples in "Notice sur les travaux scientifiques" [24], where he corrects even Poincaré (page 93 of the report, page 107 in the Complete Works).

Cartan took a paper by Study on systems of complex numbers (*Ältere und neuere Untersuchungen uber Systeme complexer Zahlen*) and multiplied it by four in size [6]. This paper is a tremendous piece of work on algebraic developments in the nineteenth century. If you want to understand the scope of Grassmann's work, read that Cartan paper.

Cartan had the genial ability of thinking of some way to avoid brute force methods and compute in a few lines what others that for which others would take pages.

A few examples are his derivation of the equations of structure of Euclidean space in a few lines as a simple exercise of change of coordinates, his computation of the Lie derivative of differential forms or of the primitives of closed differentials, etc., etc. [9]

Just a little step taken by him in his study of Lie groups [22] suggested our diagonalization of the second fundamental form of the theory of surfaces in the previous appendix, which results in a tremendous simplification of that theory, as we have shown in the previous appendix.

All this speaks of (to use Dieudonné's words [33]) "his uncanny algebraic and geometric insight that has baffled two generations of mathematicians".

B.1.8 Cartan as writer

For some reason, Cartan's writings are viewed as difficult to understand. I guess that much of it has to do with the extremely formal way in which modern mathematicians are trained, where the trees (definitions, theorems, proofs, lemmas, corollaries, remarks, etc.) and symbolism impede a vision of the forest (the leading story).

Some testimonials about the difficulty in understanding Cartan, by Gardner about himself and others:

"After thinking about this method for another twenty years and talking periodically with S.S. Chern and my students, especially Robert Bryant, I realized ... One may wonder why anyone would need twenty years to understand a paper. However, I was not alone in experiencing difficulty with parts of the theory of Repere Mobile and the method of equivalence, as the following two citations testify."

The first of those two citations is due to Hermann Weyl in his review of a Cartan book [22]. He said:

"Nevertheless I must admit that I found the book, like most of Cartan's papers, hard reading..."

Weyl goes on to wonder whether this difficulty has to do with not having been trained in the French system. The second occurs in the middle of Singer and Sternberg's paper [1965]:

"We now resume some of the principal formulae in coordinate notation with the idea of providing a partial guide to some of the writings of E. Cartan on the infinite groups and on the equivalence problem. We must confess that we find most of these papers extremely rough going and we cannot follow all the arguments in detail."

Cartan had the ability to come with some "trick" to compute in a few lines what would take others several pages to do. Gardner (again in the introduction to his book on the method of equivalence) makes a point similar to the one by Dieudonné mentioned in the previous subsection about Cartan unparalleled computational abilities:

"... working most of the examples worked out by Cartan made you feel that special tricks and brilliant observations were part of the method."

The style of Cartan, like Kähler's, is the style of physicists. He develops ideas as if he were telling a story. His style should not be difficult for physicists if given some guidance by the mathematicians who now claim to understand him. For that, mathematicians should explain to physicists what is going on rather than translate Cartan's stories into a collection of definitions, theorems, remarks, proofs, corollaries, lemmas, etc., which we physicists so much dislike.

B.1.9   Summary

Differential geometry, topology and analysis are now connected through the common language of differential forms, which Cartan created. He was unique in the theory of exterior differential systems, because he was so great in algebra. In turn his awesome contributions to classical differential geometry are made possible by his profound knowledge of the theory of exterior differential systems. And it was on the shoulders of Cartan that Kähler built his work on the theory of partial differential equations and his exterior-interior calculus.

Cartan's almost complete works comprises more than 46 hundred pages. From 1893 to 1949, both inclusive, Cartan had at least one publication each year, except for 1900 (but he published three in 1901), 1903 (but he had published three in 1902), 1906 (but he published two in 1907), 1921 (but he had four publications in 1920 and nine in 1922), and 1948 (but remember he died in 1951, of a long illness, at the age of 82).

Cartan the scientist/mathematician sprinted almost to the end of the scientific marathon he run, when force majeure (sickness) may have caused his abandoning the race at about kilometer 40. He was not from this planet. Any classification of geniuses in the mathematics/physics field only shows the ignorance of those doing the classification if they do not put him at least in competition for the first place. No other mathematician sprinted all his life. Euler was another marathoner, but his sprinting was not quite the same.

The top algebraist, the top expert on differential equations and the top differential geometer, an extraordinary mathematical technician and a physicist whose work on this subject may one day be appreciated when understood. That summarizes who Cartan was. Reportedly, he also was a great human being and family man. That is the cherry in the cocktail.

B.2 Hermann Grassmann (1808–1877)

B.2.1 Mini biography

In his day job, Hermann Grassmann was a high school teacher. In his spare time, he became one of the greatest mathematicians of all time. In addition, he was a superb linguist who translated the Hindu Rig Veda (literature and religion) from Sanskrit into German verse, and produced an extensive dictionary in the process. He was an accomplished folklorist, musician and natural scientist (botany, crystallography, color mixing, statics, electricity, acoustics, theory of tides, invented a heliostat, etc.). All that is stated in the book *"A New Branch of Mathematics"* [43], which is the main translation of much of his work without, however, constituting his complete works.

Grassmann undertook six semesters of university studies in Berlin, his subject being philology, theology, philosophy and psychology. Reportedly, he did not take any mathematics courses there, but read books in mathematics written by his father, Justus G. Grassmann, another high school teacher. In fact, the idea of purely geometric products, which are the basis of H. Grassmann's mathematical work, was his father's. Hermann developed Justus'es basic concepts into a full blown system. He passed several state examinations which qualified him to teach a variety of subjects in high school. He was eventually granted a Doctor of Philosophy Honoris Causa by the University of Tübingen. He was elected a corresponding member of the Göttingen scientific society. After his death, he was finally recognized by Felix Klein and Sophus Lie, who was instrumental in the publication of his Complete Works.

In mathematics, he is credited with having discovered exterior algebra, but he also virtually discovered Clifford algebra and advanced the study of projective algebraic varieties while working as a high school teacher. And there is much more, as we shall now intimate.

B.2.2 Multiplications galore

Grassmann did not use the vector product, which is a contraption peculiar to three dimensions, also to seven dimensions in a more artificial way. As we have seen in chapter 3, the vector product in dimension three comprises the exterior product, but in combination with Hodge duality. It did not exist at the time, but, had it existed, Grassmann would have viewed it as just a footnote to his system, as can be inferred from his comments on Hamilton's quaternions. As in vector algebra, only grades zero and one are present.

His extensive use of associative, distributive, commutative and anticommutative laws (some times in muddled ways) was totally uncharacteristic in his time. By 1844, he was dealing with mathematical objects of arbitrary dimensions (hyperplanes) living in n-dimensional space, for arbitrary natural number n. This is presently known as graded algebra, where there are scalars, vectors, bivectors (kind of antisymmetric tensors of grade two), trivectors, etc. He did that ten years before Riemann's famous lecture to the Göttingen faculty on the spaces that now bear his name, at a time where one barely had addition of vectors and multiplication by scalars. Highly recognized nowadays also is his implicit contribution to projective geometry, with implications for differential topology and algebraic geometry, and where the concept of Grassmannian honors his name.

The exterior product was only the best known of the many different products that he considered. His paper "On the Various Types of Multiplication" (included in [43]) reports three such types, namely symmetric, circular and linear. One would think that is by itself quite an achievement, considering that this is in mid nineteenth century. But then we are for a surprise. There are sixteen different species of symmetric multiplications, eight different species of circular multiplication and two different species of linear multiplication, which he calls algebraic and outer respectively. Not all of them are equally important. In this regard, we follow Cartan's report of Grassmann's products in his paper on complex numbers [6]. He speaks of the complex and the interior as the most interesting among the circular products. The algebraic and the exterior multiplications are the linear ones.

Cartan continues his report speaking of two systems of non-algebraic multiplication, or rather a system with two different multiplications. These are the progressive and regressive products, the combinatorial ones been among the progressive ones. Intimately related to the progressive product, there is the regressive product, both of great importance in modern algebra. Cartan also speaks in the same report of interior multiplication involving either progressive or regressive products.

If you are overwhelmed, well, that was my purpose. You have experienced what you would have felt if you had immersed himself into his work.

B.2.3   Tensor and quotient algebras

Not everything in Grassmann work is correct or final. Thus, for example, no attention is explicitly given to whether a product is associative or not. At one point, he may be assuming it and later drops it almost without notice. There is also an issue as to the limitations of his constrains, which will be better understood after we speak of his informal invention of the general tensor algebra and its quotients algebras.

In the first half of page 454 of [43] (this is in the paper "On the Various Types of Multiplication"), Grassmann has already introduced the essentials of the concept of general tensor algebra. If there is a limitation in this definition, it is the absence of what we called property (c) in our definition of tensor product (section 7 of chapter 3). The difference between the general tensor algebras and their quotients algebras lies precisely in that the latter ones do not comply with that property. View these comments and those to follow in the context that we are talking about a 1844 publication, four decades before the advent of the vector calculus.

Quotient algebras were introduced by Grassmann as constrains on the products of his general (tensor) algebras. Be aware that we are speaking of tensors, not tensors fields. These became part of structured mathematics with the work of Ricci and Levi-Civita [60].

As pointed out in an editorial note in the translation of the paper "*On the Various Types of Multiplication*", there is a real limitation in the restriction of constraints by Grassmann to two factors (Recall that the Jacobi identity is not of this type). It is important to notice that the defining relations of Clifford algebra (see for instance Equations (8.6) and (8.8) of chapter 3 are of this type). For further discussion of the limitation of Grassmann constraints see that editorial note.

B.2.4   Impact and historical context

The impact of Grassmann in mathematics goes far beyond algebra and calculus. Thus, for instance, Yaglom reports [93] that Dedekind developed the axiomatic definition of the natural numbers from Grassmann's constructions.

The dust has not yet settled on what was his vision on projective geometry. The paper by Dieudonné *The Tragedy of Grassmann* [32] lauds his contribution to that geometry. It was eventually followed by Rota and collaborators paper [1] who claimed that the Bourbakists (Dieudonné is one of them) had not understood the full import of Grassmann's ideas on the subject. That is ironic since both of these parties might have made claim to being the greatest modern admirers of Grassmann.

Just before dying, Grassmann came to the threshold of Clifford algebra. He had virtually abandoned mathematics during the previous ten years because of frustration with the lack of reception of his ideas. The Clifford product of two vectors is the sum of their exterior and inner products, which Grassmann created.

Reportedly, Cartan's work is difficult to understand. Grassmann work is much more so. This should not be surprising. In addition to being ahead of his time and working in total isolation from academia, he was not trained in the prevailing mathematical traditions. Top mathematicians like Cauchy and Möbius were put off by his style, thus failing to capture the tremendous depth of his ideas.

One has to wonder what might have been Grassmann's legacy if he had been offered a professorship at Göttingen to develop and explain his ideas in a more conventional way. A terrific cross-fertilization with Gauss and specially Riemann might have ensued before the "horrible vector calculus" (Dieudonné's expression), tensor calculus and gamma matrices emerged. Calculus and geometry might have been at the beginning of the 20th century better than they are even today.

Appendix C

PUBLICATIONS BY THE AUTHOR

One main motivation of the author to write this book is that his work on unification —which is now entering its high energy phase after having recently entered the more general quantum mechanical phase— will not be understood without knowledge of Clifford algebra, Kähler calculus, Finsler bundles and his temporarily abandoned work on the geometrization of classical electrodynamics with those bundles. The list of publications that follow may give an idea of where those topics may have been discussed in greater detail.

51. "$U(1) \times SU(2)$ from the Tangent Bundle", J. Physics: Conference Series 474, 012032 (2013).

50. "Real Units Imaginary in Kähler's Quantum Mechanics". ArXiv-1207.5718vI.

49. "Real Version of Calculus of Complex Variable: Cauchy's Point of view". ArXiv 1205.4256.

48. "Real Version of Calculus of Complex Variable: Weierstrass Point of View". ArXiv 1205.4657.

47. Book "Differential Forms for Cartan-Klein Geometry", Abramis, London (2012).

46. "Opera's Neutrinos and the Robertson Test Theory of the Lorentz Transformations", ArXiv 1111.2271v2.

45. "From Clifford through Cartan to Kähler", Hypercomplex numbers in geometry and physics, #1 (13) Tom 7, Moscow, pp. 165-180 (2010). It can be found in: http://hypercomplex.xpsweb.com/page.php?lang=ru&id=569. Click *Russian pdf*. Scroll down to page 165. Papers are in Russian, except mine, which is in English.

44. "The Foundations of Quantum Mechanics and the Evolution of the Cartan-Kähler Calculus". Found. Phys. 38, 610-647 (2008).

43. "Klein Geometries, Lie Differentiation and Spin", Differential Geometry – Dynamical Systems 10, 300-310 (2008).

42. "The Kaehler-Dirac Equation with Non-Scalar-Valued Differential Form". Adv. Appl. Clifford Alg. 18, 1007-1021 (2008).

41. "New Perspectives on the Kaehler Calculus and Wave Functions". Adv. Appl. Clifford alg. 18, 993-1006 (2008).

40. "Recent Developments on the Foundations of Classical Differential Geometry with Implications for the Testing and Understanding of Flat Spacetime Physics", Proceedings of XIII International Scientific Meeting on Physical Interpretations of Relativity Theory, Baumann State Technical University, Moscow (2007).

39. "Affine torsion à la Cartan", Annales de la Fondation Louis de Broglie 32, 409-423 (2007).

38. "The Idiosyncrasies of Anticipation in Demiurgic Physsical Unification with Teleparallelism", with D. G. Torr. International Journal of Computing Anticipatory Systems 19, 210-225 (2006).

37. "Anticipation at the Unification of Geometry and Calculus", with D. G. Torr. International Journal of Computing Anticipatory Systems 19, 194-209 (2006).

36. "Of Finsler Fiber Bundles and the Evolution of the Calculus", with D. G. Torr. Balkan Geometry Society Proceedings, Geometry Balkan Press, Bucharest, 183-191 (2006). www.mathem.pub.ro/dept/confer05/M-VAO.PDF.

35. "A Different Line of Evolution of Evolution of Geometry on Manifolds Endowed with Pseudo-Riemannian Metrics of Lorentzian Signature", with D. G. Torr. Balkan Geometry Society Proceedings, Geometry Balkan Press, Bucharest, 173-182 (2006). www.mathem.pub.ro/dept/confer05/M-VAA.PDF.

34. "A Gravitational Experiment Involving Inhomogeneous Electric Fields", with T. Datta and M. Yin. Proceedings of the 2004 Space Technology and applications International Forum (STAIF), 1214-1221 (2004).

33. "Is Electromagnetic Gravity Control Possible?", with D. G. Torr, Proceedings of the 2004 Space Technology and applications International Forum (STAIF), 206-1213 (2004).

32. "Synchronizations versus Simultaneity Relations", with Implications for Interpretations of Quantum Measurements", with D. G. Torr. In "Gravitation and Cosmology: From the Hubble Radius to the Planck Scale", 367-376. Editors: R. L. Amoroso, G. Hunter, M. Kafatos and J. P. Vigier, Kluwer, Boston (2002).

31. "From the Cosmological Term to the Planck Constant", with D. G. Torr. In "Gravitation and Cosmology: From the Hubble Radius to the Planck Scale", Kluwer, Boston, 1-10 (2002).

30. "Quantum Clifford Algebra from Classical Differential Geometry", with D. G. Torr. J. Math. Phys. 43 (3), 1353-64 (2002).

29. "The Cartan-Clifton Method of the Moving Frame: Finslerian Bundles on Riemannian Distances", with D. G. Torr. Algebras, Groups and Geometries, 17(3), 361-374 (2001).

28. "Marriage of Clifford Algebra and Finsler Geometry: a Lineage for Unification", with D. G. Torr. Int. J. Theor. Phys. 40(1), 273-296 (2001).

27. "Clifford-Valued Clifforms: a Geometric Language for Dirac Equations". With D. G. Torr. In Clifford Algebras and Their Applications in Mathematical Physics.; Vol. 1 (Algebra and Physics), pp. 143-162. Editors: R. Ablamowicz and B. Fauser. Birkhäuser, Boston (2000).

26. "The Theory of Acceleration within its Context of Differential Invariants: The Root of the Problem with Cosmological Models?", with D. G. Torr. Found. Phys. 29, 1543-1580 (1999).

25. "The Cartan-Einstein Unification with Teleparallelism and the Discrepant Measurements of Newton's Constant G", with D. G. Torr. Found. Phys. 29, 145-200 (1999).

24. "Teleparallel Kähler Calculus for Spacetime", with D. G. Torr. Found. Phys. 28, 931-958 (1998).

23. "The Construction of Teleparallel Finsler Connections and the Emergence of an Alternative Concept of Metric Compatibility", with D. G. Torr. Found. Phys. 27, 825-843 (1997).

22. "The Emergence of a Kaluza-Klein Microgeometry from the Invariants of Optimally Euclidean Lorentzian Connections", with D. G. Torr. Found. Phys. 27, 533-558 (1997).

21. "Elementary Geometries Underlying the Theory of Non-linear Euclidean Connections", with D. G. Torr, Cont. Math., 196, 301-309 (1996).

20. "Canonical Connections of Finsler Metrics and Finslerian Connections on Riemannian Metrics", with D. G. Torr, Gen. Rel. Grav. 28, 451-469 (1996).

19. "The Cornerstone Role of the Torsion in Finslerian Physical Theories", with D. G. Torr, Gen. Rel. Grav. 27, 629-644 (1995).

18. "Finslerian Structures: The Cartan-Clifton Method of the Moving Frame", with D. G. Torr. J. Math. Phys. 34 (10), 4898-4913 (1993).

17. "Geometrization of the Physics with Teleparallelism (II): Towards a Fully Geometric Dirac Equation", with D. G. Torr and A. Lecompte. Found. Phys. 22, 527-547 (1992).

16. "Geometrization of the Physics with Teleparallelism (I): The Classical Interactions". Found. Phys. 22, 507-526 (1992).

15. "Conservation of Vector-Valued Forms and the Question of the Existence of Gravitational Energy-Momentum in General Relativity", with D. G. Torr. Gen. Rel. Grav. 23, 713-732 (1991).

14. "On the Geometrization of Electrodynamics". Found. Phys. 21, 379-401 (1991).

13. "The Breaking of the Lorentz transformations and the Geometrization of the Physics", with D. G. Torr. Nucl. Phys. B. (Proc. Suppl.) 6, 115 (1989).

12. "On Testing the Line Element of Special Relativity with Rotating Systems", with D. G. Torr. Phys. Rev. A 39, 2878 (1989).

11. "Electrodynamics of the Maxwell-Lorentz Type in the Ten-Dimensional Space of the Testing of Special Relativity: A Case for Finsler Connections", with D. G. Torr. Found. Phys. 19, 269 (1989).

10. "Revised Robertson's Test Theory of Special Relativity: Supergroups and Superspace". Found. Phys. 16, 1231 (1986).

9. "Revised Robertson's Rest Theory of Special Relativity: Space-Time structure and Dynamics", with D. G. Torr. Found. Phys. 16, 1089 (1986).

8. "Kinematical and Gravitational Analysis of the Rocket-Borne Clock Experiment by Vessot and Levine Using the Revised Robertson's Test Theory of Special Relativity". Found. Phys. 16, 1003 (1986).

7. "Revised Robertson's Test Theory of Special Relativity". Found. Phys. 14, 625 (1984).

6. "Problems of Synchronization in Special Relativity: (A Reply to G. Cavalleri and G. Spinelli)", J. G. Vargas. Found. Phys. 13, 1231, (1983).

5. "Nonrelativistic Para-Maxwellian Electrodynamics with Preferred Reference Frame in the Universe". Found. Phys. 12, 889 (1982).

4. "Spontaneous ParaLorentzian Conserved-Vector and Nonconserved-Axial Weak Currents". Found. Phys. 12, 765 (1982).

3. "Nonrelativistic Para-Lorentzian Mechanics". Found. Phys. 11, 235 (1981).

2. "Relativistic Experiments with Signals on a Closed Path (Reply to Podlaha)". Lett. Nuovo Cimento 28, 289 (1980).

1. "Comment on Lorentz Transformations from the First Postulate". Am. J. Phys. 44, 999 (1976).

References

[1] Barnabei, M., Brini., A. and Rota G.-C. (1985): On the Exterior Calculus of Invariant Theory, J. of Algebra **96**, 120-160.

[2] Brandt, H. E. (2005): Finslerian Quantum Theory, Nonlinear Analysis **63**, e119-e130.

[3] Bryant, R., Chern, S.-S., Gardner, R., Goldschmidt, H. and Griffiths, P. (1990), Exterior Differential Systems, MSRI Publications, Springer.

[4] Cartan, É. J. (1899): Sur certains expressions différentielles et le problème de Pfaff, Ann. École Norm, **16**, 239-332.

[5] Cartan, É. J. (1908): Nombres complexes, French edition of Encyclop. Sc. math., **I5**, 329-468. This is a very extended version of an article originally by E. Study.

[6] Cartan, É. J. (1908): Les sous-groupes des groups continus de transformations, Ann. École Norm. **25**, 57-194.

[7] Cartan, É. J. (1913): Les groupes projectives qui ne laisse invariante aucune multiplicité plane, Bull. Soc. Math. de France, **41**, 53-96.

[8] Cartan, É. J. (1922); Sur les équations de la gravitation d'Einstein, J. Math. Pures et appliquées, **1**, 141-203.

[9] Cartan, É. J. (1922): Leçons sur les invariants integraux, Hermann, Paris.

[10] Cartan, É. J. (1922): Sur les equations de structure des espaces généralisés et l'expression analytique du tenseur d'Einstein, Comptes Rendus.

[11] Cartan, É. J. (1923): Sur les variétés à connexion affine et la théorie de la relativité généralisé, Ann. École Norm. **40**, 325-412; **41**, 325-412.

[12] Cartan, É. J. (1924): Sur les variétés à connexion affine et la théorie de la relativité généralisé, Ann. École Norm. **41**, 1-25.

[13] Cartan, É. J. (1924): Les récentes généralizations de la notion d'espace, Bull. Sc. Math. **48**, 294-320.

[14] Cartan, É. J. (1925): Sur les variétés à connexion affine et la théorie de la relativité généralisé, Ann. École Norm. **42**, 17-88.

[15] Cartan, É. J. (1925): La theorie des groups et les recherches récentes de géometrie différentielle, L'Ensign. math., **24**, 1-18.

[16] Cartan, É. J. (1925): La géométrie des espaces de Riemann (Mémorial Sc. Math., IX). Reprinted (1963) Gauthier-Villars, Paris.

[17] Cartan, É. J. (1930): Notice historique sur la notion de parallélism absolu, Annalen **102**, 698-706.

[18] Cartan, É. J. (1931): Le parallélism absolu et la théorie unitaire du champ, Revue Métaph. Morale, 13-28; Actual. scient. et ind., **44** (1932).

[19] Cartan, É. J. (1934): Les espaces de Finsler, Exposés de Géometrie, II. Reprinted: Hermann, Paris, 1971.

[20] Cartan, É. J. (1935): La méthode du repère mobile, la théorie des groupes continus et les espaces généralisés, Exposés de Géométrie, **V**, Hermann, Paris.

[21] Cartan, É. J. (1936): Le rôle de la théorie des groupes de Lie dans l'évolution de la géometrie moderne, C. R. Congrés intern. Oslo, **I**, 92-103.

[22] Cartan, É. J. (1937): The theory of finite continuous groups and differential geometry treated by the method of the moving frames, Gauthier-Villars, Paris. Reprinted (1992) Éditions Jacques Gabay, Sceaux.

[23] Cartan, É. J. (1938): Leçons sur la théorie des spineurs, Exposés de Géométrie, **XI**, Hermann, Paris.

[24] Cartan, É. J. (1984): Oeuvres Complètes, Editions du CNRS, Paris.

[25] Chern, S-S. (1948): Local Equivalence and Euclidean Connections in Finsler Spaces, Science Reports Tsing Hua Univ. 4, 95-121.

[26] Clifford, W. K. (1882): Mathematical Papers by William Kingdon Clifford (R. Tucker, Ed.), Macmillan, London. Reprinted by Chelsea (1968), New York.

[27] Clifton, Y. H. (1966): On the Completeness of Cartan Connections, J. of Math. and Mechanics, **16**, 569-576.

[28] Crowe, M. J. (1967): A History of Vector Analysis, Dover Publications, New York.

[29] Debever, R. Editor (1979): Elie Cartan-Albert Einstein, Letters on Absolute Parallelism: 1929-1932, Princeton University Press, Princeton.

[30] Deicke, A. (1953): Über die Darstellung von Finsler-Räumen durch nichtholonome Mannigfaltigkeiten in Riemannschen Räumen, Archiv der Math. **4**, 234-238.

[31] Deicke, A. (1954): Finsler Spaces as Non-Holonomic Subspaces of Riemannian Spaces, J. London Math. Soc. **30**, 53-58.

[32] Dieudonne, J. (1979): The Tragedy of Grassmann, Linear and Multilinear Algebra, **8**, 1-14.

[33] Dieudonne, J. (1970-1990): Biography in Dictionary of Scientific Biography, New York.

[34] Dirac, P.A.M. (1958): The Principles of Quantum Mechanics, Clarendon Press, Oxford.

[35] Donaldson, S. K., Kronheimer, B. P. (1990): The Geometry of Four-Manifolds, Clarendon, Oxford.

[36] Ehresmann, C. (1950): Les connexions infinitésimales dans un espace fibré différentiable, Colloque de Toplogie, Bruxelles, 29–55.

[37] Einstein, A. (1916): Die Grundlagen der allgemeinen Relativitätstheorie, Ann. Phys., Lpz., **49**, 769. Reprinted in English under the title "The Foundation of the General Theory of Relativity" as an article in the book The Principle of Relativity by Methuen and Company, book republished under the same title by Dover (1952), New York, pp. 109-164.

[38] Einstein, A. (1933): On the Method of Theoretical Physics (The Herbert Spencer lecture, delivered at Oxford, June 10). Published in Mein Weltbild (1934), Querido Verlag, Amsterdam. Also collected in Ideas and Opinions (1954), Bonanza Books, New York.

[39] Einstein, A. (1930): Théorie unitaire du champ physique, Ann. Ins. Henri Poincaré, **1**, 1-24.

[40] Feynman, R. P., Leighton, R. B., and Sands, M. (1964): The Feynman Lectures on Physics, Addison-Wesley, Reading, Massachusetts.

[41] Flanders, H. (1953): Development of an Extended Exterior Differential Calculus, Trans. American Math. Soc. **75**, 311-326.

[42] Gardner, R. (1989): The Method of Equivalence and its Applications, SIAM, Philadelphia.

[43] Grassmann, H. (1995): The Ausdehnungslehre of 1844 and Other Works of Grassmann, English translation by L. C. Kannenberg titled "A New Branch of Mathematics", Open Court, Chicago.

[44] Hestenes, D. (1992): Mathematical Viruses in A. Micali et al., Clifford Algebras and their Applications in Mathematical Physics, 3-16 (1992). Kluwer Academic Publishers.

[45] Kähler, E. (1934): Einführung in die Theorie der Systeme von Differentialgleichungen, Teubner. Reprinted: Chelsea Publishing Company, New York, 1949.

[46] Kähler, E. (1960): Innerer und äusserer Differentialkalkül, Abh. Dtsch. Akad. Wiss. Berlin, Kl. Math., Phy. Tech., **4**, 1-32.

[47] Kähler, E. (1961): Die Dirac-Gleichung, Abh. Dtsch. Akad. Wiss. Berlin, Kl. Math., Phy. Tech **1**, 1-38.

[48] Kähler, E. (1962): Der innere Differentialkalkül, Rendiconti di Matematica **21**, 425-523.

[49] Klein, F. (1872): Vergleichende Betrachtungen über neuere geometrische Forschungen, Reprinted in Math. Annalen **43**, 63-100 (1893).

[50] Landau, L.D., Lifshitz E.M. (1978): Mechanics, Elsevier, Burlington, Massachusetts.

[51] Landau, L.D., Lifshitz E.M. (1975): The Classical Theory of Fields, Pergamon Press, Oxford.

[52] Levi-Civita, T. (1917): Nozione di parallelismo, Rend. Circ. Mat. Di Palermo **42**, 173-205.

[53] Lichnerowicz, A. (1962): Elements of Tensor Calculus. John Wiley, Somerset, NJ.

[54] Mason, L. J. and Woodhouse, N. M. J. (1996): Integrability, Self-Duality and Twistor Theory, Clarendon Press, Oxford.

[55] Mead, C. A. (2000): Collective Electrodynamics: Quantum Foundations of Electromagnetism, The MIT Press, Cambridge, Massachusetts.

[56] O'Connor, J. J. and Robertson, E. F. (1970-1990): Élie Joseph Cartan in the MacTutor History of Mathematics Archive., http://www-history.mcs.st-and.ac.uk/Biographies/Cartan.html

[57] Pauli, W. (1921): Relativitätstheorie in Encyklopädie der mathematischen Wissenschaften **V19**, Teubner, Leipzig. English translation by G. Field titled Theory of Relativity, Pergamon Press, Oxford.

[58] Peano, G. (1888): Calcolo geometrico secondo l'Ausdehnungslehre di H. Grassmann, Fratelli Bocca Editori, Torino.

[59] Pommaret, J. F. (1988): Lie Pseudogroups and Mechanics, Gordon and Breach, New York.

[60] Ricci, G. and Levi-Civita, T. (1901), Méthodes de calcul différentiel absolu et leur applications, Math. Ann. **54**, 125-201.

[61] Riemann, B. (1854): Über die Hypothesen welche der Geometrie zu Grunde liegen, Habilitationschrift, Göttingen.

[62] Riemann, B. (1861): Comentatio mathematica, qua respondere tentatur quaestioni ab Illma Academia Parisiensi propositae: "Trouver quel doit être l'état calorifique d'un corps solide homogène indéfini pour qu'un système de courbes isothermes, à un instant donné, restent isothermes après un temps quelconque, de tel sorte que la température d'un point puisse s'exprimer en fonction du temps et de deux autres variables indépendantes". This Prize Essay was first published in 1876, in the first edition of Riemann's Collected Works, thus later than the related papers by Christoffel and Lipshitz.

[63] Riemann, B. (1953): Collected Works, Ed. Heinrich Weber, Dover, New York.

[64] Ringermacher, H.I. (1994): An Electrodynamic Connection, Class. Quantum Grav. **11**, 2383-2394.

[65] Rudin, W. (1976): Principles of Mathematical Analysis, McGraw-Hill, New York.

[66] Schmeikal, B. (2004): Transposition in Clifford Algebra. In: Clifford Algebras - Applications to Mathematics, Physics and Engineering, Ablamowicz, R. (ed.), Birkhäuser, Boston-Basel-Berlin, pp. 355-377.

[67] Sharpe, R. W. (1996): Differential Geometry, Springer, New York.

[68] Slebodzinski, W. (1931): Sur les equations canoniques de Hamilton, Bulletin de l'Academie Royale de Belgique, Classe des Sciences, Series 5, **17**, 864-870.

[69] Slebodzinski, W. (1963): Formes exterieures et leurs applications, Vol. 2, Polska Akademia Nauk, Warsaw.

[70] Struik, D. J. (1961): Classical Differential Geometry, Dover, New York.

[71] Van Dantzig (1934): The Fundamental Equations of Electromagnetism, Independent of Metric Geometry, Proceedings of the Cambridge Philosophical Society, Cambridge **30** (4), pp 421-427.

[72] Vargas, J. G. (1991): On the Geometrization of Electrodynamics, Found. Physics **21**, 610-647

[73] Vargas, J. G. and Torr, D. G. (1993): Finslerian Structures: The Cartan-Clifton Method of the Moving Frame", J. Math. Phys. **34**, pp. 4898-4913.

[74] Vargas, J. G. and Torr, D. G. (1995): The Cornerstone Role of the Torsion in Finslerian Physical Worlds, Gen. Rel. Grav. **27**, 629-644.

[75] Vargas, J. G. and Torr, D. G. (1996): Canonical Connections of Finsler Metrics and Finslerian Connections on Riemannian Metrics, Gen. Rel. Grav. **28**, 451-469.

[76] Vargas, J. G. and Torr, D. G. (1996): Elementary Geometries Underlying the Theory of Euclidean Connections on Finsler Metrics of Lorentzian Signature, Contemporary Mathematics **196**, 301-310.

[77] Vargas, J. G. and Torr, D. G. (1997): The Emergence of a Kaluza-Klein Microgeometry from the Invariants of Optimally Euclidean Lorentzian Spaces, Found. Phys. **27**, 533-558.

[78] Vargas, J. G. and Torr, D. G. (1998): Teleparallel Kähler Calculus for Spacetime", Found. Phys. **28**, 931-958.

[79] Vargas, J. G. and Torr, D. G. (1999): The Theory of Acceleration within its Context of Differential Invariants: The Root of the Problem with Cosmological Models? Found. Phys. **29**, 1543-1580.

[80] Vargas, J. G. and Torr, D. G. (2000): Clifford-valued Clifforms: A Geometric Language for Dirac Equations, in R. Ablamowicz and B. Fauser, Eds., Clifford Algebras and their Applications in Mathematical Physics, vol 1, Algebra and Physics, Birkhäuser, Boston, pp. 135-154.

[81] Vargas, J. G. and Torr, D. G. (2006): The Idiosyncrasies of Anticipation in Demiurgic Physical Unification with Teleparallelism, Int. J. of Computing Anticipatory Systems **19**, 210-225.

[82] Vargas, J. G. and Torr, D. G. (2008): Klein Geometries, Lie differentiation and Spin, Differ. Geom. Dyn. Syst. **10**, 300-311. www.mathem.pub.ro/dgds/v10/D10-VA.pdf.

[83] Vargas, J. G. (2008): The Foundations of Quantum Mechanics and the Evolution of the Cartan-Kähler Calculus, Found Phys **38**, 610-647. DOI 10.1007/S10701-008-9223-3.

[84] Vargas, J. G. (2011): "Opera's Neutrinos and the Robertson Test Theory of the Lorentz Transformations" arXiv:1111.2271 [hep-ex].

[85] Vargas, J. G. (2012): Differential Forms for Cartan-Klein Geometry (Erlangen Program with Moving Frames), Abramis, London.

[86] Vargas, J. G. (2012): Real Calculus of Complex Variable: Weierstrass Point of View, arXiv 1205.4657.

[87] Vargas, J. G. (2012): Real Calculus of Complex Variable: Cauchy's Point of View, arXiv 1205.4256.

[88] Vargas, J. G. (2012): Real Units Imaginary in Kähler's Quantum Mechanics, arXiv 1207.5718v1.

[89] Vargas, J. G. (2013): $U(1) \times SU(2)$ from the Tangent Bundle. Presented at the XXIst International Conference on Integrable Systems and Quantum symmetries (ISQS-21), Department of Mathematics, Faculty of Nuclear Sciences and Physical Engineering, Czech Technical University Prague, Prague June 11-16.

[90] Weitzenbock, R. (1922): Neuere Arbeiten der algebraischen Invariantentheorie, Differential Invarianten, Enzyklopädie **3** (3), Heft 6.

[91] Weitzenbock, R. (1923): Invariantentheorie, Groningen, Noordhoff.

[92] Whitney, H. (1936): Differentiable Manifolds **37** (3), 645-680.

[93] Yaglom, I. M. (1988): Felix Klein and Sophus Lie: Evolution of the Idea of Symmetry in the Nineteenth Century, Birkhäuser, Boston-Basel.

Index